The Financialization of Agri-Food Systems

Financialization is the increased influence of financial actors and logics on social and economic life, and is one of the key drivers transforming food systems and rural economies around the world. The premise of this book is that the actions of financial actors, and their financial logics, are transforming agri-food systems in profound ways. It is shown that although financialization is a powerful dynamic, some recent developments suggest that the rollout of financialization is contradictory and uneven in different spaces and markets. The book examines cases in which state regulation or re-regulation and social movement resistance are setting roadblocks or speed bumps in the path of financialization, resulting in a 'cooling off' of investment, as well as the other side of the argument where there is evidence of a 'heating up'. The authors address not only the limits to financialization, but also the mechanisms through which financial entities are able to penetrate and re-shape agri-food industries.

This book provides both a comparative analysis of financialization blending, and empirical findings with conceptual insights. It explores the connection between financialization, food systems, and rural transformation by critically examining: the concept of financialization and how food and farming are being financialized; the impacts of financialization in the food industry; and financialization in farming and forestry – along with the impacts this has on rural people and communities. This is a timely book, bringing together concrete case studies, from around the globe, to reveal the operations and impacts of finance capital in the 'space' of agri-food.

Hilde Bjørkhaug is Research Professor at Ruralis – Institute for Rural and Regional Research, Trondheim, Norway.

André Magnan is Associate Professor in the Department of Sociology and Social Studies at the University of Regina, Canada.

Geoffrey Lawrence is Emeritus Professor of Sociology in the School of Social Science at The University of Queensland, Australia.

Other books in the Earthscan Food and Agriculture Series

Environmental Justice and Soy Agribusiness
Robert Hafner

Farmers' Cooperatives and Sustainable Food Systems in Europe
Raquel Ajates Gonzalez

The New Peasantries
Rural Development in Times of Globalization
(Second Edition)
Jan Douwe van der Ploeg

Sustainable Intensification of Agriculture
Greening the World's Food Economy
Jules N. Pretty and Zareen Pervez Bharucha

Contested Sustainability Discourses in the Agrifood System
Edited by Douglas H. Constance, Jason Konefal and Maki Hatanaka

The Financialization of Agri-Food Systems
Contested Transformations
Edited by Hilde Bjørkhaug, André Magnan and Geoffrey Lawrence

Redesigning the Global Seed Commons
Law and Policy for Agrobiodiversity and Food Security
Christine Frison

Organic Food and Farming in China
Top-down and Bottom-up Ecological Initiatives
Steffanie Scott, Zhenzhong Si, Theresa Schumilas and Aijuan Chen

For further details please visit the series page on the Routledge website: www.routledge.com/books/series/ECEFA/

The Financialization of Agri-Food Systems

Contested Transformations

Edited by Hilde Bjørkhaug, André Magnan and Geoffrey Lawrence

Routledge
Taylor & Francis Group
LONDON AND NEW YORK

earthscan
from Routledge

First published 2018
by Routledge
2 Park Square, Milton Park, Abingdon, Oxon OX14 4RN

and by Routledge
52 Vanderbilt Avenue, New York, NY 10017, USA

First issued in paperback 2020

Routledge is an imprint of the Taylor & Francis Group, an informa business

British Library Cataloguing-in-Publication Data
A catalogue record for this book is available from the British Library

Library of Congress Cataloging-in-Publication Data
Names: Bjørkhaug, Hilde, editor. | Magnan, André, editor. | Lawrence, Geoffrey, editor.
Title: The financialization of agri-food systems : contested transformations / edited by Hilde Bjørkhaug, André Magnan and Geoffrey Lawrence.
Description: New York, NY : Routledge, 2018. | Includes bibliographical references and index.
Identifiers: LCCN 2018022563| ISBN 9781138068513 (hardback) | ISBN 9781315157887 (ebook)
Subjects: LCSH: Food industry and trade. | Financialization. | Agriculture–Economic aspects.
Classification: LCC HD9000.5 .F5162 2018 | DDC 338.1/3–dc23
LC record available at https://lccn.loc.gov/2018022563

ISBN 13: 978-0-367-58627-0 (pbk)
ISBN 13: 978-1-138-06851-3 (hbk)

Typeset in Baskerville
by Wearset Ltd, Boldon, Tyne and Wear

Contents

Contributors

Jonathan Almanza holds a Bachelor of Agricultural Economics degree from Antonio Narro Autonomous Agrarian University in Saltillo, Mexico. He also has a Masters of Economics and Innovation Management from the Polytechnic University of Madrid, Spain. His main research topics include globalization, transnational corporations and technological change in the agri-food sector.

Reidar Almås is Emeritus Professor of Rural Sociology and Regional Policy at the Norwegian University of Science and Technology in Trondheim, Norway. He is currently senior researcher at RURALIS – Institute for Rural and Regional Research, previously the Centre for Rural Research – of which he was the founder. He has published and co-published 20 books and his scientific works spans the areas of agri-food restructuring, agricultural policy, work accidents and work environment in agriculture, agricultural history, globalization and localization, and rural and regional development. He was President of the International Rural Sociology Association from 2008 to 2012.

Hilde Bjørkhaug is research professor at RURALIS, Institute for Rural and Regional Research in Trondheim, Norway. Her research interests are on aspects of agricultural restructuring, family farming and the role of small and medium farms on food and nutrition security. She also studies changes in the agri-food value chain including questions on ethics, power relations, policy and financialization. Her current project portfolio examines the transition to the bio-economy. She is an active member of the European Society for Rural Sociology and member of the executive board of the International Sociological Association's Research Committee 40 (the sociology of agriculture and food).

Michaela Böhme is a doctoral candidate at the Collaborative Research Centre 'Processes of Spatialization under the Global Condition' at Leipzig University, where she also completed an MA in Global Studies with an area focus on China. Set against the background of China's emerging global food security strategy, her PhD research explores the entangled interests and motive forces shaping Chinese investment in Australian farmland and agricultural resources. She is also interested in the political economy of China's development, the broader issues of multi-polarity, and the globalization of food and agriculture.

Jostein Brobakk is a political scientist and researcher at RURALIS, Institute for Rural and Regional Research and PhD candidate at The Norwegian University of Science and Technology, Trondheim Norway. He has been a visiting scholar at the University of Waterloo, Ontario, Canada, and has worked on several projects funded by the Norwegian Research Council on agriculture and climate change, and the valuation of farmland in a global context. His main research interests are within policy regime development, governance of the agri-food sector, and the effects of climate change on agricultural regimes and policy.

Douglas H. Constance is Professor of Sociology at Sam Houston State University in Huntsville, Texas, USA. His current research examines the impacts of the globalization of the conventional agri-food system on rural communities and alternative agri-food systems. He has numerous journal articles, book chapters, and books on these topics. His most recent co-edited book is *Alternative Agrifood Movements: Patterns of Convergence and Divergence* (Emerald Press, 2014). He is past president of the Southern Rural Sociological Association (2003) and the Agriculture, Food, and Human Values Society (2008).

Marion W. Dixon is Professorial Lecturer in the School of International Service at American University in Washington, DC. Her research centres on agriculture and food system change, with a focus on North Africa. She is currently writing a book, *Food and Revolt: Egypt, Ecology, Frontiers*, that investigates the development of a corporate agri-food system and the implications for food security and political stability in Egypt. In another project, she is researching the socioecological history of chemical fertilizer and the role of phosphate rock frontiers in that history. Her recently published and forthcoming publications are in the journals *Comparative Studies of South Asia*, *Geoforum*, and *Journal of Agrarian Change*.

Cynthia Gharios is a PhD candidate at the Graduate School Global and Area Studies at the University of Leipzig. Her research focuses on the reorientation of food security strategies of oil-rich countries from the Gulf Cooperation Council (GCC), and on understanding the multidimensional strategies of food investors from the GCC in other Arab World countries. She has previously worked on agrarian transformation in Lebanon, and is currently involved in a project on agricultural war damage in Yemen.

Midori Hiraga is a PhD candidate for International Political Economy of Food in the Graduate School of Economics, Kyoto University, Japan.

She achieved an MSc (Food and Nutrition Policy) from the Centre for Food Policy, City University London, UK. Her research is focused upon the political economy on provisioning vegetable oils for an industrial mass diet. She is now researching the historical case of the transformation, in Japan, of vegetable oils into everyday foodstuffs, including the developing trajectory, along with the current structure, of the global vegetable oil complex.

Geoffrey Lawrence is Emeritus Professor of Sociology at The University of Queensland, Australia. His work spans the areas of agri-food restructuring, neoliberal globalization, financialization, and food security. During his academic career Geoff has raised some $15 million in research grants and has published 25 books and over 500 journal articles and book chapters. Books include the co-edited *Food Security, Nutrition and Sustainability* and the co-authored *THINK Sociology*. He was President of the International Rural Sociology Association (2012–2016). He is a Life Member of the Australian Sociological Association and an elected Fellow of the Academy of Social Sciences in Australia.

Jana Lindbloom is a researcher at the Institute of Sociology at the Slovak Academy of Sciences in Bratislava. Her research focuses on economic sociology, organization studies, alternative organizations, property relations, post-socialist transformation, and public debates on agriculture and land. She has also written about qualitative research methods, mainly ethnomethodology.

André Magnan is Associate Professor in the Department of Sociology and Social Studies at the University of Regina, Canada. His interests include the sociology of agri-food systems, globalization and development, and sociological theory. André's research examines how social, economic, and political changes are affecting agricultural sectors and rural communities. His 2016 book, *When Wheat Was King*, examined the history and politics of grain marketing on the Canadian prairies. He has published several journal articles and book chapters on the financialization of farmland in Canada and Australia. He is currently involved in a four-year research project on financialization and land politics in Western Canada, funded by the Social Science and Humanities Research Council.

Francisco Martínez is Researcher and Professor at the Centre of Socio-economic Research at the Autonomous University of Coahuila and Professor in the Department of Agricultural Economics at the Antonio Narro Autonomous Agrarian University in Saltillo, Coahuila. Mexico. Dr Martínez's work focuses on the following topics: Globalization of the economy and society; Globalization, Plant Genetic Resources and Rural Development; and, Globalization and the Poultry Industry. Dr Martínez is the author of numerous publications on these topics.

Bruce Muirhead is Professor of History and Associate Vice President, Oversight and Analysis, at the University of Waterloo in Ontario, Canada. He has written extensively on Canadian trade negotiations since World War II, as well as Canadian politics, diplomacy and economic development. Bruce's more recent work has focused on the evolution of Canadian agricultural policy, and especially dairy and egg supply management. He is the Egg Farmers of Canada Chair in Public Policy, where his research focuses on issues surrounding the relevance and usefulness of supply management by comparing the Canadian system with those found in comparable sectors in Australia, NZ, the UK and the US.

Jodey Nurse-Gupta is a post-doctoral fellow in the Department of History at the University of Waterloo. She is currently researching the history of Canada's supply-managed dairy, egg, and poultry industries. Central to her work is the examination of asymmetrical power relations among stakeholders in agricultural marketing systems and the polarized agricultural policies that have emerged in the twentieth and twenty-first centuries. She has also written on topics of gender and women's history in agricultural organizations, and she recently completed a study of women's leadership roles in the Canadian egg farming industry.

Stefan Ouma is Assistant Professor in Economic Geography at Goethe University, Frankfurt. His research explores the socialities, materialities, and practicalities of global economic connections. It combines interests in global political economy with the ambition to explore these 'macro issues' in the minutiae of everyday economic life. Stefan has been working on agricultural commodity chains and marketization processes in Kenya and Ghana, logistical infrastructures in West Africa, and the rise of farmland and agriculture as an 'alternative asset class', with a geographical focus on Tanzania and New Zealand. Recent publications on the finance-driven land rush include 'Situating global finance in the Land Rush debate: A critical review' (*Geoforum* 57, 2014) and 'From financialization to operations of capital: Historicizing and disentangling the finance–farmland-nexus' (*Geoforum* 72, 2016). He is also the author of *Assembling Export Markets: The Making and Unmaking of Global Food Connections in West Africa* (Wiley-Blackwell, 2015).

Sarah Ruth Sippel holds a PhD in geography and is a Senior Researcher at the University of Leipzig, Germany, and an Honorary Senior Research Fellow with the School of Social Science at the University of Queensland, Australia. She leads a four-year research project on imaginations of land (C04, SFB 1199) funded by the German Research Foundation. She has intensively worked on the interlinkages between export agriculture, rural livelihood security, and labour migration in North Africa and the Western Mediterranean. Sarah's current research explores the nexus between global food security and financialization of natural resources, especially in the Australian context, as well emerging forms of solidarities within global agri-food systems.

Elin Slätmo is a Senior Research Fellow at the research centre Nordregio, established by the Nordic Council of Ministers. With a background and PhD in human geography her key research interests are social-environmental studies of governance of natural resources, sustainability, and the actors involved in negotiations around the use of land. With the ambition to develop conceptual and applied knowledge about the relationship between humans and the environment concerning norms, practices and power, she primarily conducts research in contexts of the global North.

Kiah Smith, PhD, is a sociologist and Research Fellow in the School of Social Science, The University of Queensland, Australia. Her research spans food security and food justice, climate change, resilience, ethical trade, gender, livelihoods and the green economy. She currently works on an ARC Discovery project exploring financialization, agribusiness, land and food security in Australia. Previously, Kiah was a research analyst with the United Nations Research Institute for Social Development. She is also a *Future Earth* Fellow, member of the ISA RC40 and RC09, TASA, and the Australasian Agrifood Research Network, and co-convenor of the Brisbane Fair Food Alliance (a civic food collective). She is the author of *Ethical Trade, Gender and Sustainable Livelihoods.*

Melanie Sommerville is a PhD Candidate in the Department of Geography at the University of British Columbia, Vancouver, Canada. Her research examines the social and ecological transformation of agriculture in (post)colonial settings. Melanie's dissertation, *Agrarian Repair*, explores new alliances between white-dominated finance and agribusiness capital and racialized, marginalized, communities in Canada (First Nation people) and South Africa (rural black populations). She has written widely on agricultural financialization and related topics, has a background in policy, continues to consult in the agricultural sector, and has recently commenced a postdoctoral fellowship in the Department of Geography and Planning, University of Toronto.

Steven Wolf is Associate Professor in the Department of Natural Resources at Cornell University, New York State. He studies the political economy of land and the environment with a particular focus on the challenges of securing public goods from landscapes characterized by strong private property rights claims. Current analyses focus on carbon offsetting from cropping systems and habitat exchanges that provide offsets to oil/gas and mining interests to conserve populations of at-risk species. Steven recently co-edited – with Alessandro Bonanno – *The Neoliberal Regime in Agri-Food: Crisis, Resilience and Restructuring* (2014) and, more recently, *Resistance to the Neoliberal Agri-Food Regime: A Critical Analysis* (2017).

Preface

This book developed out of working groups on financialization held during the Congresses of the European Society for Rural Sociology and the International Rural Sociology Association. In 2016, the editors of this book co-hosted sessions on the financialization of agri-food systems at the IRSA World Congress in Toronto. Scholars from this and previous meetings were invited to submit papers to this collection.

Hilde Bjørkhaug, André Magnan and Geoffrey Lawrence have appreciated the collaboration of the authors who have contributed to the realization of this book. We have benefited from Professor Hugh Campbell's advice and involvement at various times during the book's preparation. We also recognize the advice and assistance of Amy Johnston, Editorial Assistant, and Tim Hardwick, Senior Commissioning Editor at Earthscan Food and Agriculture Series from Routledge.

We are grateful for research funding from the Australian Research Council (DP 160101318), the National Research Foundation of Korea (NRF-2016S1A3A2924243) and the Norwegian Research Council (FORFOOD No. 220691).

1 Introduction: the financialization of agri-food

Hilde Bjørkhaug, André Magnan and Geoffrey Lawrence

In December, 2017, a headline in the *Australian Financial Review* trumpeted 'Agricultural cycle on fire: Australian farmland returns closing in on 17 per cent' (Cranston, 2017). The article described a recent boom in the returns on financial investments in Australian farm properties, as land values and livestock prices rose considerably. The figures are courtesy of the National Council of Real Estate Investment Fiduciaries (NCREIF), which has developed a new instrument called the Australian Farmland Index, compiled by tracking the performance of a set of benchmark properties. By collecting and publishing these data, the organization is providing valuable information to global investors. The figures allow pension funds and other investors to, for instance, compare the returns of Australian farmland to US farmland, and allocate their funds accordingly.

This story provides one example of how the agri-food system is becoming 'financialized' as actors, such as specialized farmland investment managers, and intermediaries, such as the NCREIF, are introducing new logics into the rural land market. For decades, agriculture was considered a poor investment – certainly, it was rare for institutional investors to be interested in buying farmland. This began to change in the mid-2000s, not only in Australia, but in many countries around the world, as investors began cobbling together agricultural portfolios by buying farm properties and other agricultural assets. Once considered somewhat exotic, the 'farmland asset class' has become increasingly mainstream thanks to a number of factors, including global economic conditions, the emergence of farmland investment experts and boosters, the development of myriad new investment models, and the growing influence of benchmarking tools like the 'farmland index' (Conrad, 2018).

For some years now, scholars have been documenting the financialization of agriculture and food at national and international scales and all along the 'food chain' – from agricultural inputs, to farmland and agricultural production, processing, the international trade in food commodities, and retail. Financialization can be defined as the process through which financial actors, logics, and processes exert increasing influence over economic and social life (see Epstein, 2005). This basic definition is necessarily broad since financialization can, and has, occurred in many social and historical contexts. In order to 'unpack' what financialization means in any given instance, we must understand the wider social context in which these processes unfold and focus on what real people are doing and why. The concept of financialization is a useful shorthand term for a range of related transformations affecting agriculture and food (Fairbairn, 2015a). However, as the contributors to this volume show, financialization is far from a monolithic or homogeneous process, with many variations across time, space, and sub-sectors.

In this introductory chapter, we provide a 'big picture' framework for understanding the financialization of agri-food. We first describe the historical and social context in which the world of finance has come to exert more influence over agriculture and food. We then survey the state of knowledge on the financialization of agri-food and suggest some of the most important intellectual and substantive questions that have yet to be adequately addressed. Finally, we provide an overview of the contributions to this volume and their potential for advancing our knowledge of food, agriculture, and finance.

Historical and social context

Finance is a fundamental feature of the capitalist political economy. Historically, finance was most closely associated with the banking sector and its role in providing credit to capitalist enterprises. Arrighi (1994) argues that the importance of finance to the capitalist world economy has ebbed and flowed through a series of historical phases. Periods of material expansion – in which production of commodities is the main engine of accumulation – have been followed by periods during which major capitalist powers have turned increasingly to financial activities to increase their wealth. Arrighi (1994) as well as thinkers such as Harry Magdoff and Paul Sweezy (see Foster, 2007 for a review), Epstein (2005), and Krippner (2005) have argued that the global economy has been undergoing a period of financialization since the 1970s. In the last few decades, the size of the financial sector relative to the rest of the economy has grown dramatically and there has been an explosion in the number and variety of financial intermediaries, instruments, and products. There has also been a cultural shift towards 'shareholder value' in capitalist enterprises, that is, businesses prioritizing the return on capital for shareholders above all other considerations. Financialization has touched all sectors of the global economy, and agri-food is no exception.

As noted, there is a long history of interaction between finance and food. Beginning in the eighteenth century, food traders established the first commodity exchanges, and in the mid-1800s new instruments such as futures contracts for grain were introduced (Cronon, 1991; Clapp, 2016). Through most of the twentieth century, the state played a key role in mediating the relationship between food and finance (Martin and Clapp, 2015). Governments provided cheap credit to farmers and regulated commodity exchanges to prevent destructive speculation by private actors. These measures stabilized agriculture, providing a framework for its industrialization and commercialization. Since the 1980s, however, deregulatory moves led by the US opened more space for financial actors to participate in key nodes of the agri-food chain. Most importantly, the US gradually loosened the rules restricting the activities of non-commercial speculators in commodity exchanges, partly in response to financial innovations such as commodity index funds (CIFs)[1] (Clapp, 2016; Schmidt, 2016). The Commodity Futures Modernization Act of 2000 consolidated this trend by exempting firms selling 'over the counter' derivatives from regulatory oversight. This meant that financial firms could participate in these markets without facing the strict 'position limits' that had traditionally restricted the role of speculators in the market. Throughout the 1990s and 2000s, financial firms massively expanded their involvement in the food commodities trade. By 2011, non-commercial traders represented 61 per cent of the trade in wheat futures, compared to only 12 per cent in 1990, and agricultural assets held by financial investors reached US$126 billion (Clapp, 2016, p. 151).

The financialization of the food commodity trade became the object of great controversy during the global 'food crisis' of 2007–8. As international prices for basic food staples like wheat, maize, and rice reached record highs, some economists and market insiders suggested that financial speculation had contributed to the phenomenon. While it is difficult to disentangle the many factors that led to the food crisis, there is strong evidence to suggest that financial speculation did contribute to the price spike and to food price volatility (Field, 2016; Schmidt, 2016, pp. 92–104). As Schmidt (2016, p. 3) explains, financial actors had succeeded, through a mix of innovation and political influence, in turning food commodities into a financial asset. This subsequently allowed financial players to 'dominate markets and determine the direction of prices in the short term'. The food crisis had devastating consequences for vulnerable people in many countries, and the episode led to much closer scrutiny of CIFs.

It was around the time of the global food crisis that activists and scholars began to document the role of financial actors in large-scale land acquisitions, referred to by critics as 'land grabs' (GRAIN, 2008). Like investment in food commodities, financial investment in farmland is driven by a narrative that suggests that, with a growing world population and increasing affluence in places like China, global demand for food will increase sharply in coming decades. In turn, this demand will drive up land prices, providing an opportunity for investors to realise healthy financial returns by purchasing farms. Despite the perception that states and state-owned entities are behind 'land grabbing', the vast majority of documented *deals* involve private or public investors (Nolte *et al.*, 2016).

A range of investors including hedge funds, pensions, wealthy individuals, and other institutional money managers, have acquired large tracts of farmland (Daniel, 2012; Fairbairn, 2014). Between 2006 and 2016, financial actors invested approximately US$45 billion in farmland globally, and that this figure has been growing by 8–10 per cent per year (de Lapérousse, 2016). By another measure, it is estimated that overseas investors concluded deals totalling 42.4 million hectares in low and middle-income countries between 2000 and 2016 (Nolte *et al.*, 2016).

There is evidence that financial interest in agri-food has waned to some extent in recent years. For instance, investment in commodity derivatives, which include exposure to food prices, peaked in 2011 at US$448 billion and had declined to US$276 billion by the end of 2014 (Clapp, 2016, pp. 151–152). In terms of large-scale land investments, Nolte *et al.* (2016, pp. 12–13) report that the number of concluded deals tracked in the Land Matrix database levelled off between 2012 and 2016.[2] Despite these indications of a slowdown in the rate of financialization, it is clear that financial entities have become prominent players in the agri-food sector. There has been a qualitative shift in the organization of the agri-food sector, which will persist through the ebb and flow of investor activity.

The transformations highlighted above have profound implications for consumers, farmers, and food workers. As we have seen, financial flows can dramatically alter the price of food, with far-reaching impacts on food security. Farmland investment flows are changing the structure of land ownership, with control over land passing from family-owned operations to corporations, investment funds, and specialized farmland managers. The financialization of agriculture influences production patterns, as investment logics shape the types of crops grown, land management practices, and farming practices. Private equity takeovers of supermarket chains can lead to layoffs and other forms of corporate restructuring that affect workers. For these reasons, and many more, agri-food scholars are compelled to better understand the financialization of food.

The state of scholarly knowledge on agri-food financialization

The ways in which financial actors were making forays into various parts of the global agri-food sector was outlined by Burch and Lawrence (2009). They did so by drawing together early reports of the financialization of farmland, farm inputs and logistics, the food commodities trade, and food retailing. They likewise provided evidence of how agri-food companies were leveraging their own resources to engage in financial activities, a phenomenon called 'financialization in reverse'. The theoretical contribution was to interpret these trends in terms of the restructuring of the global food system under a third 'food regime'. In doing so, they connected financialization to the broader dynamics of the capitalist world economy.

Since Burch and Lawrence's article, scholars have done a substantial amount of work to broaden the scope of analysis, integrate other intellectual traditions, explore case studies, and improve the methodologies used in understanding financialization. Several scholars have contributed to refining how the financialization of agri-food is occurring. Williams (2014) argued that it is wrong to think of finance 'colonizing' agriculture and food and that it is more appropriate to consider financialization as a complex social process involving actors from many backgrounds developing new practices, relations, and technologies. Here, he suggests, certain ideas from social studies of finance can be used to provide a more nuanced understanding of financialization. In a related vein, Ouma (2015) urged scholars to avoid overly abstract portrayals of financialization and focus instead on the real actors and everyday dimensions that accomplish these social and economic changes. Williams, Ouma, and others thus caution against an overly structuralist approach to financialization, one that suggests that the process is monolithic or inevitable.

A number of scholars have helped think through the nature of land as a financial asset. Li (2014, p. 591) argues that, because of its unique characteristics, land is an 'awkward, resistant or incomplete commodity'. As a result, the process of turning land into a financial asset requires 'cultural work', which is accomplished by financial actors who develop discourses and technologies that make land seem like an appealing investment. Because of the political sensitivities around farmland and food production, farmland investment is an inherently risky activity (Li, 2015). Similarly, Ouma (2015, p. 226) has argued that farmland 'cannot *easily* be turned into just another asset class' (our emphasis), and in subsequent work has sought to open various 'black boxes' that help us understand the on-the-ground work involved in joining the worlds of finance and agriculture (see Chapter 5, this volume). For her part, Fairbairn (2014) has shown that investors are well aware of the dual nature of farmland as a 'productive asset' (as the foundation for food production) and a 'financial asset' (as a means of capturing income streams and capital returns). These arguments could reasonably be extended more generally to food, which Winson (1994) called an 'intimate commodity' in his classic work. Although food is bought, sold, and traded like other goods, its fundamental role in social reproduction, culture, and human well-being means that its commodification is always incomplete, always contested.

Financialization takes different forms in various social contexts and subsectors – depending on the motivations, knowledge, and identity of key actors. Scholars have been at pains to explore the different models under which agri-food resources are harnessed to financial accumulation. Burch and Lawrence (2013) documented the strategies of private equity firms in the retail sector. Scholars such as Isakson (2015) and Breger Bush (2012) have looked at how financial firms and development agencies have introduced financial derivatives intended to protect small producers in the global South from weather and price shocks. In the literature on the financialization of farmland, scholars recognize a whole range of models including 'own operate', 'own lease-out', joint venture models, publicly traded companies, and farmland real estate investment trusts that give investors direct or indirect access to land, food production, or both (Fairbairn, 2014; Magnan, 2015; Kuns *et al.*, 2016). Knuth (2015) explains how investors can diverge quite significantly in their tolerance for risk, time horizons, and strategies which, in turn, affects where and how they invest in agriculture. Sippel (2015) has examined the behaviour of sovereign wealth funds in the financialization of agri-food, concluding that they blend strategic objectives like food security with commercial goals in complex and often opaque ways. Others have examined the role of intermediaries such as farmland investment fund managers in forging the link between finance and farming (Ducastel and Anseeuw, 2016; Anseeuw *et al.*, 2017).

Taking into account its complexity and specificity, scholars have come to understand financialization as a political process. For instance, Clapp's (2014, 2016) work has helped theorize how financialization shapes power relations in the global food system. As she has argued, financialization is only the latest in a series of developments that contribute to 'distancing' in the food system. Distancing occurs when social structures increase the social and geographical distance between actors, in turn affecting how and to what extent actors can influence those social structures. For Clapp (2014, p. 798), financialization 'has added a layer of complexity to agri-food markets' by introducing new actors and new instruments that have a great deal of influence over food and farming, but may be far removed from its on-the-ground realities. This, in turn influences the political context in which competing actors seek to shape food systems. Schmidt (2016, p. 26) describes financialization as a process in which 'finance breaks down barriers to penetrate new markets in order to increase profits'. This generally proceeds in two phases. During the first, financial actors innovate in order to bypass existing barriers to their activities or create altogether new products. In the second phase, finance uses its political clout to legitimize and justify its activities (Schmidt, 2016, p. 26). Indeed, legitimizing discourses have played a key role in the rollout of new financial vehicles involved in agriculture and food (Larder *et al.*, 2015).

Given their impacts, it is no wonder that financialization processes have caused some degree of anxiety – in certain cases, resistance – among governments, local communities, and food system actors in a wide range of social and geographical contexts. In the US, regulators tightened Dodd-Frank Act rules around speculating in food commodities in 2010, only to see these reforms promptly challenged by industry lobbying (Clapp, 2016, pp. 156–157). Governments in Canada (Magnan, Chapter 6, this volume), Australia (Sippel, 2015), Brazil (Fairbairn, 2015b), and New Zealand (Kirk, 2017) have increased oversight over farmland investment, and in some cases tightened ownership rules, often in response to public backlash. In some countries, particularly in the developing world, grass-roots opposition to large-scale land deals have derailed investment projects (Nolte *et al.*, 2016). Resistance and contestation have thus shaped the course of financialization – ensuring that its rollout has been partial and uneven.

This admittedly selective review of the literature on the financialization of the agri-food sector provides a glimpse of an increasingly mature, sophisticated, and diversified body of knowledge. There are nonetheless a number of substantive and theoretical questions that deserve more attention. First, the financialization of farmland has received much more attention than that of the financialization of other segments of the food chain like agricultural inputs, risk management, food service, retail, and restaurants. Second, though scholars have begun to do so (see for example, Sippel *et al.*, 2017), there is a need for more studies of the concrete impacts of financialization on local communities, farm sectors, and businesses. These must be sensitive to the historical and social contexts in which financialization is occurring. Third, there are some broad conceptual and empirical questions that have yet to be adequately addressed. What are the limits and contradictions of financialization in the agri-food sector? How does financialization relate to macro structures like neoliberalism, globalization, and capitalism? How does financialization create new identities, practices, and norms? Bridging the macro and the micro, the theoretical and the empirical, will provide a more robust conceptual framework for understanding the on-going transformations that are occurring.

Purpose and scope of this book

The premise of this book is that the actions of financial actors, and their financial logics, is transforming agri-food systems and rural economies around the world, and in profound ways. In recent years, scholars have examined how financialization is restructuring different 'nodes' in the food chain and the 'food regime' at large. In the global South, large-scale land purchases have been implicated in 'de-peasantization' as capital seeks land for biofuel, rubber, and oil palm production and similar forces are at work in the global North where private equity firms and developers, in purchasing land for investment, are threatening the future livelihoods of family farmers. Along the agri-food value chain, products of both Southern and Northern financialization enter global trade and markets. Chapters in this book explore the connection between financialization, food systems, and rural transformation by critically examining: the concept of financialization and how food and farming are being financialized; the impacts of financialization in the food industry; and, financialization in farming and forestry – along with the impacts this has on rural people and communities.

Although financialization is a powerful dynamic, some recent developments suggest that the rollout of financialization is contradictory and uneven in different spaces and markets. In some cases, state regulation or re-regulation and social movement resistance are setting roadblocks or speed bumps in the path of financialization, resulting in a 'cooling off' of investment. But, in other instances, there is evidence of a 'heating up'. The book examines such cases, addressing not only the limits to financialization, but also the mechanisms through which financial entities are able to penetrate and re-shape agri-food industries.

While a number of important, single-authored and single-focused, books – such as Frederick Kaufman's *Bet the Farm* (2012), Luigi Russi's *Hungry Capital* (2013) and Ted Schmidt's *The Political Economy of Food and Finance* (2016) – have provided critical insights into the operation of finance in the arenas of farming and food production, there is none, to date, that has brought together concrete case studies, from around the globe, to reveal the operations and impacts of finance capital in the 'space' of agri-food. This book provides a comparative analysis of financialization blending, as it does, empirical findings with conceptual insights. Contributors reveal which national and international forces are promoting financialization, which policy settings support its growth, and what effects it is having at the local level.

The themes of the book

The book comprises three parts. In Part I 'How are food and farming being "financialized"', contributors trace the dynamics of change as financial actors identify and act upon profit-making opportunities in the agri-food sector. In Part II 'Financialization of the food industry' selected case studies demonstrate the high level of corporate control in agri-food globalization and the extent to which finance has altered food production and has contributed to the 'massification' of diets. Part III 'Financialization of farming and forestry' provides further case studies on the motives and impacts of finance's move into the rural sector. Part III's case studies indicate how farm and forestry restructuring is occurring under the influence of the finance sector, the extent to which institutional investment is reshaping rural space, along with the consequences of such transformations. While the three parts deal with separate concerns in relation to financialization, many of the chapters throughout the book have clear overlaps. There are concerns, for example, about how globalization and neoliberalism have played their role in fostering financial penetration of the food and farming sectors, of how the state's actions have encouraged (yet, in some cases, restricted) the ambitions of financial entities, and the actions of people to oppose the excesses of changes wrought by global finance.

This introductory chapter sets out to identify the reasons why social science has begun to provide a critical understanding of the process of financialization. In Part I, Lawrence and Smith examine the concept of 'financialization' dealing, in particular, with the criticisms it has engendered. Financialization stands accused of being: a nebulous term, devoid of conceptual clarity; a catch-all concept that is incapable of revealing the highly differentiated practices and processes that are part of the world of finance; a totalizing notion which, taken to its logical conclusion, predicts the 'financialization of everything'; a concept that lacks both historical significance and empirical specificity; and, in not capturing the mathematical intricacies of financial transactions, actually obscures what is going on in the world of finance. In response to these, and other claims, the authors argue that, conceptually, financialization has utility in helping to 'ground' somewhat abstract notions like value, risk and financial flows as well as to uncover the internal dynamics of finance, itself. A focus upon new actors (such as hedge funds, private equity groups, and sovereign wealth funds), new investment vehicles (derivatives, index funds and the like) and new outcomes (for example, changes to position limits and the conversion of farmland into a liquid asset) provides insights on contemporary change over-and-above what can be achieved by the use of such concepts as globalization or neoliberalism/neoliberalization. The authors conclude that – while it remains underdeveloped – the concept of financialization has efficacy in explaining how new forms of capital accumulation are emerging, along with their impacts.

In Chapter 3, Bjørkhaug and Brobakk identify the difficulties faced by public investment funds as they seek to generate strong financial returns while, at the same time, seeking to maximize social and environmental benefit. The authors track the investment history of two Norwegian funds – the Norwegian Government Pension Fund – Global (GPFG) and the Norwegian Investment Fund for Developing Countries (Norfund). While GPFG evinces a desire to invest for the financial benefit of future generations, its charter precludes it from giving consideration to such important issues as climate change or food security in its decision-making. It is supposed to be a non-political actor that concentrates upon profit generation – moving in and out of investments to maximize income. In contrast, Norfund co-invests in local start-up companies and funds activities aimed at spurring business growth in the developing world – in an effort (in a political sense) to counter traditional, but often unsuccessful, development aid programmes. It acts as a 'responsible' investor, respecting labour rights, environmental regulations and societal norms and values. Yet, GPFG is hardly politically neutral pushing, as it does, for Norway's broader geopolitical interests. Likewise, Norfund has been accused of using tax havens and of orchestrating land grabs. As Bjørkhaug and Brobakk argue, both funds are part of a financialized global economy and, as such, have found that their desires for profit maximization clash with what would be ethically, socially, and environmentally sound investment strategies.

As with Chapter 3, Chapter 4 examines the growing presence and influence of one of today's key actors in global investments – sovereign wealth funds (SWFs). According to Sippel, Böhme and Gharios, concerns about food availability in a number of nations has resulted in a new form of agro-security mercantilism, that is, the acquisition of farmlands in offshore locations as a means of producing foods for eventual repatriation to the investing nation. The most obvious form of this 'strategic financialization' is intervention in the purchase of farms by many SWFs. These funds are also making strategic purchases of firms along agri-food chains which vertically integrate their activities and assist in fulfilling domestic food policy goals. Importantly, the SWFs often join forces with other forms of capital to co-invest in agri-food industries, making the issues of ownership and control complex and the purposes of investment sometimes difficult to determine. Nevertheless, through the activities of SWFs, states are increasing their influence in global food chains even though resistance and backlash from 'host' nations has been prevalent and has limited investment opportunities. SWFs from China and the Gulf States are chosen as exemplars of state-led overseas investment in food and farming.

The 'black box' of finance-gone-farming is opened by Ouma in Chapter 5. Ouma employs the term 'assetization', rather than financialization, to capture the process whereby farmland is turned into an asset class that can yield an income stream for finance. Eschewing what he considers to be 'abstract and disembodied' narratives about farmland purchases, Ouma identifies institutional investors (pension funds and insurance companies) as the key global actors, followed by investment banks, SWFs, transnational agribusiness corporations and high net worth individuals. The investment vehicles they employ are sophisticated and there are multiple investment rationalities that guide their investments. Not surprisingly, therefore, talk of some generalized 'land-grab' is overdrawn. While 'frontier' regions in Africa, South America and Central and Eastern Europe (known for cheap land) were a favoured target in an earlier phase, today it is more expensive farmland in nations such as Canada, Brazil, Australia and New Zealand. In all cases, investors must be cognizant of risk/returns, the existence and enforcement of property rights, the capital liquidity requirements of investors, opportunities for assetization, and investment timelines – including possible timetables for shifting capital out of the investment. Investment is a complicated business, with intermediaries playing an important role in delivering intelligence to guide purchases. It is also risky, and evidence is provided that points to a less than enthusiastic effort on the part of many institutional investors to fully embrace farmlands as a new asset class. Ouma concludes by positioning current-day attempts at understanding the assetization of farmland as part of a more general project of (re)politicizing the operations of the finance sector and, therefore, of ensuring the activities of finance receive closer greater public scrutiny.

In recognition of the recent attractiveness of farmland investment by the finance sector in the global North, Magnan, in Chapter 6, selects Canada and Australia to examine media and public discourses about the costs/benefits of investment, the state's role in regulating activity, investor motives, and the future of farming. In his analysis of the discussions and debates between 2008 and 2014, Magnan reveals that while there were national similarities in the so-called 'farmland controversy', there were also differences relating to the socio-cultural – and economic and political – makeup of each nation. The sanctity of domestic ownership of land differed, for example. While the law does not prohibit foreign ownership of land in Australia, in Saskatchewan (as the quintessential farming province in Canada) foreign ownership of land is banned. Yet, the media in both nations generally agreed with the proposition that non-farm capital is required if farming is to become more productive – which was construed, by the finance sector, to be a 'win-win' for the investors and for the farmers whose land they acquired. In response, a prominent farmer-based discourse in Canada was that institutional ownership would result in the control of agriculture by wealthy corporations, to the detriment of the local farming community, and rural society more generally. This outcome was much less of a concern in Australia. What was of concern in Australia was the lack of transparency relating to farm purchases by overseas entities, particularly Chinese state-owned enterprises. Public opposition led to the tightening of rules relating to disclosure of sales and to the lowering of the financial threshold used to judge whether an investment was in the public interest. The tightening of rules was not something backed by financial stakeholders in Saskatchewan who pointed to the already rigid regime and who predicted that capital would be driven elsewhere in the face of increased regulation. In both Canada and Australia, those supporting the continued financialization of agriculture drew upon discourses of productivity improvements, 'feeding the world' and helping family farmers to increase their flexibility. In contrast, those opposing financialization identified lack of transparency in farmland transactions, the differential power relations between large-scale investors and farmers whose land was being targeted, and the undisclosed motivations of 'foreigners' as some of the reasons to scrutinize – and potentially limit – new investments.

In Part II, 'Financialization of the food industry', Dixon's chapter examines finance, food and political tumult in Egypt. In Chapter 7 she takes, as a starting point, the food-inspired uprisings of 2007–8 and 2011 and traces them back to the increasing power of finance and agribusiness during the 1990s. Economic power became consolidated in the hands of the corporations and family and business groups, whose priorities and actions were at odds with the food needs of the majority of Egypt's citizens. Dixon identifies finance capital as a political class whose global activities impact directly on domestic class relations. In Egypt's case, family businesses and those running the state were tied closely together by economic interests, education and intermarriage. Their wealth was obtained through global sources but, as it accumulated, so their domestic hold on power weakened. Why did this happen? According to Dixon, following Egypt's membership of the WTO in 1995, the economic reforms pushed through in the early 1990s – including the privatization of banks, the floating of the exchange rate, the re-opening of the stock exchange and the establishment of various trade agreements – led to a rush of foreign investment into Egypt. Private equity firms increased their influence. The outcome was the entrenchment of corporate-style agriculture at the expense of the production of locally available staples. Dependency upon imported foods increased dramatically. And, although exports of agro-food commodities increased (in line with Egypt's policy of improving external income), imports grew at a higher rate, leading to terms-of-trade vulnerabilities. When global food prices increased in the 'crisis' years, food insecurity could not be easily curtailed by domestic government intervention. As a consequence, thousands of people took to the streets to vent their anger. Importantly, Dixon indicates that food costs as a percentage of income continue to remain high and the corporate food model remains entrenched. She predicts, in these circumstances, that should food prices again escalate, another series of protests is likely to eventuate.

Hiraga, in Chapter 8, discusses the rise in the popularity of 'flex' crops – such as soy and palm oil – as investment targets for Japanese and foreign transnationals. She traces the role of the Japanese *sogo-shosha* (general trading companies) in investing and trading in vegetable oils throughout Asia. Aided by Japan's financial conglomerates (*zaibatsu*), the borrowing and investment strategies of the *sogo-shosha* have promoted the financialization of the global vegetable oil complex. In particular, they are very active in South–South trading as part of the so-called corporate (or third) food regime. Hiraga contends that *zaibatsu* and *sogo-shosha* – originally established to foster industrial development in Japan – have had particular success in the agri-food industries. They were active investors in the Manchurian soybean industry before World War I, consolidated their position post-World War II, and – in a more liberal economic era – are now pursuing global strategies in quest of profit-making in the provision of a range of foods and food inputs for the industrial diet. The state's unswerving role in promoting the activities of the *zaibatsu* and *sogo-shosha* is seen to be a significant factor in their success and has been of positive assistance in the financialization of food in Japan.

In Chapter 9 Almanza, Constance and Martínez re-work the idea of 'land grabs' in their discussion of 'poultry grabs'. Land grabs represent an attempt by capital to identify and purchase so-called 'underdeveloped' farmlands and to profit from food price increases, the growing demand for agro-fuels, and from land speculation. Forced eviction has been a feature, but so too has the incorporation of local labour via waged work, contract farming and other mechanisms aimed at harnessing local resources in the profit-making circuits of capital driven by the new owners. Poultry grabs display this latter feature. Through vertical and horizontal integration, concentration in the industry has led to the creation of few, very large, firms with these firms incorporating smaller players via production contracts. The large firms profit not through the ownership of land and chicken-raising facilities, but through grower contracts and the exploitation of cheap non-union labour. They also strongly embrace financialization. JBS, for example, has grown to become the largest meat company in the world through its affiliation with J&F Investimentos, a private investment holding company, and BNDES, the Brazilian Development Bank. The authors describe the tactics of JBS in agri-food acquisitions but also detail its recent fall from grace linked to business and political scandals in Brazil. The lesson is that state power has been crucial in creating the neoliberal conditions under which mergers, acquisitions and portfolio diversification could occur within Brazilian agribusiness, allowing JBS to expand its operations both nationally and globally. Financialization has facilitated the acquisitions that have been crucial to JBS's success.

One of the main financial actors in food-sector restructuring has been private equity. Private equity firms are renowned for 'discovering' underperforming firms, purchasing them, reorganizing their management and asset base and worker profile to increase profits and/or reduce costs, and then to re-sell those firms for a higher price than the original purchase. The general aim is to turn over a profit from the sale in a relatively short period of time. In Chapter 10 Almås examines the role of private equity in the food sector of Norway. While private equity has operated in the wider global economy with limited scrutiny, its presence in Norway has been given careful attention by the state. Norway, like other Nordic nations, has had a long history of farmer-owned cooperatives, labour unionism, and a tradition of strong state intervention. When private equity sought to enter the chicken industry it found itself somewhat shackled – having to work closely with a Norwegian farmer-owned cooperative. Instead of implementing the usual tactic of purchase/restructure/sell, firms were operating in a partially deregulated environment where their hands were tied. They had to abide by stringent Norwegian regulations and to respect labour laws and environmental regulations which limited the extent to which they could profit from a quick move in/move out strategy. In other words, the Norwegian food companies were not easy targets for private equity firms, leading Almås to posit that strong regulation at the national level can be a fetter on the activities of finance capital, and to the processes which 'financialize' the food industry.

Sommerville's chapter on recent agri-food financialization in Canada's prairie provinces begins Part III of the book. In Chapter 11, Sommerville addresses two points raised by Brett Christophers (2015) in his critique of the concept of financialization – that it has limited power in explaining socio-economic change, and that financial involvement in the economy has strong historic roots and is not something 'new', as its adherents claim. In examining the agrarian development of the province of Saskatchewan, Sommerville outlines the role that finance has played, revealing the continuities from older to newer forms of investment. In recent times, farms have been under pressure to consolidate and new financial entities have recognized the opportunities to promote restructuring. The latter have provided crop insurance, equipment financing, forward contracts and derivatives – as part of a growing array of financial products that are at one-and-the-same time assisting farmers to stay in business, while making them more financially vulnerable. She also notes the disconnection between the new investors' desire for relatively quick financial gains and the need in agriculture for 'patient capital' to help to ride the bumps of weather and markets. In a similar vein to other contributors to this book, Sommerville warns of viewing financialization as a 'steamroller' that is on an unstoppable path of changing the nature of farming: by understanding local actions and practices it can be seen that farmers are 'key agents' in determining the trajectories of finance in farming. She believes that a more comprehensive story about prairie agrarian transformation will be achieved by researchers who study the cultural aspects of financialization, specifically the ways finance has increased its influence over the identities, and lived aspects of daily life, of farmers and others in the provinces.

In Chapter 12, Muirhead and Nurse-Gupta argue that Canada's governance of dairy, egg and poultry production through the supply management system is hindering financialization in these sectors. The model discourages foreign ownership and agribusiness takeovers of production primarily the result of not yielding to what critics would call appropriate economies of scale. It achieves this as a result of production controls managed through quotas, import restrictions and the annual negotiation of price. The numbers tell the story; an average dairy herd in Canada is about 80 cows, and for eggs, laying flocks are approximately 22,000 hens, numbers regarded as too small-scale in the Western corporate faming perspective. This enables these farms to be defined as family farms in the purest sense, and this regulated system delivers acceptable prices for both producers and consumers that are negotiated among stakeholders and that result in a sustainable income for the former. The system was implemented in the 1960s and 1970s to provide the state with assurances that it would not need to offer costly subsidies to producers who now had stability of income to make astute long-term investments in their operations. Elsewhere, the period was characterized by overproduction, price volatility and price collapse, none of which affected this Canadian paradigm. Muirhead and Nurse-Gupta demonstrate that Canadian governments over time have appreciated the supply-managed sectors' ability to self-manage without subsidies.

The system is presently facing intense pressure in trade negotiations given the high tariffs that protect its functioning and the surfeit of, for example, global dairy product especially in the United States which supports price but not production control. As well, as Muirhead and Nurse-Gupta argue, opposition comes from critics for ideological reasons – those opposing any regulation and who believe in the sanctity of the 'free market'. The authors conclude that the passive blockage of foreign takeovers and the maintenance of a family farming system has served the country well and is a critical bulwark against the financialization of the Canadian dairy, egg and poultry sectors in that it asserts a Canadian concern for the country's food security and sovereignty.

Chapter 13 examines the reshaping of farmland ownership and use in Slovakia. The former socialist state has three types of farms – large agribusiness enterprises established after the privatization of socialist state farms and collective farms, individual and family farms, and cooperatives. Two types of new investors have entered farming in recent years – those wanting to farm, and those wanting to acquire land for its potential capital gain. Those purchasing farms are a mix of foreign and local investors. Lindbloom contends that farmland has become attractive to investors because of Slovakia's entry to the EU which, through concessions, subsidies and payments, has placed a platform under rural investment. There are four 'pathways' for entry of capital – incorporation into a farm in financial difficulty, direct purchase, 'hostile takeover' and involvement through invitation. Lindbloom details the 'messiness' that occurs when outside 'intangible' capital seeks to invest in a tangible resource such as farmland. She reports on the complexity, relationality and contingency in the various deals that have occurred, suggesting that the variety of pathways and actors (especially intermediaries) creates a multifaceted picture of the financialization of farmland.

In Chapter 14, Slätmo reveals how the Swedish state has sought to alter the Swedish Land Acquisition Act in ways that will allow corporate capital to increase its investment in farmland. In past decades Swedish farmland has been primarily owned by individuals, many of whom do not work the land (only 30 per cent of the land is owned by 'active' farmers). In other words, some 70 per cent of the land is controlled by absentee owners who are not directly connected to the land and its output. The Swedish government received a report in 2015 that the current Land Acquisition Act hindered competitiveness in the agricultural sector and has sought to introduce that a more 'competitive', neoliberal, approach to drive productivity in the farm sector. This entails freeing up land for corporate investment, on the premise that more capital needs to be injected into Swedish agriculture. Slätmo interviewed farmers, politicians, representatives of NGOs, researchers, consultants and an array of rural and urban residents in an effort to establish the extent of support for the proposed changes to the Act. While it was generally acknowledged that capital was, indeed, a scarce resource and was much needed in farming, there were different views as to how that capital could be utilized. Unlike those in the policy and business spheres who readily supported changes to the Act to encourage corporate investment, farmers (and their affiliated organizations) did not think external capital should be employed to change land ownership structures to 'corporatize' farming but, rather, to meet seasonal difficulties, to enable farm succession, to support farming culture, and to ensure supply of healthy and affordable foods. The Act is currently under investigation and it is not known whether the proposed changes will come to fruition.

The final chapter in the book provides a case study of the role institutional investors played in helping to halt the ecologically-damaging process of forestland 'parcelization'. Institutional investors' ambitions to continue to pursue logging and papermaking were tempered by the activities of the state and demands of forest communities as they sought novel governance arrangements that would both allow for economic development, while preserving biodiversity and environmental integrity. Wolf's case study, in Chapter 15, demonstrates that cooperation between a variety of actors (state agencies, civil society groups, local communities and institutional investors) helped to stem the process of landscape fragmentation and community displacement, thereby providing a positive development outcome for the Adirondacks. The State of New York required funds from investors interested in owning timberland and the arrangements that were concluded resulted in a multifunctional forest landscape. In this case, 'financialization' cannot be construed as having the negative outcomes that are reported in much of the pre-existing 'land grab' and 'green grab' literature. Wolf reminds us that the role of finance in land development is complex: in the case of the Adirondacks institutional investors helped to place an economic platform under forest communities, while addressing concerns of state agencies and NGOs about ecological risks and problems.

Conclusion

The (neo)liberalization of business and trade, the privatization of state assets and the watering down of finance and banking regulations have, in concert, promoted capitalist development. As the chapters in this book reveal, the state is increasingly involved in facilitating financialization across the world. Financialized economic activities are not only being deployed by financial institutions, but also by national and private investment funds (SWFs) and state-backed organizations, for commercial purposes, for development, and for aid, among a host of other purposes. The 2008 food crisis changed agri-food policy rhetoric, turning what was previously considered 'hidden' protectionist policies of some Northern states into 'food security' discourse, where attempts were made to ensure a secure food supply in the face of global uncertainties. Some states have actively appropriated farmland abroad, while countries with food surpluses have supported financialization – within neoliberal globalization – as a mechanism to sell greater volumes of food in the global marketplace. Either way, the farming and food sectors have remained very attractive sites for investment.

Previous research has shown, and studies in this book show, that financialization of the agri-food system raises many ethical and moral questions. When nation-states help to facilitate financialization it is important that they weigh up impacts upon small and medium scale farms and food businesses and those who occupy lands (either legally or illegally). Throughout the book, authors shed light on some of the least-explored financialized parts of the food system, including the investors, their tactics, and the communities that are affected. Many contributors demonstrate that the poor and least powerful can readily become more marginalized as financialization proceeds. It is hoped that the book will contribute to a more nuanced understanding of the contours and impacts of the current, financialized, global agri-food system.

Notes

1 Commodity index funds are financial products that bundle together the value of a basket of commodities, typically including oil, metals, and agricultural products.

2 The Land Matrix is an open-source database of large-scale land deals (www.landmatrix.org/en/). The authors of the report caution that the levelling off of land deals concluded could be explained by a lag in reporting.

References

Anseeuw, W., Roda, J., Ducastel, A. and Kamaruddin, N. (2017) 'Global strategies of firms and the financialization of agriculture', in E. Biénabe, A. Rival, and D. Loeillet (eds) *Sustainable Development and Tropical Agri-chains*, Springer, Dordrecht, pp. 321–337

Arrighi, G. (1994) *The Long Twentieth Century: Money, Power, and the Origins of Our Times*, Verso, London

Breger Bush, S. (2012) *Derivatives and Development: A Political Economy of Global Finance, Poverty and Farming*, Palgrave Macmillan, New York

Burch, D. and Lawrence, G. (2009) 'Towards a third food regime: Behind the transformation', *Agriculture and Human Values*, vol 26, no 4, pp. 267–279

Burch, D. and Lawrence, G. (2013) 'Financialization in agri-food supply chains: Private equity and the transformation of the retail sector', *Agriculture and Human Values*, vol 30, no 2, pp. 247–258

Christophers, B. (2015) 'The limits to financialization', *Dialogues in Human Geography*, vol 5, no 2, pp. 183–200

Clapp, J. (2014) 'Financialization, distance and global food politics', *Journal of Peasant Studies*, vol 41, no 5, pp. 797–814

Clapp, J. (2016) *Food*, 2nd edition, Polity Press, Cambridge, UK

Conrad, J. (2018) 'A decade in the farmland asset class evolution', *Global AgInvesting*, 17 January, available at: www.globalaginvesting.com/decade-farmland-asset-class-evolution/, accessed 5 February 2018

Cranston, M. (2017) 'Agricultural cycle on fire: Australian farmland returns closing in on 17 per cent', available at: www.afr.com/real-estate/agricultural-cycle-on-fire-australian-farmland-returns-closing-in-on-17-per-cent-20171214-h04t9k, accessed 7 February 2018

Cronon, W. (1991) *Nature's Metropolis*, W. W. Norton and Company, New York

Daniel, S. (2012) 'Situating private equity capital in the land grab debate', *Journal of Peasant Studies*, vol 39, no 3–4, pp. 703–729

de Lapérousse, P. (2016) 'Agriculture: A new asset class presents opportunities for institutional investors', Global AgInvesting, available at: www.globalaginvesting.com/agriculture-a-new-asset-class-presents-opportunities-for-institutional-investors/, accessed 11 January 2018

Ducastel, A. and Anseeuw, W. (2016) 'Agriculture as an asset class: reshaping the South African farming sector', *Agriculture and Human Values*, vol 34, no 1, pp. 199–209

Epstein, G. (2005) *Financialization and the World Economy*, Edward Elgar, Cheltenham, UK

Fairbairn, M. (2014) '"Like gold with yield": evolving intersections between farmland and finance', *Journal of Peasant Studies*, vol 41, no 5, pp. 777–795

Fairbairn, M. (2015a) 'Reinventing the wheel? Or adding new air to old tires?', *Dialogues in Human Geography*, vol 5, no 2, pp. 210–213

Fairbairn, M. (2015b) 'Foreignization, financialization and land grab regulation', *Journal of Agrarian Change*, vol 15, pp. 581–591

Field, S. (2016) 'The financialization of food and the 2008–2011 food price spikes', *Environment and Planning A*, published online DOI: 10.1177/0308518X16658476

Foster, J. (2007) 'The financialization of capitalism', available at: https://monthlyreview.org/2007/04/01/the-financialization-of-capitalism/, accessed 5 February 2018

GRAIN. (2008) *Seized: The 2008 Landgrab for Food and Financial Security*, available at: www.grain.org/article/entries/93-seized-the-2008-landgrab-for-food-and-financial-security, accessed 5 February 2018

Isakson, S. (2015) 'Derivatives for development? Vulnerability and the financialisation of climate risk management', *Journal of Agrarian Change*, vol 15, no 4, pp. 569–580

Kaufman, F. (2012) *Bet the Farm: How Food Stopped being Food*, John Wiley and Sons, New Jersey

Kirk, S. (2017) 'Government to tighten rules around selling land to overseas buyers', NZFarmer, available at: www.stuff.co.nz/business/farming/99349185/government-to-tighten-rules-around-foreign-farm-buyers-from-december-15, accessed 17 January 2018

Knuth, S. (2015) 'Global finance and the land grab: Mapping twenty-first century strategies', *Canadian Journal of Development Studies*, vol 36, no 2, pp. 163–178

Krippner, G. (2005) 'The financialization of the American economy', *Socio-Economic Review*, vol 3, pp. 173–208

Kuns, B., Visser, O. and Wastfelt, A. (2016) 'The stock market and the steppe: The challenge faced by stock-market financed, Nordic farming ventures in Russia and Ukraine', *Journal of Rural Studies*, vol 45, pp. 199–217

Larder, N., Sippel, S.R. and Lawrence, G. (2015) 'Finance capital, food security narratives and Australian agricultural land', *Journal of Agrarian Change*, vol 15, no 4, pp. 592–603

Li, T. (2014) 'What is land? Assembling a resource for global investment', *Transactions of the Institute of British Geographers*, vol 39, no 4, pp. 589–602

Li, T. (2015) 'Transnational farmland investment: A risky business', *Journal of Agrarian Change*, vol 15, no 4, pp. 560–568

Magnan, A. (2015) 'The financialization of agri-food in Canada and Australia: Corporate farmland and farm ownership in the grains and oilseed sector', *Journal of Rural Studies*, vol 41, pp. 1–12

Martin, S. and Clapp, J. (2015) 'Finance for agriculture or agriculture for finance?', *Journal of Agrarian Change*, vol 5, no 4, pp. 549–559

Nolte, K., Wytske, C. and Giger, M. (2016) *International Land Deals for Agriculture: Fresh Insights from the Land Matrix: Analytical Report II*, Bern Open Publishing, Bern

Ouma, S. (2015) 'Getting in between M and M' or: How farmland further debunks financialization', *Dialogues in Human Geography*, vol 5, no 2, pp. 225–228

Russi, L. (2013) *Hungry Capital: The Financialization of Food*, Zero Books, UK

Schmidt, T. (2016) *The Political Economy of Food and Finance*, Routledge, London

Sippel, S.R. (2015) 'Food security or commercial business? Gulf State investments in Australian agriculture', *Journal of Peasant Studies*, vol 42, no 5, pp. 981–1001

Sippel, S.R., Larder, N. and Lawrence, G. (2017) 'Grounding the financialization of farmland: perspectives on financial actors as new land owners in rural Australia', *Agriculture and Human Values*, vol 34, no 2, pp. 251–265

Williams, J. (2014) 'Feeding finance: a critical account of the shifting relationships between finance, food and farming', *Economy and Society*, vol 43, no 3, pp. 401–431.

Winson, A. (1994) *The Intimate Commodity: Food and the Development of the Agro-Industrial Complex in Canada*. Toronto: University of Toronto Press

Part I

How are food and farming being ‘financialized’?

2 The concept of 'financialization'

Criticisms and insights

Geoffrey Lawrence and Kiah Smith

Introduction

The term 'financialization' has been employed to describe and explain the quite significant changes occurring under capitalism in the latter's purported move from a production-based system to one dominated by finance. For Foster (2007) there were signs as early as the 1960s – but especially during the final three decades of the twentieth century – that finance was re-shaping the capitalist economy. This was occurring because multinational corporations – which had achieved spectacular success through the sale of products to eager consumers, globally – were beginning to face conditions of oversupply. Weakening consumption led corporate decision-makers to withhold investment in productive capacity and, instead, seek to expand their money capital as a means of maintaining profit margins (Foster, 2007, pp. 1–4; Sweezy and Magdoff, 2009). From 1980 to 2003, the finance sector in the US tripled its profits as a proportion of GDP, while the manufacturing sector recorded no growth in the same period (Holton, 2012, p. 45). The primary mechanisms through which money capital was expanded were new finance industry-created products which included credit default swaps, derivatives, commodity index funds, and other 'special purpose vehicles' (Konzelmann *et al.*, 2013). Despite their being implicated in the Global Financial Crisis (GFC) of 2007–2008, these vehicles continue to underpin trade in today's financial markets (Lapavitsas, 2012; Clapp and Isakson, 2018).

While some argue that financialization is one of the three key macro-sociological concepts that can be harnessed to understand the contemporary dynamics of change within the capitalist economy (the other two being globalization and neoliberalism [Foster, 2007; Lawrence, 2015]), others have viewed the term with suspicion (Christophers, 2015; Poovey, 2015). The five most important criticisms of financialization are that the concept: is nebulous and thus lacks clarity; tells us nothing new; is a-spatial and a-historical; is 'totalizing'; and, does not help us to understand the mathematics of financial transactions.

In this chapter, the criticisms of the concept of financialization are presented in detail, together with responses to those criticisms. We argue that while financialization currently means many things to many writers, it nevertheless provides important insights into the structures and processes that define neoliberal globalization. Specifically, the financialization concept has helped to connect abstract notions of capital flows, value, and risk to (a) the socio-political construction, and consequences, of new patterns of ownership and control over land and agri-food industries, as well as to (b) the ideas and practices of financial actors seeking to construct financial value in the context of multiple food and economic crises, and the internal logics of finance itself. As Gunnoe (2016, p. 1083) has pointed out, 'the structural processes of financialization are mediated by a host of organizational and cultural factors', which include neoliberalism and globalization. Some examples are given to demonstrate its utility in understanding changes in food, agriculture, and farmland investment across scales. Recent research from agri-food studies has provided especially rich insights into the emerging patterns of actors and investment vehicles channelling capital into farmland and agribusiness, the mechanisms used to turn agriculture into a new financial asset class, the internal logics and motivations of financial actors involved, and the social-ecological implications of these processes in different country contexts. To abandon financialization – as some critics demand – would be to lose a concept providing much explanatory power. The promise of financialization lies in its capacity to provide new insights into the contours, drivers, and outcomes of neoliberal globalization.

Criticisms of 'financialization'

In 2015 the journal *Dialogues in Human Geography* featured a penetrating critique of the concept of financialization. Penned by human geographer, Brett Christophers, 'The Limits to Financialization' issued a strong note of caution to scholars who uncritically apply the concept. A series of responses to Christophers' paper were also featured in the journal. This section draws heavily from the *Dialogues in Human Geography* discussions. Although not described in the journal in the manner presented below, five major criticisms were levelled at 'financialization'.

The concept is nebulous and 'chameleon-like' – lacking conceptual clarity

While the finance system has always been crucial to an understanding of capitalist development (see, for example, Hilferding, 2006 [1910]), it was not until the late 1990s and into the early twenty-first century that the concept of financialization emerged to explain the accelerated move to a finance-led economy (van der Zwan, 2014). Christophers (2015, pp. 184–187) insists that the exuberance in scholarly use of the term – which has occurred in a 'breathless and boosterish' manner over the last decade – has resulted in obfuscation rather than clarity. He argues (p. 184) that to view financialization in a blanket way as 'the (increased) contemporary importance of finance' is to say very little, when what is required is a clear and unambiguous analytical concept that provides insights into capitalism's supposed new phase. While financialization has 'begun to crumble before our eyes ... as a chaotic, motley idea' (p. 186), more robust concepts like globalization and neoliberalism continue to have a 'core conceptual thread' (p. 187). Christophers (2015, p. 187) asks: 'Which of the 17 going-on 27 meanings of financialization (does) the writer (have) in mind.... Would it not be preferable for the sake of analytical and communicative clarity, to dispense with the term altogether?' Christophers (2015, p. 191) considers the concept of globalization as a much more potent means of understanding socio-economic change and that – in contrast to financialization – 'intellectually, it works'. Yet, globalization is hardly a coherent process or an agreed-upon concept, with Beerkens (2006) reporting that there were over 90 definitions/perspectives of globalization in use up to 2006. There are 'dozens' of definitions/understandings of neoliberalism (Aalbers, 2015, p. 215). The main point here is that macro-sociological terms arise from different theoretical perspectives (economics, sociology, geography, for example) and will therefore place stress on different elements or attributes; it is therefore not surprising that there are multiple definitions of a phenomenon like globalization or financialization. While the concept of financialization might be 'raw and undeveloped', there is no denying its power in explaining how financial entities are transforming capitalist economies (Lapavitsas, 2013, p. 55; Murphy, 2015, p. 208).

Financialization tells us nothing new

According to Christophers (2015, p. 187) financialization is essentially a re-embodiment of ideas that have been in circulation for decades and asks us to ponder whether financialization simply 'dress(es) up existing theoretical claims in new terminological clothes'. He notes that a powerful concept like shareholder value arose independently of the term financialization, throwing doubt on the latter's novelty and usefulness (Christophers, 2015, p. 188). It is 're-inventing the wheel' to adopt a new concept when presently-existing ones provide us with the insights we require (Christophers, 2015, p. 189).

In contrast, Fairbairn (2015, pp. 211–212) has defended financialization for its ability to provide new insights. Drawing upon her work on farmland investment, she demonstrates how new financial actors (hedge funds, sovereign wealth funds, pension funds) have risen in prominence as land has become a new component in portfolio diversification. For

Fairbairn (2015, p. 212), financialization helps us to understand the growing influence of financial investors, the focus upon financial profits rather than productive profits, and the growing moral legitimacy of finance in all aspects of life. While the shareholder value concept sought to explain the embeddedness of corporate managers' decision-making in relation to a macro-economic environment of economic crisis, this is just one discourse used to legitimate strategies that were/are integral to the financialization process more broadly (see Gunnoe, 2016). In addition to this, financialization draws attention to finance-sector instruments and vehicles – including leveraged buy-outs, junk bonds, derivatives, securitization, among a host of others – which now form part of the language of finance trading, but were little known before the 1980s (Lawrence, 2015). Its novelty lies in its focus upon new investors and new financial products and processes that elevate finance, and particularly its role in the roll-out of neoliberal globalization, as important objects of research.

Financialization is a-spatial and a-historical

Spatially, much of the writing about financialization is limited to 'exemplars' such as the US and UK, with this restricted lens preventing global connections from being explored and understood (Christophers, 2015, p. 192). Furthermore, the tendency to view financialization as a phenomenon explicitly associated with the current wave of capitalist development is to deny its historical roots (Christophers, 2015). Arrighi (1994) in his classic text *The Long Twentieth Century* has identified three historical periods where finance capital has asserted itself in re-shaping the economy. Similarly, it is disingenuous to suggest that shareholder value and the penetration of everyday life by finance are contemporary phenomena. Christophers (2015, pp. 192–193) argues that Keynes was writing about them in the 1930s and suggests that only during the mid-1930s to the mid-1970s was finance 'truly shackled' in the Western nations. This leads him to assert (p. 193) that if a wider historic optic is adopted, we can read change in the twentieth century as being 'a financialized capitalist *norm* punctuated ... by four decades of exceptional *de*financialization'. This is a provocative assessment. If the past is not understood and we adopt an exclusive 'presentist lens' there is a strong chance we will not only misread history but also be building new theory which creates a 'false separation' between the past and the present (Christophers, 2015, p. 194).

There are a number of responses to these criticisms. Directly confronting the concern that financialization is a-spatial, Aalbers (2015, pp. 214–216) argues that it is an 'inherently spatial phenomenon' which allows for the linkage of various scales of analysis. He cites the work of writers who examine finance in daily life, those who discuss accumulation at the level of the nation-state, and those adopting a global focus. The challenge is to connect these spaces to understand the inter-relationships of financial drivers and processes, along with its impacts across space. Furthermore, to suggest that most of the writing about financialization is focused upon the US and UK is to overlook a vast literature on the financialization of farmlands in the global South (the large-scale land acquisitions, or so-called 'land grabs' – such as the special collections of papers in *Agriculture and Human Values* 34(1); *Development and Change* 44(2); *Globalizations* 10(1), *Journal of Agrarian Change* 15(4), *Journal of Peasant Studies* 39(2) and 42(3–4), and *Third World Quarterly* 34(9)) and of the financial penetration of food and farming systems in nations in the global North lying beyond the US and UK. Indeed, a recent systematic review[1] of the financialization literature reveals that 80 per cent of papers reported case studies from a wide variety of contexts, in the following regional groupings:

- Global (55), US (20), and UK (14)
- South Africa (6), Africa (general, 3), Kenya (4), Tanzania (2), Ethiopia (2), Morocco (2), Sudan (2), DRC, Lesotho, and Uganda (1 each)
- Canada (6), Australia (7), and New Zealand (4)
- Brazil (2), Latin America (general), Argentina, Chile, Dominican Republic, Costa Rica, Guatemala, and the Caribbean (1 each)
- India (5), China (2), Malaysia (2), Vietnam (2), Asia (general), Cambodia, Hong Kong, Indonesia, Singapore, and the Philippines (1 each)
- Europe (general, 3), Spain (4), Ireland (2), Norway (3), France (3), Germany (2), Ukraine (2), Turkey (2), Italy, Greece, Portugal, and Iceland (1 each)
- Egypt, Bahrain, Palestine, Israel, and Qatar (1 each).

In other words, the criticism that financialization is missing a spatial dimension is unfounded – researchers around the globe are attempting to 'ground' financialization and show its local-level features, helping – through case studies – to build a globally-composite picture of its contours, specificities and impacts. Some of these studies are discussed in more depth later. Space is essential to an erudite understanding of social processes. As Massey (2005, p. 130) argues 'places are not ... points or areas on maps, but ... integrations of space and time ... (they are) spatio-temporal events'. Financialization is not a linear process and appears in different ways in different spaces, which is obvious from its contrasting forms and operations in the global North and global South (Bonizzi, 2014; Clapp and Isakson, 2018).

Is financialization a-historical? As Gertel and Sippel (2016, p. 215, and see Russi, 2013, p. 29) have argued, 'financialization has not appeared out of "nowhere"'. It is a process that relies upon certain techniques and metrics which have been created over several centuries'. Ouma (2015 and 2016, p. 91) argues that 'the coupling of finance and farmland/agriculture has a long history', with many writers drawing upon historical examples in their analyses. For example, to explain finance capital's current interest in food and farming Russi (2013) invokes food regime theory, which requires an understanding of British-based settler colonial farming systems, crisis periods in the 1930s and 1970s, through to corporate expansion from the 1980s. Similarly, investigating present-day financialization in the US, Krippner (2005) writes of capital accumulation difficulties since the 1950s. To explain recent foreign interest in land in Mozambique Fairbairn (2013) discusses earlier land distribution policy – from the Portuguese prazo system reaching back to the 1700s, to the nationalization of land in 1975. In explaining the rise of derivatives markets Breger Bush (2012) discusses the colonization of Africa, along with the neoliberal (free market) ideologies of Thatcher and Reagan from the 1980s. Desmarais *et al.* (2017) discuss various state-based land acts from the 1870s in detailing farm ownership changes in Saskatchewan. In fact, it is difficult to discover any published case studies of financialization that do not explicitly set the current activities of finance against the backdrop of history. In contradistinction to Christophers, Aalbers (2015, p. 216) finds a strong historical presence in the financialization literature. As van der Zwan (2014, p. 106) warns us: 'In the absence of historical contingency, the specific mechanisms underlying financialization's development remain unclear. Financialization is rendered abstract, anonymous and teleological: 'finance' becomes the driving force behind its own expansion'. It is thus incumbent upon researchers to provide historical context – and virtually all appear to do so, discrediting the view that financialization is largely 'a-historic'.

Financialization is a totalizing concept

Is financialization an unstoppable force in a world increasingly driven by the policies and practices of neoliberal globalization? Researchers have employed the concept of financialization in the study of land acquisitions, food industry investments, the purchase of ports, in disaster management, and in everyday life, with the criticism emerging that nothing will be left *un*-financialized: thus, financialization is a totalizing concept (Christophers, 2015, and see Ouma, 2016). It is true that the underlying logic of financialization – which is based on the notion that its role in capitalism is to turn everything illiquid into a liquid, tradeable, commodity – suggests inevitability. But is this so?

According to Aalbers (2015) there are many areas of life that finance has been unable to penetrate, while Bayliss *et al.* (2016) acknowledge that financialization is often 'far removed from the direct experience of everyday life'. And, where financialization's presence has created inequality and social disadvantage, there has been considerable resistance and protest – as has been demonstrated by peasant fightback over lands illegally or surreptitiously acquired by foreign capital (McKeon, 2014) and street demonstrations as part of the Occupy Wall St movement (Chomsky, 2012). Financialization does not presuppose that finance capital will be the dominant form of capital as some end-game that views bankers and investment firms controlling the world. Rather, it might also be seen as an element in the 'double movement' within capitalism, whereby the activities and motivations of corporate food actors are continually involved in a dialectic process of struggle and redefinition with food movement actors concerned with social justice, environmental sustainability, and economic transformation (Holt-Gimenez and Shattuck, 2011, and see Polanyi, 1944). In a broad sense, it is a framework for identifying how investors employ a variety of different calculative practices – over space and time – in seeking to maximise economic gains (Murphy, 2015). But financialization's trajectory is not unstoppable: that would be to deny social agency. There is no financial juggernaut but, rather, a growing presence of finance and financial entities under a neoliberal form of capital accumulation (Schmidt, 2016, p. 3).

The concept does nothing to help us penetrate the 'black box' of mathematics that underpins all modern financial transactions

While the concept of financialization draws attention to the importance of finance in everyday life, it nevertheless obscures what is 'really' going on. According to Poovey (2015, p. 221) most scholars writing about financialization fail to discuss 'the institutions, legal framework, and theoretical formulations that underwrite financial transactions'. It is the infrastructure that allows financial transactions to occur that should be the focus of scholarly attention. Her argument is that sociologists, historians, anthropologists, and other 'nonmathematical' writers approach financialization largely ignorant of the technical specificities that underpin modern financial transactions. She asserts: 'Unable to follow or participate in mathematical accounts of financial relationships and properties, scholars have resorted to the languages we have, and the result has been vague talk of "cultures of finance"' (Poovey, 2015, p. 221).

Finance has become 'enmeshed' in 'arcane mathematical language' – a 'black box' that prevents the non-mathematically trained from understanding what is occurring, leaving them to describe finance in imprecise and oversimplified ways (Poovey, 2015, pp. 221–222, and see Christophers, 2015, p. 191 and Ouma, 2015, p. 227). To put it another way, those involved in financial trading create mathematical models, with advanced statistical techniques used to price financial derivatives, predict futures markets, and guide global trading. The models and tools are 'simultaneously incomprehensible and unassailable' (Poovey, 2015, p. 222), yet they are at the heart of financialization and to leave these in a 'black box' is to ignore the very mechanisms that drive financialization: any theorizations arising must therefore be partial, at best, erroneous, at worst.

Who *can* open the 'black box'? Should social scientists retrain in finance before writing about financialization? Should they avoid being involved at all and leave any investigations to the financially literate? If so, is it only Wall Street bankers who can really understand and comment on the practices and dynamics of Wall Street banking? Many social scientists (particularly sociologists, anthropologists, and economists) have made their academic careers researching 'the professions', applying rigorous research methods tailored to investigating complex social, economic, and political processes from a wide range of theoretical perspectives. In fact, it could readily be asserted that it is preferable *not* to be so intimately bound to the object of study, if what one is attempting is a critical analysis: it is often the 'outsider' who is best at deconstructing what are taken-for-granted ideas, practices and motives (Latour, 1981). With financial discourse so embedded in everyday culture, and in light of increasing public critique of the power of financial elites to dictate public policy (Epstein and Jayadev, 2005), alternative perspectives are needed now more than ever. A key challenge for financialization researchers is to 'make sense of practices of financial economization' alongside 'thinking about the emancipatory power of our approach(es)' (Ouma, 2015, p. 228).

Financialization – in the context of globalization and neoliberalism

One of the recommendations made by Christophers is to dispense with the concept of financialization. Financialization is 'fundamentally fragmented' and an 'unhelpful signifier' (Christophers, 2015, p. 187), while concepts such as globalization and neoliberalism have more conceptual coherence (despite, too, having their problems). In contrast, in this section, it is argued that financialization cannot be subsumed within either globalization or neoliberalism: it has a dynamic of its own and influences, and is influenced by, those two broader processes.

In its most basic sense, globalization refers to the spatio-temporal processes of change which drive the transformation of social, economic, and political organization across space (Held *et al.*, 1999, p. 15). It is characterised by growing connectedness, integration, and interdependence with events and decisions in one location of the world being conveyed quickly to – and often having impacts within – other locations (Martell, 2010). Facilitated by technologies such as the personal computer, smartphone, internet, and satellites, information flows almost instantaneously across national boundaries, helping to foster transnational and transcultural interactions. Transnational corporate activity is an important feature of globalization, with businesses able to organise labour and resources in a manner which increases international trade and maximizes profits on a global scale (Kotz, 2002; McMichael, 2017). Driven by technology, time/space compression is being experienced by most economies throughout the world (Bauman, 1999). And yet, while being described as the 'big idea' of the twentieth century (Held *et al.*, 1999), globalization cannot explain all that is occurring to economies and societies.

The term neoliberalism has been employed by scholars to describe the belief system that claims 'free' markets are preferable to regulated markets, that government activity crowds out private investment, and that individual freedom is enhanced when state activities are reduced/redirected (Kotz, 2002; McMichael, 2013). The sanctity of the marketplace and of individual responsibility are key ideas, and those ideas are translated into policies by governments (Springer *et al.*, 2016, p. 2). They do so through what is described as 'economic rationalism' – encompassing policies such as deregulating the economy (and reregulating it to stimulate the 'free' market), privatizing state assets, reducing welfare expenditure, endorsing 'user pays', lowering corporate taxes, and, imposing fiscal austerity (Stilwell, 2002). The assumption is that these measures will promote entrepreneurship, open up markets, and increase trade, thereby stimulating economic development and wealth creation (Springer *et al.*, 2016) even if, in practice, development is skewed and most of the wealth created is distributed unevenly (Monbiot, 2016).

In contrast to both globalization (which stresses global interconnections) and neoliberalism (which promotes free markets), financialization is a term which encapsulates the growing presence and influence of financial actors, financial markets, and financial motives in the operation of national and global economies and institutions (Epstein, 2005, p. 3), with financial traders now dominating strategic markets (Schmidt, 2016; Clapp and Isakson, 2018). Financialization has three key elements. The first is the new actors who have appeared during the past two or three decades. While commercial and investment banks have centuries-old histories, the presence of private equity firms, hedge funds, pension funds, and sovereign wealth funds has only risen to significance in the global economy since the 1990s (Epstein, 2005). The second is the creation and use by the financial sector of a number of new products and instruments, including leveraged buyouts, credit default swaps, junk bonds, derivatives and 'securitization', which have altered the dynamics of finance in the economy (Holton, 2012). The third is a set of new outcomes: commercial and industrial enterprises are relying less on banks and are performing their own financial transactions; banks do less lending to business and are relying more on household borrowings; households have become financialized, being increasingly exposed to new financial products; and, financialization of the non-finance sector has reduced growth in that sector, leading to lower levels of employment, to stagnant wages, to income transfers from employees to employers, and to the growth of social and economic inequality (Lapavitsas, 2012; Lin and Tomaskovic-Devey, 2013; Isakson, 2014; Kornrich and Hicks, 2015; Tori and Onaran, 2015). Peasant dispossession via 'land grabbing', homelessness and destitution after the collapse of the sub-prime mortgage market, targeted social protest, and demands for reregulation of the finance sector, have been other consequences (Glaister and Bruce-Lockhart, 2008; Chomsky, 2012; Konzelmann and Fovargue-Davies, 2013; Margulis *et al.*, 2014).

If globalization, neoliberalism, and financialization are separate concepts and processes, how are they related? Figure 2.1 provides a schematic view of the relationships between the three concepts. Globalization has primacy: it is occurring independently of neoliberalism and financialization but influences, and is influenced by, them. Globalization is a macro-level force that has encouraged the growth of neoliberal ideology, neoliberal (economic rationalist) practices, along with financialization (particularly foreign direct investment – see Kotz, 2002, p. 71). Neoliberalism, as an ideology, justifies and promotes neoliberal practices which, in turn, help to foster the ideology of free markets, less government intervention, and so forth. Neoliberalism stimulates globalization via growing business interconnections, and via the support for global regulatory bodies (such as the World Trade Organization) which promote free market policies on a global level. Neoliberal ideology justifies the settings that foster financialization, encouraging new actors, creating new mechanisms/techniques of profit-making, and producing new outcomes. Neoliberal practices (particularly financial deregulation) have been key to the growing importance of finance in the global arena. In turn, financialization reinforces neoliberal ideologies and practices, as well as promoting globalization through the rapid movement of funds on a global scale, and the speeding up of financial transactions through computerization and the use of sophisticated algorithms (Gertel and Sippel, 2016). These interrelations are captured in the two-directional arrows in Figure 2.1.

There is nothing functionally beneficial or inevitable about the relationships in Figure 2.1, and all three processes can be viewed as potentially containing the seeds of their own destruction – ultimately challenging the current trajectory of neoliberal globalization. For example, while neoliberalism is expected to 'liberate' capital and so improve capital accumulation, it can also undermine that process. Lowering the wages of workers and reducing government expenditure can slow demand in the economy, reducing profits and resulting in economic stagnation (Kotz, 2002). The inequality it generates can also promote class conflict, which discourages investment and can initiate the flight of capital (Kotz, 2002). Reducing public regulation of the finance sector makes capitalism more vulnerable to financial crises and recessions (Kotz, 2002), as was demonstrated in the GFC of 2008 (Konzelmann and Fovargue-Davies, 2013). As Schmidt (2016, p. 3) notes, financialization transforms commodities into assets, and asset prices are quite volatile leading, in some instances, to speculative 'bubbles'. Globalization can 'come unstuck' if governments of powerful nations reject its supposed benefits and reinstate protectionist policies (German, 2000; *Spiegel Online*, 2017).

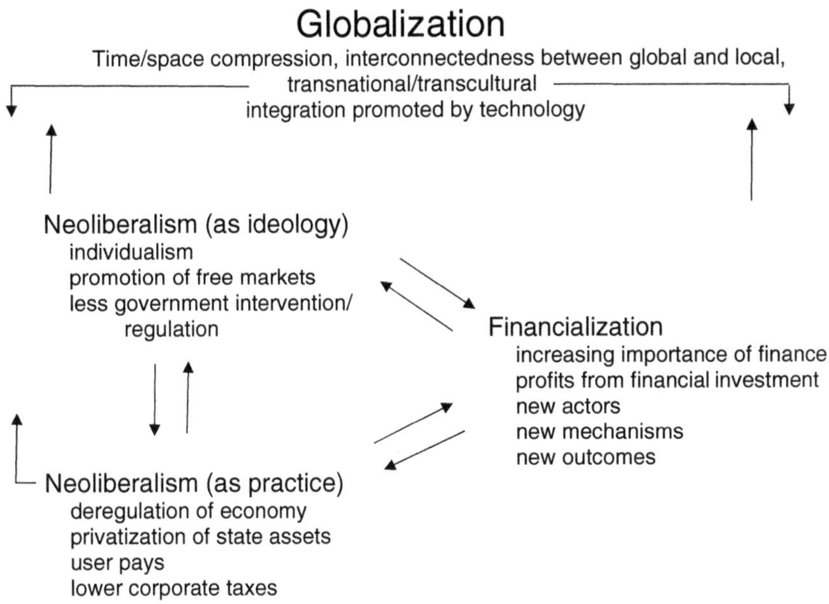

Figure 2.1 Schematic relationship between globalization, neoliberalism and financialization.

Source: authors.

The importance of Figure 2.1 is to demonstrate that financialization has its own set of characteristics and is a process *in its own right* – albeit, one closely associated with globalization and neoliberalism. Much is lost in understanding the dynamics of the world economy if financialization is treated simply as an 'effect' of globalization and neoliberalism. Furthermore, the growing power of financial actors and financial logics to define and redefine the political parameters of globalization and neoliberalism is seriously underplayed if financialization is not treated as being a leading force, alongside those concepts. In the next section, the current-day significance of financialization is demonstrated by grounding the concept in contemporary agri-food research.

Insights from agri-food applications of financialization

Following the world food and economic crises of 2008, scholars have sought to theorise the ways in which food-system restructuring is connected with the world of finance, with financialization becoming an explanatory concept in agri-food studies. The link between food and finance is not new – there has been a long history of research on farm debt, institutional land ownership, agricultural risk and capital return, and commodities markets (see, for example: Buttel and Newby, 1980; Goodman and Redclift, 1989; Harl, 1990; Dudley, 2000; Ouma, 2015). Most of this work did not refer to financialization, but to various 'crises' which generated farm debt and bank foreclosure. Today – with the term financialization being increasingly employed by writers – two key issues have emerged within agri-food research. They are those of:

- attributing causality to recent food and economic crises (to what extent are food price spikes a result of speculation in commodity derivatives markets, for example?) (see Fuchs *et al.*, 2013; Schmidt, 2016; Clapp and Isakson, 2018); and
- understanding the changing distribution of power and resources between finance and other aspects of food supply chains, usually either within a 'land grabs' (in the global South) or 'large scale land acquisitions' (global North), framework.

The above draws attention to the complexity of financialization as both a concept and as a strategic project. The remainder of this section focuses on empirical agri-food research that has advanced understandings of financialization and its relationship with neoliberalism and/or globalization.

Agri-food research on financialization has generally taken two paths. The first – and most common – applies the concept to explain spatially and temporally-specific patterns of capital accumulation, land ownership, and agricultural transformation (and its impacts). Writing on the political economy of the food crisis and subsequent food provisioning, Isakson (2014) has described how traditional finance actors have increasingly moved into agriculture, and *vice versa*, and how agri-food enterprises (particularly retailers) have increased their financial activities, with significant implications for labour and smallholder farmer livelihoods. This argument informs much of the financialization research that seeks to explain the intersection between national contexts (including histories) with current-day differentials in power and control between financial and other supply chain actors (for example, through changing land ownership or control of supply chains). The work of Magnan (2015) and Sommerville and Magnan (2015) has explored how domestic and foreign investment is affecting patterns of farm structure and ownership in Canada and Australia, as corporate and financial investment into agricultural production and farmland expand. While both countries have followed a neoliberal agricultural development paradigm – including support for free trade, the rationalization or dismantling of national wheat boards, and the near-elimination of farm/agricultural subsidies – different regulatory arrangements *vis a vis* investment has meant that the financialization of farmland looks quite different for these nations. In particular, while foreign investment is welcomed in Australia it is currently prohibited in the Canadian prairies. Desmarais *et al.* (2017) have specifically focused on the outcomes of an increasingly neoliberal approach to the regulation of land investment in Saskatchewan, Canada, finding an increase in the consolidation of land, absentee farm owners, and a shift from the values of 'social investment' (investment in labour, ingenuity, social commitments) to the values of 'financial investment' (economic gains through rents and asset appreciation). Together, these have the potential to undermine ecological sustainability and 'land sovereignty' (Desmarais *et al.*, 2017).

Other political economy studies of the 'geographies of financialization' (Williams, 2014) include: the negative effects of short term speculative strategies over longer term production-oriented strategies on the performance, agricultural and otherwise, of Nordic-owned farm companies in Russia and Ukraine (Kuns *et al.*, 2016); the emergence of farmland investment management organizations (FIMOs) and publicly-traded farmland real estate investment trusts (REITS) in the US, as a case study of a 'particular iteration of financial expansion' (Fairbairn, 2015, p. 212) and the 'first step towards the securitisation of farmland' (Fairbairn, 2014, p. 777); financialization as an historical practice of corporate consolidation and 'frontier making' between Egypt and Sudan, with implications for the end of cheap food and oil, greater food insecurity and political instability (Dixon, 2014); development of new 'corporate enclosures', together with dispossession and biodiversity loss, resulting from Norwegian land acquisitions in Uganda (Richards and Lyons, 2016); Leguizamon's (2016) study of the distance and abstraction of food from farms, and its socioecological contradictions, in Argentine soybean production (see also Clapp, 2014); and, the important role of the state in encouraging – through neoliberal policy and regulatory settings – the financialization of food, farmland, and agribusiness industries (Martin and Clapp, 2015; Clapp and Isakson, 2018).

The above studies support the conceptualization of financialization as a force separate from neoliberalism, and illustrate the role that globalization plays in constructing narratives around rising demand for agricultural or farmland investment. This demand can be framed in terms of opportunities for income generation and hedging risk (such as Salerno's [2014] study of corporate land investment through private equity acquisitions in the Philippines) or narratives of food security (including Sippel's [2015] work on Gulf State investments in Australia). According to Desmarais *et al.* (2017, p. 151) the political economy research demonstrates that: '(f)inancialization is both a contribution of and a deepening of the neoliberal restructuring of agriculture ... agri-food financialization is driven in part by the broader transformation of the global economy under neoliberalism, but also by changes specific to the agri-food sector'. A second theme in the empirical research takes a 'social study of finance' approach to set its gaze on the internal workings of financial logics. Ouma (2016, p. 83) uses the concept of 'operations of capital' from finance studies to: 'attun[e] us to the situated modes, processes and practices of financial economization that have reworked organizations, economic relations, labour and nature at specific historical conjunctures'.

Drawing on research into institutional investment in Australia, Ouma's work addresses the critique that financialization research has little to say about finance itself, by focusing on the 'practical accomplishment' of turning agriculture into an alternative asset class through the practical and abstract activities of capital, resources, property, and value. Ducastel and Anseeuw (2017) have examined the strategies and instruments used by financial intermediaries (sub-financial industry actors such as agricultural funds managers and consultants) to construct the 'assetization process' of farmlands in South Africa. Their work contributes insights on how financial value, liquidity, and cash flow are created out of a traditionally illiquid asset (the farm), the importance of standardised metrics for measuring risk and returns, and the de-politicization of farmland as a key task for financial actors. Other examples of empirical research investigating the 'work of financialization' include: Williams' (2014) study of farmer risk management, investment funds and technologies of valuation and commodities re-engineering in Canada; and, Gunnoe's (2016) study of the ways in which the pursuit of shareholder value reshaped the US forest products sector in the interests of the financial community – while undermining long-term stability of the industry and the people who were dependent upon it. This body of research has contributed important insights into how the 'collective logics' or 'moral economy' of neoliberalism, free markets, enhanced productivity, bureaucracy, entrepreneurialism and technology underpin and influence the activities of finance actors (and others) who are actively rethinking and reshaping the relationship between finance and farming (see Williams, 2014).

Conclusion

Social science research into financialization is a productive and growing field of endeavour. New concepts, theories and methods are emerging over time to help researchers grasp and explain changes that are occurring within the economy, society and polity. Globalization and neoliberalism are key concepts, in this regard, with financialization having recently been added to help explain the dominance of finance capital within neoliberal globalization. Not everyone is in agreement with its 'elevation' in status. Financialization has been viewed as lacking in conceptual clarity and power, of being incapable of telling us something new, of lacking spatial and historical specificity, of being 'totalizing' (with the 'financialization of everything' being an outcome of its supposedly unstoppable trajectory) and being unable to engage with the mathematical intricacies at the heart of finance. The pre-existing concepts of globalization and neoliberalism are seen as being perfectly capable of capturing changes wrought by finance capital, without recourse to 'financialization'.

In this chapter each of these criticisms has been examined and challenged, with the conclusion that many of these charges are either overly drawn or without foundation. Furthermore, when the features of financialization are outlined, it is clear that there is something new going on – particularly in relation to actors, financial products, and outcomes. Figure 2.1 demonstrates that – as a concept and process in its own right – it can be placed within a schematic framework, alongside globalization and neoliberalism. It is clear that there are strong linkages between all three concepts/processes. As Fuchs *et al.* (2013, p. 220) write:

> the politics of financialization is not limited to what actors 'do', whether they are state or non-state actors, rather it includes a structural component deriving from material and ideational contexts, allowing or precluding, advantaging or disadvantaging certain political, economic or social options.

Many of the agri-food examples given in this chapter demonstrate how financialization is a key concept in explaining spatially and temporally specific patterns of capital accumulation, land ownership and agricultural transformation (and its impacts) – that is, of the politics and practices of financialization. This process is summed up neatly by Leguizamon (2016, p. 315), who explains how distance (between and within corporate finance and land/agribusiness assets themselves): 'transforms agriculture from a food-producing activity into a commodity and, eventually, to a stock in the financial market... agriculture goes *beyond* the commoditization of food (and land) toward the utter dematerialization, decontextualization and financialization of nature'. Other agri-food research has focused on 'making sense of the motivations and practices of the actors involved' (Magnan, 2015, p. 3), an approach which is particularly useful for investigating the relationship between cultural values and practices connected with neoliberalism, globalization, and/or their contestation and critique.

If there is real problem for financialization then it is the same as that for concepts like globalization and neoliberalism: it is amorphous and messy (Schmidt, 2016). According to Aalbers (2015, p. 215), financialization is:

> A very loosely defined concept that covers many processes, structures, practices, and outcomes at different scales and in different time frames ... this is part of its strength. In some studies financialization is the *explanandum* (the phenomenon to be explained), in others the *explanans* (the thing that explains) ... so is not that different from ... globalization and neoliberal(ism) ... (which try) to make sense of contemporary capitalism and the way it is embedded in societies as well as changing ... those same societies.

The concept of financialization has utility because it provides insights over and above those of globalization and neoliberalism. It identifies new actors, new mechanisms, and new outcomes (including new forms of resistance). When it is utilized in on-ground agri-food research studies it provides considerable explanatory power of the role of finance in altering the trajectories of food and farming systems. It is a concept-in-the-making and deserves to be honed and perfected, rather than subsumed under globalization and neoliberalism or – as some would demand – be eliminated from the lexicon of social science.

Acknowledgements

The research presented in this chapter was part-funded by the Australian Research Council (Project No. DP 160101318). Emeritus Professor Lawrence was also part-funded by the Ministry of Education of the Republic of Korea and the National Research Foundation of Korea (NRF-2016S1A3A2924243) and the Norwegian Research Council FORFOOD Project No 220691). Thanks also to Dr Nicolette Larder for her assistance with data analysis in the systematic literature review referred to in note 1.

Note

1 These data are taken from an on-going systematic literature review of over 220 academic journal articles on financialization: A paper is currently being prepared for publication.

References

Aalbers, M. (2015) 'The potential for financialization', *Dialogues in Human Geography*, vol 5, no 2, pp. 214–219

Arrighi, G. (1994) *The Long Twentieth Century: Money, Power, and the Origins of Our Times*, Verso, London

Bauman, Z. (1999) *Globalization: The Human Consequences*, Polity Press, Cambridge

Bayliss, K., Fine, B. and Robertson, M. (2016) 'Introduction to special issue on the material cultures of financialisation', *New Political Economy*, published online 28 November, DOI: 10.1080/13563467.2017.1259304

Beerkens, E. (2006) 'Globalisation: Definitions and perspectives', available at: www.beerkens.info/files/globalisation.pdf, accessed 15 March 2017

Bonizzi, B. (2014) 'Financialization in developing and emerging countries', *International Journal of Political Economy*, vol 42, no 4, pp. 83–107

Breger Bush, S. (2012) *Derivatives and Development: A Political Economy of Finance, Farming and Poverty*, Palgrave Macmillan, New York

Buttel, F. and Newby, H. (eds) (1980) *The Rural Sociology of the Advanced Societies: Critical Perspectives*, Croom Helm, London

Chomsky, N. (2012) *Occupy*, Penguin, UK

Christophers, B. (2015) 'The limits to financialization', *Dialogues in Human Geography*, vol 5, no 2, pp. 183–200

Clapp, J. (2014) 'Financialization, distance and global food politics', *Journal of Peasant Studies*, vol 41, pp. 797–814

Clapp, J. and Isakson, S. R. (2018) *Speculative Harvests: Financialization, Food and Agriculture*, Fernwood Publishing, Canada

Desmarais, A., Qualman, D., Magnan, A. and Wiebe, N. (2017) 'Investor ownership or social investment? Changing farmland ownership in Saskatchewan, Canada', *Agriculture and Human Values*, vol 34, pp. 149–166

Dixon, M. (2014) 'The land grab, finance capital, and food regime restructuring: The case of Egypt', *Review of African Political Economy*, vol 41, no 140, pp. 232–248

Ducastel, A. and Anseeuw, W. (2017) 'Agriculture as an asset class: Reshaping the South African farming sector', *Agriculture and Human Values*, vol 34, pp. 199–209

Dudley, K. (2000) *Debt and Dispossession: Farm Loss in America's Heartland*, University of Chicago Press, Chicago

Epstein, G. (2005) 'Introduction: Financialization and the world economy', in G. Epstein (ed.) *Financialization of the World Economy*, Edward Elgar, Cheltenham, pp. 3–16

Epstein, G. and Jayadev, A. (2005) 'The rise of rentier incomes in OECD countries: Financialization, central bank policy and labor solidarity', in G. Epstein (ed.) *Financialization of the World Economy*, Edward Elgar, Cheltenham, pp. 46–74

Fairbairn, M. (2013) 'Indirect dispossession: Domestic power imbalances and foreign access to land in Mozambique', *Development and Change*, vol 44, no 2, pp. 335–356

Fairbairn, M. (2014) '"Like gold with yield': Evolving intersections between farmland and finance', *Journal of Peasant Studies*, vol 41, no 5, pp. 777–795

Fairbairn, M. (2015) 'Reinventing the wheel? Or adding new air to old tires? *Dialogues in Human Geography*, vol 5, no 2, pp. 210–213

Foster, J. (2007) 'The financialization of capitalism', *Monthly Review*, vol 58, no 11, pp. 1–12

Fuchs, D., Meyer-Eppler, R. and Hamenstadt, U. (2013) 'Food for thought: The politics of financialization in the agri-food system', *Competition and Change*, vol 17, no 3, pp. 219–233

German, R. (ed.) (2000) *Globalization and its Critics: Perspectives from Political Economy*, Macmillan, Basingstoke

Gertel, J. and Sippel, S. (2016) 'The financialization of agriculture and food', in M. Shucksmith and D. Brown (eds) *Routledge International Handbook of Rural Studies*, Routledge, London, pp. 215–226

Glaister, D. and Bruce-Lockhart, A. (2008) 'Subprime crisis: US foreclosures bring homelessness to the middle class', *Guardian*, available at: www.theguardian.com/world/2008/jun/25/usa.subprimecrisis, accessed 20 March 2017

Goodman, D. and Redclift, M. (eds) (1989) *The International Farm Crisis*, Macmillan, London

Gunnoe, A. (2016) 'The financialization of the US forest products industry: Socio-economic relations, shareholder value, and the restructuring of an industry', *Social Forces*, vol 93, no 3, pp. 1075–1101

Harl, N. (1990) *The Farm Debt Crisis of the 1980s*, Iowa State University Press, Ames

Held, D., McGrew, A., Goldblatt, D. and Perraton, J. (1999) *Global Transformations: Politics, Economics and Culture*, Polity, Cambridge

Hilferding, R. (2006 [1910]) *Finance Capital: A Study in the Latest Phase of Capitalist Development*, Routledge, New York

Holt-Gimenez, E. and Shattuck, A. (2011) 'Food crises, food regimes and food movements: Rumblings of reform or tides of transformation?' *Journal of Peasant Studies*, vol 38, no 1, pp. 109–144

Holton, R. (2012) *Global Finance*, Routledge, London

Isakson, R. (2014) 'Food and finance: The financial transformation of agro-food supply chains', *Journal of Peasant Studies*, vol 41, no 5, pp. 749–775

Konzelmann, S. and Fovargue-Davies, M. (2013) 'Conclusions', in S. Konzelmann and M. Fovargue-Davies (eds) *Banking Systems in the Crisis: The Faces of Liberal Capitalism*, Routledge, London, pp. 267–284

Konzelmann, S., Fovargue-Davies, M. and Wilkinson, F. (2013) 'The return of "financialized" liberal capitalism', in S. Konzelmann and M. Fovargue-Davies (eds) *Banking Systems in the Crisis: The Faces of Liberal Capitalism*, Routledge, London, pp. 32–56

Kornrich, S. and Hicks, A. (2015) 'The rise of finance: Causes and consequences of financialization', *Socio-Economic Review*, vol 13, no 3, pp. 411–415

Kotz, D. (2002) 'Globalization and neoliberalism', *Rethinking Marxism*, vol 14, no 2, pp. 64–79

Krippner, G. (2005) 'The financialization of the American economy', *Socio-Economic Review*, vol 3, pp. 173–208

Kuns, B., Visser, O. and Wastfelt, A. (2016) 'The stock market and the steppe: The challenges faced by stock-market financed, Nordic farming ventures in Russia and Ukraine', *Journal of Peasant Studies*, vol 45, pp. 199–217

Lapavitsas, C. (ed.) (2012) *Financialization in Crisis*, Haymarket Books, Chicago

Lapavitsas, C. (2013) 'Financialization and capitalist accumulation: A structural account of the crisis of 2007–09', in K. Yagi, N. Yokokawa, H. Shinjiro and G. Dymski (eds) *Crises of Global Economy and the Future of Capitalism: Reviving Marxian Crisis Theory*, Routledge, London, pp. 55–74

Latour, B. (1981) 'Insiders and outsiders in the sociology of science: Or, how can we foster agnosticism?' *Knowledge and Society*, vol 3, pp. 199–216

Lawrence, G. (2015) 'Defending financialization', *Dialogues in Human Geography*, vol 5, no 2, pp. 201–205

Leguizamon, A. (2016) 'Disappearing nature? Agribusiness, biotechnology and distance in Argentine soybean production', *Journal of Peasant Studies*, vol 43, no 2, pp. 313–330

Lin, K. and Tomaskovic-Devey, D. (2013) 'Financialization and U.S. income inequality, 1970–2008', *American Journal of Sociology*, vol 118, no 5, pp. 1284–1329

Magnan, A. (2015) 'The financialization of agri-food in Canada and Australia: Corporate farmland and farm ownership in the grains and oilseed sector', *Journal of Rural Studies*, vol 41, 2015, pp. 1–12

Margulis, M., McKeon, N. and Borras, S. (2014) 'Land grabbing and global governance: Critical perspectives', in M. Margulis, N. McKeon and S. Borras (eds) *Land Grabbing and Global Governance*, Routledge, London, pp. 1–23

Martell, L. (2010) *The Sociology of Globalization*, Cambridge, Polity Press

Martin, S. and Clapp, J. (2015) 'Finance for agriculture or agriculture for finance?' *Journal of Agrarian Change*, vol 15, no 4, pp. 549–559

Massey, D. (2005) *For Space*, London, Sage

McKeon, N. (2014) '"One does not sell the land upon which the people walk": Land grabbing, transnational rural social movements, and global governance', in M. Margulis, N. McKeon and S. Borras (eds) (2014) *Land Grabbing and Global Governance*, Routledge, London, pp. 105–122

McMichael, P. (2013) *Food Regimes and Agrarian Questions*, Fernwood Publishing, Halifax

McMichael, P. (2017) *Development and Social Change: A Global Perspective*, Sixth Edition, Sage, California

Monbiot, G. (2016) 'Neoliberalism – the ideology at the root of our problems', *Guardian*, available at: www.theguardian.com/books/2016/apr/15/neoliberalism-ideology-problem-george-monbiot, accessed 20 March 2017

Murphy, L. (2015) 'Financialization (un)limited', *Dialogues in Human Geography*, vol 5, no 2, pp. 206–209

Ouma, S. (2015) 'Getting between M and M′ or: How farmland further debunks financialization', *Dialogues in Human Geography*, vol 5, no 2, pp. 225–228

Ouma, S. (2016) 'From financialization to operations of capital: Historicizing and disentangling the finance-farmland-nexus', *Geoforum*, vol 72, pp. 82–93

Polanyi, K. (1944) *The Great Transformation*, Boston, Beacon Press

Poovey, M. (2015) 'On "the limits to financialization"', *Dialogues in Human Geography*, vol 5, no 2, pp. 220–224

Richards, C. and Lyons, K. (2016) 'The new corporate enclosures: Plantation forestry, carbon markets and the limits of financialised solutions to the climate crisis', *Land Use Policy*, vol 56, pp. 209–216

Russi, L. (2013) *Hungry Capital: The Financialization of Food*, Zero Books, Winchester, UK

Salerno, T. (2014) 'Capitalizing on the financialization of agriculture: Cargill's land investment techniques in the Philippines', *Third World Quarterly*, vol 35, no 9, pp. 1709–1727

Schmidt, T. (2016) *The Political Economy of Food and Finance*, Routledge, London

Sippel, S. (2015) 'Food security or commercial business? Gulf State investments in Australian agriculture', *Journal of Peasant Studies*, vol 42, no 5, pp. 981–1001

Sommerville, M. and Magnan, A. (2015) 'Pinstripes on the prairies': Examining the financialization of farming systems in the Canadian prairie provinces', *Journal of Peasant Studies*, vol 42, no 1, pp. 119–144

Spiegel Online. (2017) 'The new age of protectionism: Trump's attack on Germany and the global economy', available at: www.spiegel.de/international/world/the-new-age-of-protectionism-trump-attacks-german-business-model-a-1132050.html, accessed 22 March 2017

Springer, S., Birch, K. and MacLeary, J. (2016) 'An introduction to neoliberalism', in S. Springer, K. Birch and J. MacLeary (eds) *The Handbook of Neoliberalism*, Routledge, New York, pp. 1–13

Stilwell, F. (2002) *Political Economy: The Contest of Economic Ideas*, Oxford, Melbourne

Sweezy, P. and Magdoff, H. (2009) *Dynamics of U.S. Capitalism*, Monthly Review Press, New York

Tori, D. and Onaran, O. (2015) *The Effects of Financialization on Investment: Evidence from Firm-level Data for the UK*, *Greenwich Papers in Political Economy*, University of Greenwich, Greenwich

van der Zwan, N. (2014) 'Making sense of financialization', *Socio-Economic Review*, vol 12, pp. 99–129

Williams, J. (2014) 'Feeding finance: A critical account of the shifting relationships between finance, food and farming', *Economy and Society*, vol 43, no 3, pp. 401–431

3 Profit, aid and ethics in public financialization

Hilde Bjørkhaug and Jostein Brobakk

Introduction

Through two large public investment funds, the Norwegian state is involved in land-related and other investments outside its own borders. The Norwegian Ministry of Finance funds and controls one of the largest sovereign wealth funds (SWF) in the world, the Norwegian Government Pension Fund – Global (GPFG), while the Ministry of Foreign Affairs funds and controls the Norwegian Investment Fund for Developing Countries (Norfund). In different ways, these investment funds represent a particular case of financialization and are involved in land and agri-food-related investments both directly and indirectly. In addition to bonds and unlisted property, GPFG invests in companies that might be involved in land acquisitions subject to critique because of negative impacts, locally or regionally. Norfund, co-investing directly in start-ups with local partners, has been more directly involved in issues over land and land ownership.

Paradoxes surface when large public investors involved in economic and financial activities seek to maximise profits while claiming that their activities are both environmentally and socially responsible. In this chapter, we analyse some of the most evident paradoxes and ask:

- How do these SWFs represent financialization, and what is the logic and motivation for their investments?
- How are these funds involved in land and agri-food related financialization?
- How do ethical considerations play a role in the management of GPFG and Norfund in land-related investments?

The chapter is structured as follows. First, we present an outline of the theoretical foundations of the chapter, focusing on financialization theory, theories on the role of capital, and research on the role of ethical considerations in large investment portfolios. Second, we present information about GPFG and Norfund, followed by an analysis of the case of financialization and land investments, We end the chapter with a summary and concluding remarks. Data are drawn from interviews with representatives from the two SWFs and Norwegian NGOs, policy documents, media articles and secondary literature.

Financialization and the role of capital and ethics in the agri-food chain

In an early definition of financialization, Palley (2008) points to the (1) elevated significance of the financial sector relative to the 'real' economic sector, (2) the transfer of income from this 'real' sector to the financial sector, and (3) the contribution to increased income inequality and wage stagnation in the sector in question. 'Financialization' is employed to describe a new period in capitalist development, marked by the shift from the dominance of productive capital to the dominance of finance capital (see Klimek and Bjørkhaug, 2017). While commodity production and trade has been the previous dominant economic logic, financialization constitutes new patterns of capital accumulation, with profits created through financial means (Krippner, 2005), such as investing in financial assets, derivatives trading, and private equity buyouts. There is seen to be a different logic and motivation in 'financialized' capitalism, where buying into products, industries or businesses is motivated not by the value of what is being produced, but rather by speculation in the volatility in financial value. There is also a changing horizon from long-term to short-term investments. And, there is gradual income-transfer from the 'real', commodity-production-based sector, to the financial-assets-based sector (Perez, 2002; Epstein, 2005; Palley 2008).

In the globalized economy, the financial sector, which has grown rapidly in both employment and revenue, is transferring its logic of financialization into the 'real' economy – income generation in 'traditional' companies and corporations is becoming portfolio-based rather than being a result of earnings from basic production of goods and services (Dore, 2008). This structural change affects the behaviour of companies in both sectors and how and where they place their investments. Focusing on changing firm behaviour, Krippner (2005) refer to this as a shift from an activity-centred focus to an accumulation-centred focus. Hence, financialization operates through several channels: through the structure and operation of financial markets; by influencing the behaviour of nonfinancial corporations; and, by influencing economic policy (Palley, 2008).

Financialization of the agri-food chain

The 2007–2008 food crisis was associated with reduced access to basic food commodities and rising food prices and generated considerable interest from agri-food scholars (see Lawrence *et al.*, 2010; Brobakk and Almås, 2011; Clapp, 2014). The crisis itself generated on-ground actions, from nation-states looking for agricultural land for food security purposes, to financial institutions seeking to profit from commodity future markets and from land price inflation (McMichael, 2012). Asset managers began investing increasingly in all parts of the food chain, and in areas where they were previously not involved, including farmland, agricultural inputs, storage and logistics, inspection and certification, food production and processing, commodity trading, and retail services (Burch and Lawrence, 2013; Lawrence and Smith, Chapter 2 in this volume). Their activities ultimately had a strong impact upon the full value chain – from the farm to the food served to the consumer (Burton and Bjørkhaug, 2015).

Literature on agri-food financialization has focused upon the phenomenon of investments/speculation in land, along with the 'land-grab' in the global South (McMichael, 2012; White *et al.* 2012). Further, others have focused on large-scale land accusations, food-chain speculation and private equity buyouts in the global North (Klimek and Bjørkhaug, 2017; Muirhead, forthcoming; Lawrence and Smith, Chapter 2 in this volume). Sippel *et al.* (Chapter 4, this volume) provide an analysis of a new and powerful type of financial actor on the international scene – the national Sovereign Wealth Fund, operating at the intersection between the financial sector and the state. Such 'strategic financialization' is characterized by public investments that are made in the 'national interest'. Ouma (2015) argues that there are limits to the effects of financialization within agriculture due to the sector's specific features such as weather dependency, socioecological embeddedness, moral entanglements, and political and strategic factors. He further argues that farmland is more difficult to trade financially than other assets and therefore cannot easily be turned into 'just another asset class', thus providing farmland with a level of protection from the most aggressive forms of speculation. Following the same line of argument, Fairbairn (2014) makes a distinction between the 'productive' and 'financial' logics in farm-sector investments, arguing that farmland is a 'quasi-financial' asset producing both short to medium term speculative gains as well as a long-term production-based value creation.

Land acquisition by financialized capital

Financial managers have a strong interest in entering new areas to gain financial profit. De Soto's (2001) work on capital in *The Mystery of Capital*, highlights how formalized property rights constitute the primary cause of economic growth – a key element in the rise of capitalism. De Soto's understanding of property and the role of capital is used in many developing countries as a roadmap in establishing institutions and transforming their societies in order to create growth and to seek prosperity. Hall (2013) claims De Soto's approach has been adopted by both powerful nations (the US) as well as international organizations (the UN) in their desire to catalyse economic development. De Soto (2001) advocates for the formalization of property rights and argues that developing countries would benefit from a Western style property-system and titling of land. However, the formalization of property rights can confront and undermine old agreements on land, such as a common property and 'traditional' or informal property systems that are more frequently found in indigenous management systems, in nomadic cultures, and in systems with common grazing rights. It can also create increased 'grabbing', through a mechanism of forcing all land to be part of a formalized property system (Hall, 2013). In such a development, access to land might be restricted through the amount of capital one is able to raise, effectively forcing traditional users out of business.

Moral and ethical concerns in agri-food financialization

Organizations such as the Oakland Institute (see Daniel and Mittal, 2009) and websites such as farmlandgrab.org monitor frequently post information on – foreign farmland investment and leasing activities, along with the food security and livelihood impacts experienced by those affected. There is evidence on the dispossession of local farmers and villagers with informal property rights, causing increased economic and social distress. In such instances, often when formalized investments come in conflict with non-formalized local traditions, moral and ethical challenges are likely to occur.

In addition to the concerns of the Oakland Institute and a host of similar organizations, farmer-based organization, such as La Via Campesina, have brought attention to new moral and ethical issues connected to the shortcomings of 'financializing' agricultural land and food (see McMichael 2009, forthcoming). These issues relate to food sovereignty and the right to sufficient and nutritious food. Ouma (Chapter 5, this volume) raises some crucial issues that need to be taken into account in the debate about the morals and ethics of financialized farming. These are connected to the dominance of the discourse of financial elites, the 'valuation of everything' in the pursuit of future profits, and the lack of concern for future sustainability. Enforcing a capitalistic property-rights system brings with it quite significant moral and ethical issues and concerns.

With financialization and the entry of large investors, such as the SWFs, there are many potential negative consequences for those affected by the investments. Norway's SWFs – GPFG and Norfund – have been subject to critical debate linked to their investment activities. In the following section, we describe these funds and consider the role moral and ethical issues play in their management.

The Norwegian investment funds

GPFG and Norfund are funded and controlled by the Norwegian state, with a strategy dominated by profit generation from financial investments. They are, however, quite different when it comes to size, investment strategy, risk taking, and governance.

The Norwegian Government Pension Fund – Global (GPFG)

Funded by a surplus from the oil and gas sector, GPFG was established in 1990 to give Norway room for fiscal policy manoeuvring independent of fluctuations in the oil price. One of its main goals is to '... safeguard and build financial wealth for future generations' (NBIM, 2018). By the end of 2017, GPFG was worth around US$1,100 billion (NOK8,500 billion), with a portfolio consisting of equity (66 per cent), fixed-income/bonds (32 per cent) and unlisted real estate (2 per cent). The fund invests in nearly 9,000 companies in 77 countries which is 1.3 per cent of listed companies worldwide and 2.3 per cent in Europe (with a maximum share-holding of 10 per cent in any company). The average annual return between 1998 and 2017 was 6 per cent, or 4.1 per cent when adjusted for inflation and management costs (NBIM, 2018).

The GPFG is controlled by the Ministry of Finance but day-to-day management is the responsibility of Norges Bank Investment Management (NBIM). Parts of the trading is outsourced to international investment banks such as Barclays in London. The Norwegian parliament has the ultimate say when it comes to the formal framework specified in the Government Pension Fund Act (GPFA) and approves major changes in the fund's investment strategy. The strategy describes how GPFG should undertake a long-term horizon and take moderate risk to secure the highest possible return on investment (ROI). There is broad political agreement that the fund is a financial actor and not an instrument for Norwegian foreign, or climate, policy (NBIM, 2017). Following wider Norwegian political and societal expectations, all investments are transparent and listed on the fund's webpage.

As GPFG grew in both size and global impact towards the end of the 1990s, a debate on ethics and social responsibility surfaced among Norwegian politicians and NGOs. The Bondevik government, established after the 1997 parliamentary election, set up a commission on values and ethics in order to contribute to a value-based mobilization of Norwegian society. The report from the commission underlined the importance of preventing moral indifference and promoted humanistic and democratic values and increased care for the environment (Government, 2001). One outcome of this political initiative was a parliamentary discussion on GPFG's societal responsibility, including whether parts of the fund ought to be invested on the African continent despite the likelihood of increased risk and reduced profit for the fund. Focusing on this and other value-based issues, ethical guidelines and a Council on Ethics for GPFG were introduced in 2004. The mandate of the Council on Ethics is to provide evaluation on whether or not investments in specified companies comply with the established ethical guidelines. The guidelines put an emphasis on social, economic and environmental sustainability, describing product-based and conduct-based reasons for exclusion or observation of companies in which GPFG has invested (NBIM, 2017). If a decision to invest is found to be in breach of the GPFG ethical guidelines, the Council of Ethics writes a report to NBIM, suggesting exclusion or observation of the company. The assessments are based on monitoring activity by the Council on Ethics as well as external consultants in several parts of the world, and acts as a screening process: 'We do assessments ... then decide on which companies we actually should [do] in-depth research on. If we decide to proceed [and] we need more information, then we hire consultants' (Interviewee, Council on Ethics). Following the establishment of the Council on Ethics, the number of cases under evaluation has increased rapidly. However, these reflect not only a strengthened ethical focus but also the growth in the number of GPFG investments.

The first exclusions were those of tobacco (e.g. Philip Morris), production of nuclear weapons (e.g. Lockheed Martin) and cluster munitions (e.g. Textron). Since 2016, involvement in coal production was placed on the list for product-based exclusion, due to coal's heavy GHG footprint (NBIM, 2016). Since 2013, GPFG is excluding companies involved in palm oil if they cannot demonstrate sustainability in production. This decision is based on reports showing that palm oil production is a major cause of rainforest destruction, particularly in Malaysia and Indonesia. More recently, GPFG has also received criticism for investing in Asian banks that fund palm oil plantations (Aidenvironment, 2017). Other investments in agricultural-based businesses or investments indirectly involving GPFG in land-related investments, do not qualify as a reason for exclusion or observation, unless they include severe local environmental damage or violation of human rights (Interviewee, GPFG Council on Ethics).

Reasons for conduct-based exclusion of companies include human rights violations (e.g. Wal-Mart), local environmental damage (Rio Tinto), corruption (ZTE Corp.), breaching individual rights in war and conflicts (Africa Israel Investments) or other serious violations of fundamental ethical norms (for example, the Potash Corp of Saskatchewan's purchase of phosphate from Western Sahara). If a company is excluded or put under observation, the report backing the decision is made public. The ethical decisions made by the GPFG managers do make an impact, for the involved company and for other investors: '... what we know is that the recommendations are used by other investors who enter into dialogue with the companies' (Interviewee, Council on Ethics).

Norwegian Investment Fund for Developing Countries (Norfund)

Norfund is an equity fund established in 1997, owned by the Ministry of Foreign Affairs and funded from the development budget. Unlike GPFG, Norfund is a political instrument for promoting economic growth and business development in the developing world, as an alternative to the more traditional development aid.

Following the trends in aid and development policies in Western Europe setting up Development Finance Institutions (DFI), a 1993 government appointed North–South aid commission published a report calling for a development fund that could invest in countries with low and medium welfare (NOU, 1995, p. 5). The aim of the fund is to support production, employment and business activities in the global South, through trade policy or in collaboration with the private sector. Efforts to increase environmental sustainability have high priority. The commission also recommended new arrangements for joint ventures with Norwegian investors in business development projects (NOU,1995, p. 5). While the Norwegian Agency for Development Coordination (Norad) had been able to fund local projects through loans for some years, it did not have a mandate to invest directly in businesses or companies. The proposal of the establishment of Norfund was presented to the parliament in 1996 (Ot. prp. nr 13 [1996–1997]), and the Norfund Act was adopted in 1997 (Norfund, 2017). Norfund's current priority areas are clean energy, financial institutions, food and agribusiness, and SME-funds in Sub-Sahara Africa, Central America and South-East Asia (Norfund, 2018a, 2018b).

By the end of 2016, Norfund had a portfolio of US$2 billion (NOK17 billion), and included 770 companies. The average net return since its establishment has been 4.9 per cent (Norfund, 2018a). Norfund aims at exiting their projects after a period of active ownership, typically between five and seven years. Investment profits are returned to Norfund.

Equity is a principal instrument and, according to an interview with a Norfund representative, food and agribusiness are prioritized sectors:

> We also do loans but we do equity because we think that is what is most needed. It's easier to get a loan than to get equity through these investments. We do start-ups because we think that is also something that is very hard to raise capital for. It's easier to raise capital if you already have a cash-positive business that has shown itself to be viable. But if it's just bushland and nothing, then the risk of course is much higher.
>
> (Interviewee, Norfund)

Overall, Norfund expects high returns from their investments despite taking considerably more risk than GPFG: 'So typically, we should have a return of at least between 5 and 10 per cent on our investments as a portfolio' (Interviewee, Norfund). Norfund (2018b) underlines its role as a *responsible* investor and refers to the environmental and social standards of the World Bank's International Finance Corporation (IFC) as a rule for all co-investors in a project. The IFC standards comply with International Labour Organization (ILO) standards and restrictions on forced and child labour, regulation on labour rights, social and environmental standards, and indigenous peoples' rights. On its website, Norfund states it has zero tolerance for corruption, and has established a Business Integrity Policy (BIP) as a means to avoid financial irregularities in its operations. Any irregularities are reported to the Ministry of Foreign Affairs (Norfund, 2018b).

Analysis

GPFG and Norfund as financialization agents

In this chapter we describe financialization as a new pattern of capital accumulation through financial means (Krippner, 2005; Perez, 2002; Epstein, 2005; Palley, 2008). Despite having different management systems and investment strategies, where Norfund is also doubling as an investment fund and an instrument for Norwegian development policymaking, both GPFG and Norfund can be characterized as being part of a 'financialized' capitalistic system.

In the financialization literature, there has been a debate as to whether financialization entails something new, or is simply a reinvention of the wheel (see discussion in Lawrence, 2016). Looking at the investment strategies of GPFG and Norfund, we argue that both funds fit with central elements in the financialization theory literature as presented early in this chapter, as well as represent something new in a Norwegian and global context. GPFG and Norfund are examples of the rise of new financial actors aiming for an investment-based return rather than seeking profit from trade or production. While being strategic SWFs operating in the intersection between the financial sector and the state, they are designed and managed as private investment funds of the sort that, in recent decades, have come to dominate the global economy (Sawyer, 2014). When established, they became the first of their kind within the Norwegian state apparatus.

While Norwegian oil- and gas-based revenues used to fund the social security system directly, the GPFG was set up as a way to accumulate wealth by seeking profit from financial investment abroad. In its first annual report, the head manager Svein Gjedrem repeated GPFGs mandate, to 'achieve the highest possible return and ensure that satisfactory risk management systems and control routines are in place' (NBIM, 1998). Reflecting GPFG's intended role as a non-political actor, the investors in NBIM enjoy a high degree of freedom without day-to-day interference from the

Ministry of Finance. Politicians do set general limits, when it comes to risk taking and ethical considerations but, within those limits, GPFG operates as other investment funds – both state controlled and private – in the global markets. According to the international rating company Morningstar, GPFG gained a five-star rating from its state-of-the-art financial management (Henriksen, 2014).

It is interesting to observe that both political documents and GPFG reports use the word 'investment' and not 'speculation'. Speculation might come across as something cynical or unethical. However, despite providing long-term investments, the GPFG quickly moves in and out of companies or sectors depending on the opportunity to gain profit from share price volatility. For instance, after the 2008 financial crisis when global stock values were at a low, GPFG started buying. When the tide turned, GPFG began to sell, gaining profit from its investments (Henriksen, 2014). Some might label such volatility-driven behaviour speculation.

Despite representing a different investment approach with its emphasis on ownership in local businesses, Norfund has a similar logic as GPFG built in to its investment strategy. Profit from investments is a goal. Because Norfund takes considerably higher risk than GPFG, it needs to have a more aggressive return policy in order to compensate for losses. Norfund's behaviour means that it withdraws from local investments after a while, typically after five to seven years, redirecting profit back to the capital base, ready for new investments. Investment surplus is not shared with the local business: 'They don't [circulate back through those communities]. It goes back to the owners as a profit' (Interviewee, Norfund).

For countries struggling to get access to investment capital, the growth in development activity by actors such as Norfund is considered positive, but when the profit is redirected to the investment funds, little is left to the local communities for development purposes. Norfund and Norwegian authorities argue that the basis for local development should not be based on profit from Norfund's investment, but from sustainable businesses and job creation from Norfund's financial involvement.

Norfund is also a Norwegian development policy instrument. When established, Norfund represented a different way of thinking about development in the global South, supplementing the more traditional development aid strategy that funded recipient states directly. With the establishment of Norfund, more capital was directed through projects involving the private sector, including Norwegian companies interested in setting up businesses in the global South. As an investment fund, Norfund can be labelled an agent of financialization. As a policy instrument, it represents a more value-driven foreign policy that emerged throughout the 1990s, promoting the spread of democratic capitalism and formalization of the informal economy (De Soto, 2001). Although its role as an investment fund is stated as being important, Norfund is a hybrid. Basically, it is a governmental development agency promoting democratic capitalism as the way to bring economic growth and prosperity to the global South – as well as being an investment fund. For some, the current Norwegian foreign and development policy is described as being 'neutral' or 'value-free', based on the UN declaration of universal human rights (Tvedt, 2011, 2016). However, promoting development in the global South through a Western-style capitalist or financialization logic might come across as anything but value-free.

One criticism of Norfund's activity relates to the arrangement of funding companies via smaller investment funds located in tax havens. A 2009 investigation revealed that as much as 81 per cent of Norfund's investment capital was channelled via tax havens before being invested in various projects (Vermes, 2009). A governmental committee recommended that Norfund change its practice, but so far it remains unchanged. According to Kjell Roland, chief executive officer of Norfund, to invest in the global South, one must use existing (undeveloped) infrastructure (Vårt Land, 2016). This is a negative for poor countries as they are deprived of a potential income-source from taxation, while at the same time missing the opportunity to establish their own finance sectors. The former Minister for Development in Norway, Erik Solheim, stated that 'we cannot allow Norfund to continue investing in tax havens while at the same time trying to fight them' (Government, 2009).

The actual neutrality of GPFG's investments can also be questioned. Officially being set up as a non-political investment fund, it can be interpreted as representing Norwegian foreign policy ambitions. For instance, the former Minister of Finance, Kristin Halvorsen, stated in 2007 that Norway should have an ambition to help 'create a better world'. Ensuring that GPFG's investments reflected an ideal of fairness and sustainability, and paying attention to human rights and the environment, was one way of obtaining this goal.

Financialization of land and agriculture

GPFG and Norfund are exposed to land-investments – and the food and agricultural sectors – in quite different ways. GPFG never engages directly in land or agriculture as such, but is indirectly involved through some of its investments. For instance, GPFG's second largest investment – worth US$6.2 billion (NOK51 billion) – is Nestlé, which is involved throughout the whole food value chain. Other large agribusiness investments are Bayer, Monsanto and Syngenta, multi-nationals controlling large parts of the value chain. A much-debated GPFG investment is linked to companies involved in the palm oil industry. In 2012, several Norwegian NGOs started a campaign to reduce the use of palm oil produced in a non-sustainable manner, and raised public awareness about loss of rainforest to palm oil plantations. For instance, the Norwegian Rainforest Foundation pushed for a ban on palm oil from companies that were not members of the

Roundtable on Sustainable Palm Oil. GPFG's involvement in the palm oil industry created political pressure to put the involved companies on the exclusion list. In 2013, GPFG withdrew from 23 companies, including Wilmar, which was named by *Newsweek* 'the worst company in the world' (GRAIN, 2016), and GPFG's palm oil investments were down 40 per cent (Rainforest Foundation, 2013). Nils Ranum (Food Navigator, 2013) of the Norwegian Rainforest Foundation stated that: 'this is a major step forward and a victory for the remaining rainforests. One of the world's biggest investors sends a clear signal to the oil palm industry that its destructive practices are unacceptable'. A main pillar in the Norwegian climate policy is preserving rainforest through the UN Programme on Reducing Emissions from Deforestation and Forest Degradation (REDD+) (Kasa, 2013). Accordingly, it is counter-productive to have GPFG heavily involved in the palm oil industry. Therefore, non-sustainable palm oil production is now a reason for exclusion of companies from the GPFG portfolio. However, GPFG continues to own shares in Sime Darby, a Malaysian based conglomerate involved in palm oil plantations in Asia. GPFG's exclusion of palm oil companies shows the difficulty in separating the fund's investments as purely based on financial assessments from broader domestic political processes. This example also shows how Norwegian interests as expressed through the GPFG investments can be in direct conflict with official foreign policy interests.

Malaysian-based Sime Darby also charters 850 square miles of palm oil land in Liberia (Hinshaw, 2011), making GPFG indirectly involved in land deals. Another example of GPFG's indirect involvement in land deals is its investments in South-East Asian banks that lend money to companies buying land to produce palm oil or similar plantations. International NGOs report that these activities often involve grabbing land from indigenous people or land without formalized property rights (Rainforest Foundation, 2017). If land-related activity results in negative impacts for local people and/or the environment, it does not automatically qualify for exclusion or observation by the GPFG, according to the Council on Ethics. A manager in the Council on Ethics explains:

> We do not look at the land acquisition as such and the impact that has for security ... [This is] not on our radar at all. Because that is more a political approach and we do not – I mean political approaches are for the Ministry of Foreign Affairs. The most of those which I am familiar with is that the company has acquired a concession from the Government and then has a legal right to exploit the land.
>
> (Interviewee, Council on Ethics)

From this point of view, land grabs and their consequences should be dealt with by Norwegian politicians: they are not a local matter with which the GPFG should interfere. Only if there was a possible link between the violation of people's rights – such as forced displacement – and the actual company GPFG has invested in would the Council on Ethics would intervene.

Norfund is heavily involved in the agricultural sector in sub-Sahara Africa. In particular, its investments in plantations have spurred public debate and have resulted in criticisms from aid and environmental NGOs. Norfund has been accused of contributing to land grabs, overthrowing local informal land use practices, and forcefully displacing people. Norfund says it is sensitive to these accusations (Norfund, 2011) but considers them a part of the risk it takes as an investor in African agribusiness. A Norfund representative explains how risk is evaluated: 'We consider risk along four parameters ... financial, implementation risk and, social and environmental. (R)eputational risk is also very important because of the land issue' (Interviewee, Norfund).

NGOs, such as Norwegian People's Aid (NPA), are particularly concerned about the rights of smallholders and changes in informal land-use practices, expressing concern over the ideology guiding Norfund's activities as part of the Norwegian development policy: 'So, Hernan De Soto, he has this theory, that if you formalize the informal economy, then actually you could make huge, huge, steps for the development of a country' (Interviewee, NPA manager). Part of this ideological driven development approach, adopted by successive Norwegian development ministers, as well as Norad, implies that funds invested locally should be more sensitive to so-called under-invested sectors and local demand. Norfund engages in sectors or projects they believe are lacking in investment, such as the African food and agricultural sector, and investment requests from local business representatives play a large role (Interviewee, Norfund). Norfund does not view their involvement in companies buying up or leasing land as speculation, and the word 'development' is often used instead of 'investment'. These are important nuances, as they might indicate the fund's approach to ethically challenging cases. A Norfund manager explains its approach to the question on (land) speculation and investor interests:

> from my point of view [it is not] speculation. It's to get more of the agricultural potential in Africa to open up and running, in a way. So it's not to speculate in land. That is not the purpose at all but, of course, it gives some kind of extra comfort to the investors that they expect some kind of land [value] increase.
>
> (Interviewee, Norfund)

Two Norfund partners accused of grabbing land are the British owned New Forests Company and the Norwegian Green Resources. The establishment of the New Forests Company's (NFC) tree plantation in Uganda – timber products supported by the local government, and energy from biomass – resulted in a forceful displacement of 22,500 people (Oxfam,

2011). Oxfam and the Uganda Land Coalition filed a complaint to the World Bank complaint mechanism, CAO. CAO found that the grievances were justified, and initiated mediation between NFC and the local community. In 2010, Norfund invested around US$6.5 million (NOK50 million) in NFC via the Mauritius-based agribusiness fund Agri-Vie (Norfund, 2018a) but did not take action when the Uganda Land Coalition started protesting (Oxfam, 2011). Despite its ethical framework, Norfund, to date, has not pulled out of Agri-Vie and NFC. Norfund has no independent complaint mechanism similar to that of the CAO.

Green Resources is the largest afforestation company in Africa, with over 40,000 ha of standing forest in Mozambique, Tanzania and Uganda, operating plantations as well as sawmills (Green Resources, 2018). Established in 1995 as a Norwegian company with 80 private shareholders, it have received loans from Norfund since 2009 (Norfund, 2018a), in addition to support from the government. Despite subscribing to responsible forest management certification under the Forest Stewardship Council, Green Resources stands accused of contributing to grabbing land, violating local rights to land access, and displacing people through its many land-lease deals (Lexterra, 2016). In 2016, after years of critique and complaints from local organizations, Norfund started a monitoring process to ensure that Green Resources operates within international standards for sustainability (Dagsavisen, 2016).

Ethical considerations

Both GPFG and Norfund operate under different investment standards and ethical frameworks. The main institution in GPFG is the Council on Ethics, while Norfund subscribes to the environmental and social standards of the World Bank's IFC and the ILO convention on labour rights (Norfund, 2018c). Whether these guidelines make a difference, and whether the fund's investments are deemed ethically sound, are constantly debated among Norwegian politicians and NGOs.

Due to its size and the high number of investment objects, it is difficult for the Council on Ethics to detect potential and actual unethical conduct by the companies in which they invest. When GPFG was established, there was a debate over the introduction of positive screening of companies, only allowing GPFG to invest in companies assessed as ethical. This approach was abandoned as it would have created an enormous amount of work given the size of GPFG's investment portfolio. Also, it would have reduced the number of companies cleared for investments, reducing GPFG's possibility to spread its investments and, thereby, its risk (Ramberg, 2011). The current arrangements mean that the Council on Ethics, guided by the mandate given by the Ministry of Finance, can only monitor fractions of the GPFG portfolio, often by conducting media searches throughout the world with assistance from consultants filing monthly reports on between 30 and 40 companies. A member of the Council on Ethics explains that: 'we are really in the lower end. We do only focus on what is severe, what is gross, what is really bad' (Interviewee, Council on Ethics). The Council on Ethics can only give advice on the ethical performance of the companies. The Ministry of Finance and the Norges Bank take the advice provided into account before deciding on exclusion or observation of the companies in question. These decisions are transparent, but the assessment result of the companies that stay is not. International observers have stated that GPFG and its transparency policy makes the fund an international benchmark when it comes to sustainable investments (NRK, 2017). The implementation of ethical guidelines does make a difference, both because GPFG's decisions are often followed by other large investment funds, and because the ethical considerations have come at a cost. The 2016 GPFG report shows that the investment return the last decade could have been 1.1 per cent higher if there were no exclusions on the basis of conduct or product (Hustad, 2018). Particularly costly for GPFG is the exclusion of tobacco companies (Kasikci and Rajasingam, 2017). When GPFG excluded tobacco companies, such as Philip Morris and British American Tobacco, it was taken as an expression of Norwegian foreign policy (Ramberg, 2011). The decision certainly reflects the view of Norwegian politicians and health authorities on tobacco, but had more to do with domestic politics than foreign policy. A GPFG position in conflict with Norwegian policy is the exclusion of Lockheed Martin due to its involvement in producing cluster munitions and nuclear weapons (Council on Ethics, 2013). In 2012, the Norwegian government decided to purchase up to 52 F-35 jetfighters produced by Lockheed Martin, meaning that the Norwegian government continues to trade with a company GPFG is not investing in for ethical reasons.

While Norfund wants to be evaluated on the various standards they subscribe to, including ethical guidelines, critics refer to the actual performance of the companies they invest in. These different positions and the lack of coherence between standards and performance, leaves Norfund open to criticism. In 2015, an evaluation report from the Norwegian Agency for Development Cooperation (Norad) found that Norfund's reporting on plans for environmental and social governance was limited and that information was disclosed mainly on an ad hoc basis (Norad, 2015). The report also stated that the degree of reporting was in line with other similar institutions, and if problems appeared, Norfund successfully engaged in a dialogue with local stakeholders. Both the ethical framework as well as compliance with the framework was judged to be satisfactory.

These findings are in conflict with the criticism Norfund has experienced throughout its 20 years of existence. Norfund has been accused of choosing business over ethics, deviating from its own ethical standards, and neglecting warnings about environmental damage and human rights violations conducted by companies in which they have invested

(Holmqvist, 2018). Since mid-2017, Kjell Roland, Norfund's chief executive officer, has been engaged in a public debate over NGO based criticism of the fund's lack of compliance with both international investment standards as well as Norfund's own ethical standards. Roland and Norfund had to respond to these accusations to the Norwegian parliament in an open hearing. The Norfund response was to point to the demanding international standards, such as the OECD Standards for Responsible Business Conduct and Policy Framework for Investments and the IFC Performance Standards – to which Norfund subscribes (Roland, 2018). In addition, Roland argued that the UN Declaration of Human Rights is directed towards states, therefore if some of Norfund's partners are violating international standards, it is Norway and the recipient country that need to address the issue. Norfund says it welcomes fair criticism and values monitoring work done by journalists and NGOs, but that the criticism has come to resemble a smear campaign, and is counter-productive and harmful for those in need of investment capital (Roland, 2018). Norfund's position is that creating sustainable businesses and jobs is the best way out of poverty because traditional development aid has largely failed to make a difference. With continuing criticism, Norfund fears that investors are being scared away. This debate clearly shows the difficulty in implementing an ethical standard at the level expected by NGOs and the public, while at the same time conducting investments that are supposed to generate a high level of profit.

Discussion and conclusion: profit, aid and ethics in public financialization

This chapter opened with the claim that paradoxes surface when large public investors, involved in financialized activities, aim to pursue environmental and social responsibility at the same time as profit maximization. Two large, but very different, Norwegian public investment funds, the GPFG and Norfund, were investigated for the purposes of understanding how they 'fit' within financialization, the logic and motivation behind their investments, how they are involved with land- or agri-food financialization and, finally, how (if at all) ethical considerations play a role in the management of those funds.

GPFG, on the one hand, has defined its main goal in genuine sustainability terms, 'safeguarding and building financial wealth for future generations' (NBIM, 2018). However, the political rationale is purely financial and, as shown in this chapter, is not supposed to take into consideration issues relating to climate, foreign affairs or aid. Norfund, on the other hand, does embrace policy goals, and is a political instrument for promoting economic growth and business development in the developing world (Norfund, 2018c) – with development being its main purpose. We have argued that both of these funds operates in a financialized manner.

The two funds differ in their investment strategy and logic, GPFD aiming for maximum profit in long-term investments with low risk, often in large multinational companies, and Norfund investing in start-ups with high risk, equity being its principal instrument. Both funds are involved in land and agri-food financialization, with Norfund moving into this area with increased speed, as a response to the 2007–2008 food crisis. GDPF's involvement in land is indirect – when companies in which the fund invests are active in land-deals. In the agri-food chain, GPFD invests directly in companies, and often also indirectly in land, through investment in companies and other investment funds.

Financialization of land and the agri-food value chain

Land as an asset for investment or speculation has gained strong attention over the past decade, and there has been increasing activity in land and agri-food speculation in the global North as well as the global South. Supported by the approach of formalization of property rights for development within a capitalistic system (as promoted, for example, by De Soto, 2001), many nation-states in the global South, have opened up their territory for foreign investment in land that was formerly governed by rules and rights based on local customs and traditions.

Current foreign and development policy provides a blending of idealism and self-interest. Politicians and NGOs with 'good intentions' have implemented property right prerequisites in their aid and development policies. It is assumed that democratic capitalism and the 'formalization' of what were once informal economies is the best outcome for those countries, and for the world.

Yet, it should be asked: who is advantaged by 'democratic capitalism'? Our analysis has shown that the two Norwegian SWFs have received profits from their financialized investments, while providing questionable benefits to those countries and communities receiving investment. Critical voices have accused both funds for harmful effects. Given the structure of the investments, Norfund is more exposed to direct criticism, while GPFD operates through others, and its Council on Ethics does not interfere in the activities of other companies. Both funds continue to invest in conflicting and contradictory areas, such as companies developing environmentally questionable palm oil plantations, and capital transactions that lead to the accumulation of capital in tax havens.

Responsibility instruments

We have seen that different strategies are employed by the two funds regarding responsible investment instruments – with implications for potential sanctioning. The nature of the GPFG investment logic, being a minority shareholder in corporations, has been used as an argument by the fund not to adhere to any official CSR standard. With this, GPFG disclaims responsibility as an actor in the 'real' economy. Yet, GPFG has proved to respond actively – in a more democratic manner – in areas that are flagged on the political agenda (for example, its withdrawal from 23 companies involved in unsustainable palm oil production in 2013, without a specific mandate from the Council on Ethics). Norfund, also being a public fund – but operating as an independent actor – applies a business ethics logic through employing, according to their statutes, CRS and OECD principles. Our analysis indicate that the democratic model is more transparent and open for negotiation than the business ethics model.

Ethics for the present and the future

Any activity impacting upon food and nutritional security should be scrutinized along moral, ethical and sustainability lines. The major interest in the financialization of land and the agri-food value chain was arguably not the financial elite's sudden desire to enhance global food and nutritional security. For some nation-states, food security for their citizens has been a key driver, but actors such as the GPFG and Norfund have not revealed such intentions. The financialization of land and of agri-food industries is consistent with a 'valuation of everything' discourse. Nevertheless, one must ask: Who benefits from this 'valuation'? And the answer appears to be: the investing funds, and rich countries such as Norway. In this chapter, we have provided some evidence from two major funds with and without a public ethics council model. Despite its imperfections, the democratic ethics model would seem to be preferable to the business ethics model in delivering the best outcomes within the era of 'financialization'.

References

Aidenvironment. (2017) *Nordic Investments in Banks Financing Indonesian Palm Oil*, Rainforest Foundation of Norway, Oslo

Brobakk, J. and Almås, R. (2011) 'Increasing food and energy prices in 2008: What were the causes and who was to blame?', *International Journal of Sociology of Agriculture and Food*, vol 18, no 3, pp. 236–260

Burch, D. and Lawrence, G. (2013) 'Financialisation in agri-food supply chains: Private equity and the transformation of the retail sector', *Agriculture and Human Values*, vol 30, no 2, pp. 247–258

Burton, R. and Bjørkhaug, H. (2015) 'Introduksjon av ikke-mat i verdikjeden for mat' in H. Bjørkhaug, R. Almås and J. Vik (eds) *Norsk Matmakt i Endring*, Fagbokforlaget, Bergen

Clapp, J. (2014) 'Financialisation, distance and global food politics', *Journal of Peasant Studies*, vol 41, no 5, pp. 797–814

Council on Ethics. (2013) *Recommendations Concerning the Exclusion of Lockheed Martin Corp. From the Investment Universe of Government Pension Fund Global*, Council of Ethics, Oslo

Dagsavisen. (2016) 'Tree planting surveillance' (from Norwegian), *Dagsavisen*, available at: www.dagsavisen.no/innenriks/skal-overvake-treplanting-1.807040, accessed 2 March 2018

Daniel, S. and Mittal, A. (2009) *The Great Land Grab Rush for World's Farmland Threatens Food Security for the Poor*, Oakland Institute, California

De Soto, H. (2001) *The Mystery of Capital: Why Capitalism Triumphs in the West and Fails Everywhere Else*, Black Swan, London

Dore, R. (2008) 'Financialization of the global economy', *Industrial and Corporate Change*, vol 17, no 6, pp. 1097–1112

Epstein, G. (2005) *Financialization and the World Economy*, Edward Elgar, Cheltenham

Fairbairn, M. (2014) '"Like gold with yield": Evolving intersections between farmland and finance', *Journal of Peasant Studies*, vol 41, no 5, pp. 777–795

Food Navigator. (2013) 'Norway fund pulls out of unsustainable palm oil companies', available at: www.foodnavigator.com/Article/2013/03/12/Norwegian-fund-pulls-out-of-unsustainable-palm-oil-companies# accessed 30 April 2017

Government. (2001) *Report from the Government's Commission on Values and Ethics* (from Norwegian), available at: www.regjeringen.no/no/dokumenter/verdikommisjonens-sluttrapport/id2567979/, accessed15 January 2018

Government. (2009) 'Norway ought to step up the fight against tax havens' (from Norwegian), Ministry of Foreign Affairs press release, 18 June 2009, available at: www.regjeringen.no/no/aktuelt/pm_kapitalflukt/id567661/, accessed 5 March 2018

GRAIN. (2016) 'ADM's offshore links to Wilmar, world's worst environmental offender', available at: www.grain.org/article/entries/5499-adm-s-offshore-links-to-wilmar-world-s-worst-environmental-offender, accessed 26 February 2018

Green Resources. (2018) Company website, available at: http://dev.greenresources.no/, accessed 26 February 2018

Hall, D. (2013) *Land*, Cambridge, Polity Press

Henriksen, E. (2014) 'The oil-fund can do better' (from Norwegian), chronicle, *Aftenposten*, 4 April

Hinshaw, D. (2011) *The Great African Land Rush*, available at: https://farmlandgrab.org/18452, accessed 27 February 2018

Holmqvist, J. (2018) 'A possible way' (from Norwegian), chronicle, *Vårt Land* daily, 15 February

Hustad, J. (2018) 'Norwegian double standards' (from Norwegian), chronicle, *Dag og Tid* weekly, 26 January

Kasa, S. (2013) 'An elite driven climate policy', in R. Almås, R., H. Bjørkhaug, H. Campbell and C. Smedshaug (eds) *Towards a Sustainable and Climate Adapted Norwegian Agriculture* (from Norwegian), Akademika Publishers, Trondheim, pp. 137–159

Kasikci, F. and Rajasingam, S. (2017) T*he Norwegian Government Pension Fund – Global: The cost of Ethical Exclusions*, BI Norwegian Business School, Oslo

Klimek, B. and Bjørkhaug, H. (2017) 'Norwegian agro-food attracting private capital; Varieties of capitalism – varieties of financialisation?', *Sociologia Ruralis*, vol 57, no 2, pp. 171–190

Krippner, G. (2005) 'The financialization of the American economy', *Socio-Economic Review*, vol 3, no 3, pp. 173–208

Lawrence, G. (2016) 'Defending financialization', *Dialogues in Human Geography*, vol 5, no 2, pp. 201–205

Lawrence, G., Lyons, K. and Wallington, T. (eds) (2010) *Food Security, Nutrition and Sustainability*, Earthscan, London

Lexterra. (2016) *The Progress of Forest Plantations on the Farmers' Territories in the Nacala Corridor: The case of Green Resources Moçambique*, Lexterra, Maouto

McMichael, P. (2009) 'A food regime genealogy', *Journal of Peasant Studies*, vol 39, no 1, pp. 139–169

McMichael, P. (2012) 'The land grab and the corporate food regime restructuring', *Journal of Peasant Studies*, vol 39, no 3–4, pp. 681–701

McMichael, P. (forthcoming) 'The 21st century land question, and its politics', in H. Bjørkhaug, P. McMichael and B. Muirhead (eds) *Frogs, Fuel, Finance or Food? Cultures, Values, Ethics and Arguments on Agricultural Land*, Toronto University Press, Toronto

Muirhead, J. (forthcoming) '"Indirect" land grabbing, private certification and global GAP' in H. Bjørkhaug, P. McMichael and B. Muirhead (eds) *Frogs, Fuel, Finance or Food? Cultures, Values, Ethics and Arguments on Agricultural Land*, Toronto University Press, Toronto

NBIM. (1998) *Annual Report 1998*, Norges Bank Investment Management, Oslo

NBIM. (2016) 'Exclusion of coal companies from the fund', NBIM press release, 14 April, available at: www.nbim.no/en/transparency/news-list/2016/exclusion-of-coal-companies-from-the-fund/, accessed 19 January 2018

NBIM. (2017) *Strategy 2017–2019*, Norges Bank Investment Management, Oslo

NBIM. (2018) *Norges Bank Investment Management. The Fund*, available at: www.nbim.no/en/, accessed 26 February 2018

Norad. (2015) *Evaluation of the Norwegian Investment Fund for Developing Countries*, report 1/2015, Norwegian Agency for Development Cooperation, Oslo

Norfund. (2011) *Report on Operations 2011*, Norwegian Investment Fund for Developing Countries, Oslo

Norfund. (2017) *Norfund*, available at: http://intranett.norfund.no/historie/category484.html, accessed 25 April 2018

Norfund. (2018a) *Investment Key Figures* available at: www.norfund.no/, accessed 17 January 2018

Norfund. (2018b) *Strategy 2016–2020*, available at: www.norfund.no/strategy-2016-2020/category1074, accessed 17 January 2018

Norfund. (2018c) *How We Work*, available at: www.norfund.no/tjenester_2/, accessed 17 January 2018

NOU. (1995) *Norsk Sør-politikk for en Verden i Endring Rapport fra Nord-Sør/Bistandskommisjonen: Norges Offentlige Utredninger* (in Norwegian), Ministry of Foreign Affairs, Oslo

NRK. (2017) 'Interview with professor of economics, Beate Sjåfjell', University of Oslo, 13 November, Norwegian Broadcasting Corporation, Oslo

Ot.prp. nr. 13. (1996–1997) *Om lov om Statens Investeringsfond for Næringsvirksomhet i Utviklingsland* (NORFUND) (in Norwegian), Ministry of Foreign Affairs, Oslo

Ouma, S. (2015) 'Getting in between M and M′ or: How farmland further debunks financialization', *Dialogues in Human Geography*, vol 5, no 2, pp. 225–228

Oxfam. (2011) 'New Forests Company must address safety concerns of Uganda land grab community', available at: www.oxfam.org/en/pressroom/pressreleases/

2011-10-17/new-forests-company-company-must-address-safety-concerns-uganda, accessed 7 March 2018

Palley, T. (2008) 'Financialization: What it is and why it matters', in E. Hein., T. Niechoj, P. Spahn and A. Truger (eds) *Finance-led Capitalism: Macroeconomic Effects of Changes in the Financial Sector*, Marburg, Metropolis-Verlag, pp. 29–60

Perez, C. (2002) *Technological Revolutions and Financial Capital*, Edward Elgar, Cheltenham

Rainforest Foundation. (2013) 'The oil-fund pulls out of palm oil villain', (from Norwegian), The Rainforest Foundation, Oslo

Rainforest Foundation. (2017) 'Nordic banks and government pension funds linked to Indonesia's palm oil deforestation', available at: www.regnskog.no/en/news/new-report-calls-out-nordic-countries-commitment-to-forest-protection-by-linking-nordic-banks-and-government-pension-funds-to-indonesias-palm-oil-deforestation, accessed 7 April 2017

Ramberg, J. (2011) 'Ethics or profit? An evaluation of the Government Pension Fund – Global' (from Norwegian), Thesis, NTNU, Trondheim

Roland, K. (2018) 'Second best as the enemy of the good', (from Norwegian), chronicle, *Dagens Næringsliv* daily, 9 January

Sawyer, M. (2014) 'What is financialization?', *International Journal of Political Economy*, vol 42, no 4, pp. 5–18

Tvedt, T. (2011) 'Understanding Norwegian foreign policy' (from Norwegian), essay, *Dagbladet*, May 3

Tvedt, T. (2016) *Norwegian Ways of Thinking* (from Norwegian), Aschehoug, Oslo

Vermes, T. (2009) 'Here are the Norwegian tax havens', (from Norwegian), available at: www.abcnyheter.no/nyheter/2009/06/17/90761/her-er-den-norske-stats-skatteparadis, accessed 6 March 2018

Vårt Land. (2016) 'Norwegian investment companies still linked to tax havens', News article, available at: www.vl.no/nyhet/norsk-investeringsselskap-bruker-fortsatt-skatteparadiser, accessed 7 March 2017

White, B., Borras, S. M., Hall, R. Scoones, I. and Wolford, W. (2012) 'The new enclosures: Critical perspectives on corporate land deals', *Journal of Peasant Studies*, vol 39, no 3–4, pp. 619–647

4 Strategic financialization?

The emergence of sovereign wealth funds in the global food system

Sarah Ruth Sippel, Michaela Böhme and Cynthia Gharios

Introduction

For some years now scholars have been investigating what has been termed the 'financialization' of the agri-food system, namely the heightened and redefined intersections between the worlds of food and finance (Clapp, 2014; Isakson, 2014; Visser *et al.*, 2015). In addition to changes taking place within agri-food businesses and the emergence of agricultural derivatives trading, one focus is the direct investment of financial actors in primary production where farming and farmland are being constructed as alternative financial asset classes (Gertel and Sippel, 2016). In the literature, 'financial actor' is often used as an umbrella term referring to diverse groups of investors and various investment vehicles such as private equity funds, hedge funds, pension funds or sovereign wealth funds. Although they often employ the same investments models and rationales for farmland investments (Ouma, 2016), these actors and entities pursue substantially different interests, and strategies, which translate into diverging approaches to, and implementations of, agricultural investment projects. By taking a closer look at the emergence of one particular group of investors in the global food system, namely Sovereign Wealth Funds (SWF), this chapter seeks to contribute to a more differentiated picture of the category of 'finance actors'.

From the very beginning, scholars have identified SWFs as a driving force behind large-scale farmland acquisitions in foreign countries (see Burch and Lawrence, 2009; Cotula, 2012) and as an important source of demand for agricultural-based financial investment products (see Murphy *et al.*, 2012; Clapp, 2014). SWFs have been identified, in particular, with the notion of a renewed 'agro-security mercantilism', a strategy by which states seek to guarantee access to food via direct land acquisitions offshore (McMichael, 2013). That is, rather than relying on market access to food, agro-security mercantilism aims at achieving direct access to productive land for food supplies that are repatriated to the investing states. Apart from one in-depth case study (Sippel, 2015), however, there has been little attention paid to this particular type of finance actor and the intersection of investment motives and interests in the agri-food sector it represents.

Drawing on the examples of SWFs in the Arab Gulf States and China as representing two regional contexts that have seen important activities of SWFs in agri-food sectors in recent years, we make two main points. First, SWFs operate at the intersection of 'financial' and 'strategic' logics. While they need to produce financial returns for their shareholder – the state – they also follow strategic investment mandates layered on top of their financial activities. Especially in state-directed economies, SWFs function as a conduit for linking finance with the political and economic development goals of the state at both national and global levels. Second, in comparison to the scope of direct farmland investments that have been announced and anticipated in the aftermath of the conjunction of the financial, food and energy crises 2007/8, the number of effectively realized projects has remained small as several investments have either not materialized or faced considerable backlash. Yet, this does not mean that SWFs have withdrawn from the agri-food sector. Rather, they have been active in investing further up the value chain of agribusinesses and have started to act as facilitators for state-owned entities that likewise make acquisitions further up the value chain. Being at the centre of a network of closely interrelated financial and commercial state agencies, SWFs help states gain a stronger foothold in the global agricultural value chain and fulfil political goals that go well beyond security-mercantilist practices. These goals range from supporting domestic food security strategies to maintain social stability to redefining the state's position within global agribusiness. The investigation of SWFs activities in the agri-food system thus needs to move beyond the exclusive focus on primary production to identify the strategic motives and interests of the agri-food investments that are undertaken and facilitated by these hybrid finance-state entities.

In doing this, the chapter advances our understanding of the 'financialization' of the agri-food system while highlighting the redefined role of the state within agri-food politics that goes along with this development. In regard to the food system, financialization has been mostly employed to capture the growing influence and importance of financial capital, markets, and motives throughout the agri-food system. Here, SWFs represent an interesting case as they point to the new intersections between state and finance which, we tentatively suggest, is a form of 'strategic financialization', representing the coupling of the interests of capital in securing a resource and gaining from its appreciation, together with the pursuit of politico-strategic interests of nation-states. The emerging activities of SWFs in the food system since 2007/8 thus not only point to the increasing enmeshment of financial motives and rationales in agri-food but also represent a novel form of state intervention in the global food system. This intervention can be understood as a redefinition of state-driven food policies by making use of, and being linked to, financial entities and respective investment logics.

While the significant role states have assumed in accommodating land acquisitions has received some attention (Keulertz and Woertz, 2015), the role of the state in actively facilitating and promoting the linkages between finance and agri-food has been little studied so far (Larder *et al.*, 2017). Martin and Clapp (2015) outlined how in industrial contexts in North America, Europe, and Australia, the relationship between agriculture and finance has been historically mediated by the state in multiple ways, including providing state-backed agricultural credit along with explicit interventions in financial markets. Since the 1970s, the state has withdrawn from key areas of provision and safeguarding in regard to financial services but still sets the regulatory framework, thereby laying the 'groundwork' for making agriculture attractive to financial investments (Martin and Clapp, 2015, p. 556). The creation of SWFs, however, goes far beyond this 'regulatory' function as states have reasserted their authority in global finance as major investors in the market (Helleiner, 2009). The emergence of SWFs as direct players in the agri-food system in recent years indicates that states have started to adopt a new and more active role in regard to food, an active role that is often part of the broader redefinition of national food security strategies.

Following Pritchard *et al.* (2016) we emphasize that the role of the state in the food system needs to be reconsidered along simultaneous 'stepping back' and 'moving in' dynamics that constitute an amalgam of both reactionary and progressive elements, leading to confusing new political architectures. Having said this, it is important to note that states do not have a 'single face' (Pritchard *et al.*, 2016, p. 693) nor do they represent a relationship devoid of contradictions (Jessop, 2007). Instead of 'black boxing' state action we seek to identify how SWFs have designed their agricultural investments within the respective contexts of the Arab Gulf and China. In particular, we ask which actors are involved in the investment process, which agricultural asset classes are being targeted, and how financial versus strategic interests intersect and are being negotiated. Our analysis draws on available public sources (including databases, media sources and companies' websites) as well as information received during qualitative interviews Böhme and Sippel conducted with company representatives in Australia and China in 2016 and 2017.[1]

The chapter is organized as follows. We start with a brief outline of the emergence of SWFs as new powerful actors in the finance sector since 2000 and their recent interest in agri-food sectors. We then present examples of SWFs' activities in agri-food sectors from the Arab Gulf States and China. Based on that, we discuss the state-finance nexus represented by SWFs in the light of the financialization of the agri-food system and the redefined role of the state.

The emergence of SWFs in global finance and investments in agriculture

SWFs are an emerging and increasingly powerful category of state-owned investors. SWFs have a long history that dates back to the 1950s, but their growth has markedly accelerated since the 2000s, turning SWFs into key actors in global financial markets. According to UNCTAD's World Investment Report (2015), there were more than 100 SWFs managing assets in excess of US$7 trillion in 2014, accounting for one-tenth of the world's total assets under management. While various governance structures, institutional mandates, and investment strategies exist, the term generally refers to state-owned investment vehicles that manage public funds by actively investing in a variety of assets in domestic and international markets. For the purpose of this chapter, we follow Fotak *et al.*'s (2014, p. 584) definition in focusing on entities that are (1) investment funds (rather than operating companies); (2) wholly owned by a sovereign government but separately organized from central banks or finance ministries (although, in some cases, connections can be blurred[2]); and (3) invest (internationally and domestically) to realize commercial returns. This definition excludes other types of sovereign investors such as public pension funds.[3] Two main groups of SWFs can be distinguished based on their sources (Fotak *et al.*, 2014). Governments whose revenue streams depend mainly on the value of one underlying commodity have established one group of funds. These funds are mostly concentrated in countries with resource wealth such as the oil-rich Arab Gulf States. The second group of funds is financed out of accumulated foreign exchange reserves that stem from persistent and large net exports, such as in the case of China and other East Asian exporters.

With the exception of Norway, SWF investments are not usually publicly disclosed; information is difficult to obtain. Based on available data, however, some trends can be identified. First, most investments are directed to foreign markets and, with a share of between 50 and 70 per cent, mostly target OECD contexts (Bortolotti *et al.*, 2016, pp. 24–28). Second, while the banking, insurance and trading sector represented the bulk of investments undertaken during the financial crises in 2007/8 and 2011, in more recent years investments have shifted. This shift has been towards 'safe assets' (the sectors of Infrastructure and Utilities, Real Estate and Restaurants, Hotels, Motels), on the one hand, and towards natural resources including energy, agriculture/farmland, metals and mining, timberland, and water opportunities, on the other (see Bortolotti *et al.*, 2016, pp. 20–23). For instance, in 2017, 55 per cent of SWFs had allocated some proportion of their portfolio to natural resources, 7 per cent more than in 2016 (Preqin, 2017).

The shift in portfolio allocation from the banking and finance sector to real estate, infrastructure, and natural resources is underpinned by a change in investment strategy as many SWFs have begun to diversify away from fixed income and public equity to private markets. In the latter they either make direct investments (in housing, infrastructure, or resource assets, for example) or invest through unlisted funds (including private investment funds, private equity funds, and hedge funds) (Preqin, 2017). In 2014, SWF investments in private markets accounted for some 26 per cent of overall SWF asset allocation compared to 12 per cent in 2007, before the financial crisis (Hentov, 2015). This shift has not only been interpreted as a strategy of inflation-hedging in a post-crisis financial environment shaped by quantitative easing, but also as a sign of the growing expertise of SWFs, enabling them to diversify from traditional asset classes into more unconventional assets (Hentov, 2015). This development also points to the increasingly complex role of SWFs as they become enmeshed in more private networks of investment and as they work through private investment vehicles.

As resource shortages and volatile commodity prices are increasingly perceived as a constraint for long-term economic growth, a number of SWFs have begun to set up formal or implicit strategies to increase investments in the natural resources, energy, and food production sectors (Avendano, 2010). As listed in Table 4.1, with the exception of the Norwegian SWF and the Russian Reserve Fund, the ten largest SWFs by assets (see Bortolotti *et al.*, 2016) have all established dedicated agricultural investment vehicles, made direct investments through strategic agricultural investment companies, or announced co-investment initiatives.

Investments in agriculture – in particular those targeting non-listed agribusinesses and farmland assets – bear out the increasing complexity of SWF investments. Table 4.1 points to the growing interest of SWFs in agriculture and food-related investments as well as the diversity amongst SWFs and the ways they engage in agricultural investments. First, while eight out of the ten largest SWFs have expressed interest in the food production sector, publicly available information indicates that this interest has not always materialized in concrete deals. Given the paucity of information, more fieldwork-based research is necessary to unpack each individual investment project and its status of implementation and the challenges that may have caused some SWFs to rethink their investment strategies or abort deals. Second, investment strategies are multifaceted and range from acquisitions of physical farming assets to investments into agribusinesses further up the value chain. This suggests that SWFs are not merely acquiring farmland and exporting raw materials, but are pursuing more complex investment strategies that integrate sovereign interests into the food system in novel ways. Third, while little is known about the way SWF deals in agriculture are structured, the collected data suggest that many SWFs operate through a subsidiary operational company or enter into joint venture projects and co-investments with other, often private, funds. This indirect involvement of SWFs through equity participation in more

Table 4.1 Dedicated agricultural investment vehicles and investment intentions announced by the top ten SWFs by assets

SWF	*Investment details*	*Status*
Norway – Government Pension Fund – Global	No direct investments in agriculture (Norges Bank, 2017).	
UAE – Abu Dhabi Investment Authority	In 2013, Mubadala and the Russian Direct Investment Fund (RDIF) announced the launch of a 50–50 joint venture to identify long-term investment opportunities in Russia (Sovereign Wealth Centre, 2017). In 2016, state investment vehicle Mubadala Development Corporation announced plans to co-invest in a number of funds raising up to US$300 million for investments in Russia's agriculture and food production sector (Arabian Business, 2016).	Announced
China Investment Corporation (CIC) and Russian Direct Investment Fund (RDIF)	In 2012, CIC and RDIF announced the creation of a joint Russia China Investment Fund (RCIF) to invest in projects that advance bilateral economic cooperation between China and Russia. The fund has received US$2 billion in commitments from CIC and RDIF and has identified food and agriculture as one of its priority sectors (RCIF, 2017).	Deals executed
Kuwait – Kuwait Investment Authority (KIA)	In 2009, KIA signed investment deals with nine Asian countries and announced potential investment deals in agriculture and other sectors (Reuters, 2008).	Announced
Singapore – Government of Singapore Investment Corporation (GIC)	In line with its current investment strategy, GIC regularly invests in food and agriculture. In 2015, agricultural was its third largest sector of investment (Bortolotti *et al.*, 2016).	Deals executed
Qatar Investment Authority (QIA)	QIA is probably the most active SWF invested in agriculture with investments in AdecoAgro, a diversified farmland and agribusiness venture in South America, and investments in farmland and agribusiness through its subsidiary Hassad Food (Luyt *et al.*, 2013).	Deals executed
UAE – Investment Corporation of Dubai (ICD)	ICD operates subsidiary company Zeera Investment LLC as a special purpose vehicle for investments in agriculture (ICD, 2015).	Operating

continued

Table 4.1 Continued

SWF	*Investment details*	*Status*
Singapore – Temasek	In 2015, agriculture was Temasek's third largest sector of investment (Bortolotti et al, 2016). Furthermore, in 2011, Temasek Holdings set up a US$300 million private equity fund called Tana Africa Capital with a primary focus on consumer goods and the agricultural sectors across Africa (PwC, 2016).	Deals executed
Russia – Reserve Fund	No direct investments in agriculture (Russia Reserve Fund, 2017).	
Australia – Australia Future Fund	The Australian Future Fund invests in tangible assets such as timber and agriculture where an income return is generated based on an underlying physical asset. In addition, the fund holds public equity in a number of global agri-food businesses operating higher up the value chain (Australia Future Fund, 2017).	Deals executed

Source: compiled by authors from reports listed in the table.

Note

The table excludes the China National Security Fund listed as 7th largest SWF in Bortolotti *et al.* (2016). As a pension fund established with the aim to aid the country's aging population and support social security expenditure, the China National Security Fund falls outside our definition of SWF, even though it is listed by many sources as one of China's four SWFs – together with the China Investment Corporation, the SAFE Investment Corporation, and the China-Africa Development Fund.

directly engaged companies increases the complexity of investment structures and of the networks of involved actors, thereby contributing to the opacity surrounding SWF investments in agriculture. Below, we will investigate these aspects further by drawing on the examples of the Arab Gulf States and China and the activities of SWFs in the agri-food sector that can be observed in the two contexts.

Strategic repositioning post-2008: SWFs in the Arab Gulf States

The Arab Gulf States represent a context where SWF activities in agri-food sectors have been especially important to the reorientation of national food security policies since 2007/08. Given the arid climate, little available arable land, and scarce water resources, the Gulf Cooperation Council (GCC) countries have low levels of self-sufficiency in food. For all Gulf States, arable land surfaces are lower than 5 per cent of the country's landmass (Zurayk *et al.*, 2011, p. 130). They are as well among the most water-stressed countries worldwide. In light of a depleting resource base coupled with ongoing population growth, the Gulf States are thus highly dependent on food imports, posing a two-fold risk: a supply risk regarding the availability of food imports, and a price risk regarding the affordability of food imports (Bailey and Willoughby, 2013; Woertz, 2013).

The food crises in 2007/8 and 2011 made these risks particularly apparent as the external food supply of the Gulf became both increasingly expensive due to food price hikes as well as uncertain as major supplies implemented export bans. Subsequently, the Gulf States have reoriented their food security strategies by focusing on four dimensions (Bailey and Willoughby, 2013): (1) measures to stabilize and support the affordability of food (including price controls and food subsidies as well as social welfare expenditures more generally); (2) diversification of trade by developing new trade and partnership relationships; (3) increase of food storage facilities; and (4) investments in agri-food sectors both at home and abroad. With regard to the latter, GCC SWFs have played an important role as emerging and increasingly capital-strong investment entities.

While a number of SWFs, or, to a larger extent fund agencies, have existed for some decades in the Gulf,[4] historically they kept a relatively low profile, were oriented towards portfolio investments, and did not acquire more than 5 per cent in any given company to avoid disclosure requirements (Woertz, 2013, p. 219). Until the early 2000s, state-led agro-investments abroad existed but took place via international development funds that pursued a developmental strategy and left the implementation of projects to companies and governments in the host countries. Investments in agricultural projects neither exceeded 18 per cent of the funds' spending, nor involved taking equity stakes (Woertz, 2013, p. 220). Since the 2000s, and fuelled by unprecedented high oil incomes, several new SWFs were established and the assets held by existing ones grew significantly. Combined, the GCC's SWFs reach almost US$2.9 trillion in assets, representing a significant portion of global SWF funds (Naser, 2016, p. 1660). GCC SWFs have thus become 'the main state financial vehicles for influencing change and managing national wealth' (Naser, 2016, p. 1661). Their objectives are not publicly known and can vary greatly but are generally seen in storing the nation's wealth on the one hand and in implementing certain national financial and economic objectives on the other, such as the diversification of income and wealth accumulation and, most importantly, the diversification away from oil and gas incomes (Rios-Morales *et al.*, 2012; Shahandeh, 2012). As with most SWFs, the GCC's SWFs do not usually disclose information about their assets, investments, and returns. The few studies available indicate that GCC's SWFs mostly invest internationally rather than domestically while favouring 'developed markets' (North America, Europe) as well as 'safe', fixed-income assets such as US Treasury bills and bonds (Salisu and Taher, 2012). At the same time, and corresponding with the general investment trends outlined above, since some ten years there has as well been a shift towards increasing investments in 'emerging' markets and non-traditional asset classes (Shahandeh, 2012). The increased involvement in agri-food sectors can be considered as being part of this shift while being related to the particular moment of world market food insecurity in 2007/8.

Following the 2007/8 food crisis, several GCC SWFs appeared as major players in funding and implementing the new national food security strategies. They figured prominently among those actors identified as 'food security grabbers' (GRAIN, 2008) that pursue a novel way of 'agro-security mercantilism' (McMichael, 2013). Several years later, however, there was still a large gap between the projects that had been announced after 2008 and those actually realized on the ground. Woertz (2013, p. 227) identifies a number of reasons for this 'implementation gap', such as reduced funding options due to the financial crisis, issues in regard to the legal frameworks, business environments, and natural resource situation encountered in targeted countries, as well as political resistance by governments and civil society. It would be misleading though to conclude that GCC SWFs have withdrawn from considering agri-food sector investments an important element in their national food security strategies. Rather, and in line with Woertz (2013) and Keulertz and Woertz (2015), we suggest that the emphasis has shifted from primary production to investments in international and domestic agribusinesses as well as emerging areas such as digital farming technologies. Below we illustrate this by outlining the activities of two SWFs, the Qatar Investment Authority (QIA) and Bahrain's SWF Mumtalakat, both of which are exceptions with regard to the accessibility and transparency of information provided.

QIA: from farmland investments to agribusiness

QIA's involvement in the agri-food sector started after the food crisis in 2007/8 and occurs via two channels: its subsidiaries Hassad Food and Qatar Holding LLC (see Table 4.2). As Qatar's 'premier investor in the food and agri-business sectors', Hassad Food is entrusted with a double objective of supporting the needs of Qatar's growing economy while contributing to food security needs: 'Since its commencement, the company adopted a state of the art investment model that mainly focused on investing globally in viable prospects to support Qatar and the region's food requirements' (Hassad Food, 2018). Hassad Food has been leading the country's quest for investments in primary production both domestically and overseas with, according to the company's website, agricultural operations in Qatar, Australia, and Pakistan. Other primary production projects abroad that, among others, had been announced for Sudan have not been officially confirmed. Instead, Hassad has complemented its portfolio by setting up an agricultural technology company specializing in hydroponic greenhouse technology, establishing a food marketing company, as well as acquiring shares to become the majority shareholder in a large poultry

Table 4.2 QIA's Investments in agriculture and agribusiness

	Qatar Investment Authority[1] *(2005)*							
Company #1 (ownership structure)	Qatar Holding LLC (2006) Al Gharrafa Investment Company (wholly-owned subsidiaries)	Hassad Food (2008) (wholly-owned subsidiary)						
Company #2 (ownership/ investment structure/year of establishment or acquisition)	Adecoagro (shareholder)	Hassad Qatar (wholly-owned subsidiary) (2009)	Hassad Australia (wholly-owned subsidiary) (2009)	Mahaseel (wholly-owned subsidiary) (2011)	Roza Hassad (wholly-owned subsidiary) (2012)	Senwan Pakistan (wholly-owned subsidiary) (2012)	Zulal Oasis (part-ownership) (2013)	A'Saffa Foods (majority shareholder) (2015)
Investment model	Acquisition of shares (13.2%)	Direct investment in primary production	Direct investment in primary production	Direct investment in agribusiness company	Direct investment in primary production	Direct investment in primary production	Partnership with Oasis Agrotechnology (consortium led by Primaflor Group)	Acquisition of shares (33%)
Geographical scope and agri-food activity	Argentina, Brazil and Uruguay Various agricultural products	Qatar Green fodder production (8,000 tonnes/ year)	Australia Sheep/grain production (250,000 head, 179,000 tonnes/ year)	Qatar, Tunisia Food marketing, packaging and export[2]	Qatar Greenhouse production (3 M flowers, 1 M bedding plants, 100 tonnes vegetables/ year)	Pakistan Rice production (10,000 tonnes/ year)	Qatar Agricultural technology	Oman Poultry production (22,000 tonnes/ year)

Sources: SWF Institute, 2017; Hassad Food, 2018.

Notes
1 Qatar Investment Authority has a volume of US$320 billion mainly stemming from liquefied natural gas exports.
2 For example, Mahaseel holds a 51% share in the Qatari Tunisian Food Company, a packaging and export company for olive oil in Tunisia.

company. The close engagement with agribusiness is also evident for investments undertaken by Qatar Holding LLC. In 2011, its subsidiary Al Gharrafa Investment Company acquired shares of the large agribusiness company Adecoagro when the company was listed on the New York Stock Exchange. Adecoagro operates across three countries in South America and is a fully integrated food and energy crop producer and processor as well as farmland investor (Adecoagro, 2017a, 2017b).

QIA's engagement in agriculture and food has thus shifted from the original emphasis on direct 'food security investments' in primary production towards a more intense engagement in agribusiness activities more generally. This development can be related to the substantial backlash so-called 'food security investments' faced even in an assumed low-risk context such as Australia where the motive of producing for the food security of another nation became highly politicized and was effectively ruled out as a legitimate motive for future agricultural investments in a federal enquiry (Sippel, 2015). Hassad Australia's recent reorganization of its farming portfolio in Australia supports this development. In 2017, the company started to divest some of its major holdings as part of a shift from food security to 'overall returns for investors' and announced that it would focus its future attention on downstream operations such as food marketing and processing (CEO John McKillop, quoted in *Australian Financial Review*, 2017).

In sum, QIA's portfolio expansion indicates that apart from basic food needs such as rice and grain production, investments have been undertaken in strategic areas beyond 'pure food security' considerations, responding to the *meatification* of diets (Weis, 2007) that is also occurring in the Gulf States, on the one hand, and geared towards upcoming investment areas such as new agricultural technologies, on the other. Moreover, the development of domestic agriculture has been an important focus complementing its international engagement. Last, rather than further pursuing direct farmland purchases abroad the fund has been active in acquiring shares in agribusinesses, which is perceived as less controversial and, potentially, more stable and profitable.

Mumtalakat: facilitating domestic food industries

The case of Bahrain adds a further perspective on SWF activities in the Gulf, especially with regard to their role within domestic food policies and the linkages with state-owned enterprises (SOEs). In comparison to the other Gulf countries, Bahrain's natural (oil and gas) resources are only limited (Cadène and Dumortier, 2013). Since the 1960s, the country has diversified its economy and is today considered as the first post-oil economy of the GCC (Peterson, 2009). In 2006, the government established two SWFs to improve the performance and governance of its state-owned enterprises: the Noga Holding (to manage the state-owned oil and gas assets); and, the Mumtalakat Holding (to manage governmental stakes in non-oil-and-gas companies) (BMHC, 2017). Since then, Mumtalakat has invested in more than 40 companies in a variety of sectors with the objective of growing and diversifying the state's economy together with securing long-term returns and creating wealth for Bahrain's future generations – all within the framework of the government's Economic Vision 2030.

Following attempts to diversify investments abroad between 2007 and 2010, the unfolding of political unrest in the country in 2011, however, turned the focus inwards. Currently, Mumtalakat's investments concentrate exclusively on the domestic market (Sovereign Wealth Center, 2017). This also applies to the food sector. Mumtalakat's food-related portfolio consists of four companies (see Table 4.3). All these companies were established as SOEs in the 1970s and the 1980s and subsequently underwent ownership restructuring, that is, stock market listing or privatization. Their activities focus on wheat and meat (protein) production. As a result of Mumtalakat's investment in local agribusinesses, Bahrain now produces 70 per cent of consumed red meat and 25 per cent of consumed poultry. In short, Mumtalakat's investments focus on basic foodstuffs as well as the increasing demand for protein while fulfilling the strategic goal of fostering both domestic economic growth and (political) stability.

China's SWF: emerging linkages with the global agri-food system

Since the 2007/08 food crisis, Chinese companies have been identified as major actors in the controversy around states securing farmland and food in poorer countries (GRAIN, 2008). This 'China expansion' discourse (Hofman and Ho, 2012) is underpinned by the notion that domestic population growth coupled with shrinking land and water supply as well as severe environmental degradation has pushed Chinese companies, many of which are state-controlled, to lease or purchase foreign farmland in order to '[get] down to the work of crop production' (GRAIN, 2008, p. 3).

A closer look at the shifting role of the state in Chinese food politics reveals, however, a more complex strategy that positions China 'between self-sufficiency and global integration' (Hoering and Sausmikat, 2010). Since imperial times, Chinese governments have considered food security an integral part of national security and have emphasized concepts such as grain self-sufficiency and domestic production (Lin, 2017). For example, in a 1996 white paper, the State Council set a 95 per cent self-sufficiency ratio for key grains, including wheat, maize, and rice, and a minimum area of 120 million hectares of cultivatable land as benchmarks for food security (China State Council, 1996). These targets were to be achieved through increases in productivity and cropland reclamation and have been supported by state interventionist policies such as tariff rate quotas

Table 4.3 Mumtalakat's investments in agribusiness

Company's name	*Ownership structure*	*% shares owned by Mumtalakat*[1]	*Founding year*	*Activities*
General Poultry Company	State-owned enterprise	100.0	1977	Production of fresh eggs
Bahrain Flour Mills Company (al-Matahin)	State-owned enterprise and publicly listed[2]	65.7	1972	Import of wheat; production of wheat flour and related products sold to domestic and export markets
Delmon Poultry Company	State-owned enterprise and publicly listed[2]	15.0	1980	Production of chicken and animal feed for domestic and export markets
Bahrain Livestock Company	Joint venture (since privatization in 2000, between Trafco Group BSC, Delmon Poultry Company, and Mumtalakat)	27.5	1983	Import and distribution of chilled and locally slaughtered meat

Sources: Bahrainlivestock, 2017; Bfm, 2017; BMHC, 2017; Dawajen, 2017; SWF Institute, 2018.

Notes

1 Mumtalakat is a non-commodity SWF and has a volume of US$ 10.4 billion.

2 The remaining shares are publicly listed on the Bahrain stock exchange.

on grain imports, subsidized price floors for domestic farmers, and stockpiling programs run by the government (OECD/FAO, 2016). Indeed, since 1999, China has maintained more than 95 per cent self-sufficiency in staple grains (Lin, 2017) and achieved 40 per cent, 32 per cent and 100 per cent production increases in grains, meat, and dairy output, respectively, between 2003 and 2013 (Gale *et al.*, 2015).

Despite these achievements, food imports have been growing steadily, and China is now the world's second largest importer of agricultural products, totalling over US$120 billion per annum (FAO, 2017). So far, import growth has been mainly fuelled by the growing demand for animal feeds (particularly soy) and agricultural input commodities such as rubber, cotton, or palm oil (FAO, 2017) but, as consumers' demand for high-quality agricultural products is increasing, imports are diversifying to include more nuts, fruits, wine, meat, and processed foods (Gale *et al.*, 2015).

Rising food imports have prompted an adjustment of the national food security strategy that now allows food imports to supplement domestic food supply (Shen, 2014). At the same time, 'taking good control of its own food bowl' continues to be a fundamental principal of China's food security strategy (Shen, 2014). As China gears itself towards greater reliance on international markets, it also seeks to enhance its control over global supply chains. Embedded within the framework of its 'going out' policy, the Chinese government is now encouraging 'agriculture to go global and actively use foreign resources' (Hornby, 2014). By enlarging the foothold of Chinese companies in the global food system via outbound investment in overseas logistics, trading, and production assets, the aim is to establish reliable supply chains for the Chinese market, secure a larger share of profits for Chinese companies, and to gain influence over the determination of international prices (Gale *et al.*, 2015).

As Chinese companies are integrating into the global food system and are seeking to gain more control over the global flows of agricultural commodities and the profits from agricultural trade, the state's interventions in food markets are becoming more multifaceted. As a result, the state's food security strategy is increasingly situated within a network of interrelated operational agencies including central and provincial-level SOEs, privately owned 'national champions',[5] as well as state-owned financial actors such as SWFs and state-owned banks (SOBs), which have begun to align and coordinate their investments in the global food system. In this web of interrelated investment activities, China's main SWF – the China Investment Corporation (CIC) – plays a crucial role.

The role of China Investment Corporation (CIC)

CIC was established in 2007 under the direct control of the State Council with a mandate to diversify China's foreign exchange reserves. By 2015, its total assets had surged from US$200 billion to US$810 billion, turning CIC into the third largest SWF worldwide (Bortolotti *et al.*, 2016). CIC's investment principles emphasize the financial nature of its operations and its mission to 'seek maximum returns for its shareholder [the state]' (CIC, 2015, p. 7). Top-level officials of CIC have repeatedly sought to dispel ongoing international concerns over the potential geopolitical nature of CIC's investments, as exemplified by a recent interview with former CIC chairman and CEO Ding Xuedong, who emphasized that 'the decision to invest or not invest rests in our hands. I want to make it clear: We will not invest if we believe we will lose money' (Cheng, 2016). Furthermore, the initial capitalization of CIC through the issuance of bonds with a coupon rate of 4.5 per cent has put pressure on CIC managers to make profitable investments with a rate of return that, at a minimum, covers the coupon rate of the issued bonds (Shih, 2009).

Yet, the pressure to make profits does not necessarily stand in opposition to an investment strategy supporting strategic goals. Indeed, CIC's investments often appear to be closely aligned with China's development agenda. For example, since the global financial crisis 2007/8, CIC has shifted its focus to the natural resource sector, assisting China's 'going out' strategy as an increasing number of state-owned and private companies seek to acquire natural resources overseas (Backer, 2010). With respect to agriculture, CIC had previously shown little interest in the sector, despite repeated food price hikes since 2007/8. However, as China's food security strategy has begun to rely more on international markets, CIC's focus on agriculture has sharpened (Anderlini, 2014).

Since 2014, CIC has developed a strategy of indirect control to coordinate its investment activities in the agricultural sector. Direct investments in foreign land assets are largely left to private firms and provincial-level SOEs (Puel, 2012). Instead, CIC operates through co-investments and partnerships with SOEs, national agricultural champions, as well as national and international financial actors to target investments across the entire agricultural value chain. According to its annual report, CIC is actively expanding its partnerships and, in 2015, 'carried out over 10 cooperation projects with leading enterprises in key industries and actively developed new ways to promote cooperation between the real economy the financial sector' (CIC, 2015, p. 5). To support CIC's focus on co-investments, in 2015 CIC incorporated a new affiliate, CIC Capital, a specialized direct investment platform that engages in private equity-style partnerships with leading Chinese companies, primarily SOEs, to fund the acquisition of overseas alternative assets in infrastructure and natural resources (Cheng, 2016). Furthermore, through its indirect ownership of key state-owned commercial banks via its subsidiary Central Huijin, CIC is able to help SOEs secure capital for their foreign investments (Liew and He, 2010). As a central player within China's financial system, CIC is also able to establish strategic partnerships with other national and international institutional investors, including private equity funds, SWFs, as well as international financial institutions such as the International Finance Corporation (IFC), a World Bank affiliate focusing on global development.

This mode of operation has a number of advantages. First, it allows CIC to share risk across different consortium partners. In addition, by making alliances with other state-controlled entities such as SOEs, CIC can hedge against domestic political risk in case of investment failure by spreading the blame across different factions of political power (Shih, 2009). Second, diversification into agriculture requires operational expertise that SWFs generally lack. By pooling resources with strategic partners, CIC is able to tap into the industry know-how and operational capabilities required to successfully manage these investments. Third, co-investments with high-profile institutional partners such as, for example, IFC may lend additional legitimacy to the investment, while at the same time obscuring ownership structures. Last, this mode of operation allows CIC to align its investment activities with those of China's key agricultural companies, providing strategic support for increasing China's foothold in the global food system.

Understanding CIC's mode of operation: the case of COFCO

CIC's mode of operation in regard to the agri-food sector is best illustrated through its interaction with state-owned China National Cereals, Oils and Foodstuffs Corporation (COFCO) – China's largest food and agribusiness company, as well as the country's monopoly importer and exporter of key commodities such as wheat, corn, and rice. Founded in 1949, COFCO has grown into an extended network of business units with subsidiaries in the oilseed crushing, feed production, grain trade, branded foodstuffs, dairy, meat, and other sectors. While COFCO has operated solely within China for most of its existence, it has recently pooled resources with CIC to support its transformation into a corporation with a global presence and vertically integrated supply chain.

From 2014 to 2016, COFCO embarked on a series of step-by-step takeovers of two international agricultural trading companies – Hong Kong-based Noble Group and Dutch grains trader Nidera – with the potential to dramatically transform the global agro-industrial landscape. Both Noble and Nidera have significant agricultural assets and operations throughout the agricultural value chain in Brazil, Argentina, Australia, and the Black Sea region, with a special focus on the production and trade of bulk commodities. By integrating Noble's and Nidera's assets, technical expertise, and operational capacities into one international platform operating under the name of COFCO International, COFCO has grown into an integrated global agribusiness with access to some of the world's top grain, vegetable oil, sugar, and coffee producing regions. Its six-continent supply chain has been set up to compete with the dominant global grain traders

ADM, Bunge, Cargill, and Louis Dreyfus – the so-called 'ABCDs' (Terazono and Sanderson, 2015).

While the attention regarding these deals is mainly focused on COFCO, some components of the acquisitions were structured as co-investments involving a complex set of financial actors, and with CIC playing a crucial role. In fact, Chinese investment interests in Noble go back to 2009, when CIC acquired a 15 per cent share in the company. This was followed in 2014 by a joint venture under the name of Noble Agri JV, in which COFCO and a group of high-level financial partners, including Chinese private equity firm HOPU Investment, the Singaporean SWF Temasek, as well as IFC, acquired 51 per cent in Noble (World Grain, 2014). In 2015, COFCO went on to acquire a 51 per cent stake in Nidera and subsequently set up a COFCO International Holdings, as a joint venture with CIC, to restructure control over its new acquisitions. In the joint venture, COFCO has an 80 per cent stake while CIC owns the remaining 20 per cent (Reuters, 2015).

CIC's involvement in COFCO's new global agricultural platform has not only increased its exposure to the agricultural sector, but has also facilitated the global restructuring of one of China's largest food and agribusinesses. Since 2015, COFCO has further integrated its platform, buying the remaining stakes in both Noble and Nidera and, in April 2017, it officially launched its new division, COFCO International. While no concrete timeline has been announced, COFCO's ultimate goal remains to consolidate its acquisitions and turn COFCO International into a publicly listed agricultural trading powerhouse before the end of the decade (Terazono and Hume, 2016). Towards this end, COFCO has been aiming to complete what the company refers to as its 'overseas layout' and integrate its acquired assets into a global value chain. For example, through Nidera's operations in Brazil, COFCO engaged in a joint venture with United Arab Emirates firm Al Khaleej Sugar and local Brazilian partners to construct a 16 million tonne grain terminal in Santa Catarina (Oliveira, 2015). Furthermore, the company recently completed the construction of a new open-access bulk agri-terminal at Port Kembla, Australia, to enhance its exporting capacity and better integrate its origination operations in Australia with its global import markets (Locke, 2014).

COFCO's transition into a global agribusiness highlights the complexity of CIC's investments in agriculture. While SOEs, as well as large private agricultural champions, act as the public face of China's global agricultural expansion, CIC provides discreet support and facilitates expansion strategies through co-investments and partnerships. By mobilizing its financial capacities, CIC thus underwrites investments by Chinese state-owned firms and national champions into strategic sectors within the global food system, thereby seeking to challenge the oligopolistic market power of the Western world's ABCDs in agricultural commodity production and trade.

Conclusion: SWFs as new hybrid agri-food players?

As government-owned institutions investing in financial markets, SWFs straddle the line between sovereign and financial actors. While SWFs often emphasize the financial nature of their portfolio allocations, this chapter has demonstrated that many investments are also closely aligned with other politico-strategic interests. Not always fully insulated from domestic politics, SWFs link financial and strategic objectives in novel and complex ways, creating a new space for state action in global food politics. We conclude by sketching out two observations in regard to the state-finance nexus represented by SWFs.

First, SWFs are an extremely heterogeneous group of investors that operate according to fundamentally different investment rationales and logics. Exact typologies and general principles about their mode of operation are hard to derive, given the lack of public information provided and the wide range of ways in which SWFs have started to become enmeshed in agri-food sectors. This caveat notwithstanding, the cases examined in this chapter reveal significant levels of activity and rising investment flows from SWFs into the agricultural sector. However, important differences remain as to the investment scope, strategies, and goals of those SWFs that were investigated.

Second, despite international concerns about the increasing role of SWFs in farmland acquisitions, direct investments in farmland and primary production did not always materialize or, if it did, faced considerable resistance in the target country. Public backlashes and the difficulties surrounding agriculture as an 'asset class', such as the lack of large-scale investment opportunities, unstable returns, and complex operating and management processes, may well have resulted in the cessation of some announced or planned land deals. While SWFs do not emerge as the major vehicles through which governments seek to gain access to foreign agricultural land, they have been active further up the value chain. Here, they are often implicated in complex joint venture arrangements and co-investment structures that involve state-owned and private, financial and commercial, as well as domestic and international actors. These complex arrangements and webs of interrelated interests call for more specific, fieldwork-based, research that transcends the myopic focus on land-based investments and seeks to unpack the broader ways in which SWFs reshape the dynamics of the global food system. From the examples given in this chapter, there are two main ways in which dynamics are being reshaped. There is a shift in global capital flows given that powerful SWFs have been emerging in geographies that previously did not appear as important sources of capital. These new actors from former 'peripheries' thus alter, and potentially challenge, the existing power relations in the global food system. At the same time, through the engagement of SWFs in agri-food investments, states achieve a further and significant presence in global food politics, as both owners and investors. SWFs represent an important source of state capital that is being infused into global agribusinesses – a source of state capital however that is allocated towards assets that promise to deliver a return on investment.

In sum the involvement of SWFs in the food system points to the complex interplay of financial and strategic logics simultaneously at play in many SWF investments that can be described as 'strategic financialization'. On the one hand, SWFs have become a powerful tool in the hands of states to achieve food security goals and underwrite industrial policy. This process has resulted in new ways of leveraging finance to promote development and policy goals. At the same time, the increasing incorporation of financial instruments, capacities, and interests in the food security strategies of the state may well reshape the behaviour of states themselves. From this perspective, the rise of SWFs is contributing to an ongoing financialization of the state, in which the food security strategies employed by states increasingly respond to the financial logics of risk-reward analysis and returns on investment. SWFs can thus be considered hybrid players in the agri-food system, blending 'state/public' and 'financial' capital as well as 'political/strategic' and 'return seeking' objectives in novel ways. It is this hybrid and fluid nature where state and financial interests are becoming increasingly realigned that deserves further attention in agri-food studies.

Notes

1 The research presented in this chapter is part of the research project C04 'Land Imaginations: The Repositioning of Farming, Productivity, and Sovereignty in Australia' (SFB 1199) at the University of Leipzig and funded by the German Research Foundation (2016–19).
2 For example, the Qatar Investment Authority started as an investment fund of the Qatari Ministry of Finance.
3 Public pension funds have, as well, become important agricultural investors since 2007/8 and pursue distinct investment strategies that deserve attention in their own right.
4 For example, Kuwait's SWF (Kuwait Investment Authority) was established in 1953 and Saudi-Arabia's SWF (Public Investment Fund) was established in 1971.
5 National champions are lead companies the government considers crucial for the internationalization and upgrading of China's national economy. In return for furthering industrial strategy, national champions receive high-level political support as well as financial subsidies. Many national champions are amongst China's largest companies and traditionally belong to the state-owned sector, but there is also an increasing number of privately owned champions.

References

Adecoagro. (2017a) 'Company website', available at: www.adecoagro.com, accessed 10 June 2017

Adecoagro. (2017b) 'Institutional presentation', available at: http://ir.adecoagro.com/Download/20170605140321.pdf, accessed 20 June 2017

Anderlini, J. (2014) 'China's Sovereign Wealth Fund shifts focus to agriculture', available at: www.ft.com/content/64362b08-f61a-11e3-a038-00144feabdc0, accessed 24 May 2017

Arabian Business. (2016) 'Abu Dhabi's Mubadala may invest in Russian agriculture firms', available at: www.arabianbusiness.com/abu-dhabi-s-mubadala-may-invest-in-russian-agriculture-firms-652446.html, accessed 5 July 2017

Australian Financial Review. (2017) 'Qatar's sovereign wealth fund sells another Australian farm', available at: www.afr.com/real-estate/qatars-sovereign-wealth-fund-sells-another-australian-farm-20170721-gxg9ar, accessed 10 January 2018

Australia Future Fund. (2017) 'How we invest', available at: www.futurefund.gov.au/investment/how-we-invest, accessed 29 June 2017

Avendano, R. (2010) *Sovereign Wealth Fund Investments: From Firm-Level Preferences to Natural Endowments*, Paris School of Economics, Paris

Backer, L. (2010) 'Sovereign investing in times of crisis: Global regulation of Sovereign Wealth Funds, State-Owned Enterprises, and the Chinese experience', *Transnational Law and Contemporary Problems*, vol 19, pp. 3–144

Bahrainlivestock. (2017) 'Company's website', available at: www.bahrainlivestock.com/about-us, accessed 27 June 2017

Bailey, R. and Willoughby, R. (2013) *Edible Oil: Food Security in the Gulf*, Briefing paper EER BP 2013/03, Energy, Environment and Resources, Chatham House, London

Bfm. (2017) 'Company's website', available at: http://bfm.bh/about-us/, accessed 27 June 2017

BMHC. (2017) 'Company's website', available at: www.bmhc.bh/, accessed 27 June 2017

Bortolotti, B., Fotak, V. and Loss, G. (2016) 'SWF investment in 2015', in B. Bortolotti (ed.) *The Sky Did Not Fall. Sovereign Wealth Fund Annual Report 2015*. Sovereign Investment Lab, Bocconi University, Turin, pp. 15–40

Burch, D. and Lawrence, G. (2009) 'Towards a Third Food Regime: Behind the transformation', *Agriculture and Human Values*, vol 26, no 4, pp. 267–279

Cadène, P. and Dumortier, B. (2013) *Atlas of the Gulf States*, Brill, Boston

Cheng, A. (2016) 'China's CIC boosts direct investments in search of returns', available at: www.institutionalinvestor.com/article/3583237/investors-sovereign-wealth-funds/chinas-cic-boosts-direct-investments-in-search-of-returns.html?ArticleId=3583237&single=true#/.WRl6pPmGPIV, accessed 14 May 2017

China Investment Corporation. (CIC) (2015) *China Investment Corporation Annual Report*, available at: www.china-inv.cn/wps/wcm/connect/62e332fc-8ddd-4eb9-b21c-b5cc3585e82a/CICAnnualReport2015.pdf?MOD=AJPERES&CACHEID=62e332fc-8ddd-4eb9-b21c-b5cc3585e82a, accessed 2 May 2017

China State Council. (1996) *The Grain Issue in China White Paper*, available at: http://news.xinhuanet.com/zhengfu/2002-11/18/content_633192.htm, accessed 20 April 2017

Clapp, J. (2014) 'Financialization, distance and global food politics, *Journal of Peasant Studies*, vol 41, no 5, pp. 797–814

Cotula, L. (2012) 'The international political economy of the global land rush: A critical appraisal of trends, scale, geography and drivers', *Journal of Peasant Studies*, vol 39, no 3–4, pp. 649–680

Dawajen. (2017) 'Company's website', available at: www.dawajen.bh/en/content/about-usm, accessed 27 June 2017

FAO. (2017) *FAOSTAT*, available at: www.fao.org/faostat/en/#data, accessed 30 April 2017

Fotak, V., Gao, J. and Megginson, W. (2014) 'The financial role of Sovereign Wealth Funds', in M. Wright, D. Siegel and K. Keasey (eds) *The Oxford Handbook of Corporate Governance*, Oxford University Press, Oxford, pp. 581–604

Gale, F., Hansen, J. and Jewison, M. (2015) *China's Growing Demand for Agricultural Imports*, available at: www.ers.usda.gov/publications/pub-details/?pubid=43940, accessed 10 April 2017

Gertel, J. and Sippel, S.R. (2016) 'The financialisation of agriculture and food', in M. Shucksmith and D. Brown (eds) *Routledge International Handbook of Rural Studies*, Routledge, London, pp. 215–226

GRAIN. (2008) *Seized. The 2008 Landgrab for Food and Financial Security*, available at: www.grain.org/article/entries/93-seized-the-2008-landgrab-for-food-and-financial-security, accessed 14 May 2017

Hassad Food. (2018) 'Company's website', available at: www.hassad.com/English/AboutUs/Pages/default.aspx, accessed 2 January 2018

Helleiner, E. (2009) 'The geopolitics of Sovereign Wealth Funds: An introduction', *Geopolitics*, vol 14, no 2, pp. 300–304

Hentov, E. (2015) 'How do Sovereign Wealth Funds invest? A glance at SWF asset allocation', available at: www.ifswf.org/sites/default/files/Publications/How%20do%20Sovereign%20Wealth%20Funds%20Invest_0.pdf, accessed 6 May 2017

Hoering, U. and Sausmikat, N. (2010) 'Agriculture in China: Between self-sufficiency and global integration', available at: www.asienhaus.de/public/archiv/hoering2011_agriculture_china.pdf, accessed 30 November 2016

Hofman, I. and Ho, P. (2012) 'China's "developmental outsourcing": A critical examination of Chinese global "land grabs" discourse', *Journal of Peasant Studies*, vol 39, no 1, pp. 1–48

Hornby, L. (2014) 'China's COFCO takes on global trading houses', available at: www.ft.com/content/978dae32-a452-11e3-b915-00144feab7de, accessed 10 May 2017

Investment Corporation of Dubai. (ICD) (2015) 'Consolidated financial statements', available at: www.icd.gov.ae/wp-content/uploads/2016/06/ICD-Audited-Consolidated-FS-Dec-2015.pdf, accessed 8 May 2017

Isakson, S. (2014) 'Food and finance: The financial transformation of agro-food supply chains', *Journal of Peasant Studies*, vol 41, no 5, pp. 749–775

Jessop, B. (2007) *State Power*, Polity, Cambridge/Malden

Keulertz, M. and Woertz, E. (2015) 'States as actors in international agro-investments', *International Development Policy*, vol 6, no 1, pp. 30–52

Larder, N., Sippel, S.R. and Argent, N. (2017) 'The redefined role of finance in Australian agriculture', *Australian Geographer* (Published online 23 October 2017) DOI: 10.1080/00049182.2017.1388555

Liew, L. and He, L. (2010) 'Contributing to a harmonious society: China's Sovereign Wealth Fund', in Y. Xu and G. Bahgat (eds) *The Political Economy of Sovereign Wealth Funds: Implications for Global Development*, Palgrave Macmillan, Basingstoke, pp. 26–46

Lin, S. (2017) 'State capitalism and Chinese food security governance', *Japanese Journal of Political Science*, vol 18, no 1, pp. 106–138

Locke, S. (2014) '$75 million upgrade at Port Kembla will double its grain export and make handling more efficient by rail', available at: www.abc.net.au/news/rural/2014-08-20/nrn-port-kembla-upgrade/5683460, accessed 10 May 2017

Martin, S. and Clapp, J. (2015) 'Finance for agriculture or agriculture for finance?' *Journal of Agrarian Change*, vol 15, no 4, pp. 549–559

McMichael, P. (2013) 'Land grabbing as security mercantilism in international relations', *Globalizations*, vol 10, no 1, pp. 47–64

Murphy, S., Burch, D. and Clapp, J. (2012) *Cereal Secrets: The World's Largest Grain Traders and Global Agriculture*, Oxfam International, Oxford

Naser, H. (2016) 'The role of the Gulf Cooperation Council's Sovereign Wealth Funds in the new era of oil', *International Journal of Economics and Financial Issues*, vol. 6, no 4, pp. 1657–1664

Norges Bank. (2017) 'Holdings', available at www.nbim.no/en/the-fund/holdings/, accessed 8 May 2017

OECD/FAO. (2016) *OECD-FAO Agricultural Outlook 2016-2025*, available at: http://dx.doi.org/10.1787/agr_outlook-2016-en, accessed 24 April 2017

Oliveira, G. (2015) *Chinese and Other Foreign Investments in the Brazilian Soybean Complex*, available at: www.tni.org/files/download/bicas_working_paper_9_oliveira.pdf, accessed 25 July 2017

Ouma, S. (2016) 'From financialization to operations of capital: Historicizing and disentangling the finance–farmland nexus', *Geoforum*, vol 72, pp. 82–93

Peterson, J. (2009) 'Life after oil: Economic alternatives for the Arab Gulf States', *Mediterranean Quarterly*, vol 20, no 3, pp. 1–18

Preqin. (2017) *The 2017 Preqin Sovereign Wealth Fund Review. A Comprehensive Guide to Sovereign Wealth Funds*, available at: www.preqin.com/docs/reports/2017-Preqin-Sovereign-Wealth-Fund-Review-Sample-Pages.pdf, accessed 8 May 2017

Pritchard, B., Dixon, J., Hull, E. and Choithani, C. (2016) '"Stepping back and moving in": The role of the state in the contemporary food regime', *Journal of Peasant Studies*, vol 43, no 3, pp. 693–710

Puel, J-M. (2012) 'Les fonds souverains dans l'agriculture: un investissement politique?', *Études Rurales*, 190, pp. 161–176

PwC. (2016) *Sovereign Investors 2020. A Growing Force*, available at: www.pwc.com/gx/en/sovereign-wealth-investment-funds/publications/assets/sovereign-investors-2020.pdf, accessed 8 May 2017

Reuters. (2008) 'Kuwait signed $27 bln of deals in Asian tour-paper', available at: http://uk.reuters.com/article/kuwait-asia-idUKLH49515120080817, accessed 8 May 2017

Reuters. (2015) 'China's COFCO, CIC to set up venture to run agricultural businesses', available at: www.reuters.com/article/cic-cofco-jointventure-idUSL3N0Y35CU20150513, accessed 23 April 2017

Rios-Morales, R., Ramady, M.A. and Brennan, L. (2012) 'GCC Sovereign Wealth Funds: Challenges, opportunities, and issues arising from their growing presence on the global landscape', in M. Ramady (ed.) *The GCC Economies*, Springer, New York, pp. 237–250

Russia-China Investment Fund. (RCIF) (2017) 'Website', available at: www.rcif.com, accessed 8 May 2017

Russia Reserve Fund. (2017) 'Investment management', available at: http://old.minfin.ru/en/reservefund/management/, accessed 30 June 2017

Salisu, M. and Taher, N. (2012) 'Putting sovereign wealth funds to good use: Strategic options for SWFs in the GCC region', in M. Ramady (ed.) *The GCC Economies*, Springer, New York, pp. 273–288

Shahandeh, B. (2012) 'The misconceptions regarding the GCC sovereign wealth funds', in M. Ramady (ed.) *The GCC Economies*, Springer, New York, pp. 251–260
Shen, Q. (2014) *No. 1 Central Document Targets Rural Reform*, available at: http://news.xinhuanet.com/english/china/2014-01/19/c_133057121.htm, accessed 13 May 2017
Shih, V. (2009) 'Tools of survival: Sovereign Wealth Funds in Singapore and China', *Geopolitics*, vol 14, no 2, pp. 328–344
Sippel, S.R. (2015) 'Food security or commercial business? Gulf State investments in Australian agriculture', *Journal of Peasant Studies*, vol 42, no 5, pp. 981–1001
Sovereign Wealth Center. (2017) 'Bahrain Mumtalakat Holding Company', available at: www.sovereignwealthcenter.com/fund/4/bahrain-mumtalakat-holding-company.html#.WV-dMIiGPIU, accessed 6 July 2017
SWF Institute. (2017) 'Website', available at: www.swfinstitute.org/swfs/qatar-investment-authority/, accessed 14 July 2017
SWF Institute. (2018) 'Website', available at: www.swfinstitute.org/swfs/mumtalakat-holdings/, accessed 4 January 2018
Terazono, E. and Hume, N. (2016) 'COFCO Agri eyes market listing by end of the decade', available at: www.ft.com/content/ff363410-0174-11e6-99cb-83242733f755, accessed 7 May 2017
Terazono, E. and Sanderson, H. (2015) 'COFCO teams up with Chinese fund to create global grains trader', available at: www.ft.com/content/0bce9d96-fa39-11e4-b432-00144feab7de, accessed 7 May 2017
United Nations Conference on Trade and Development. (UNCTAD) (2015) *World Investment Report. Reforming International Investment Governance*, UNCTAD, Geneva
Visser, O., Clapp, J. and Isakson, S. (2015) 'Introduction to a symposium on global finance and the agri-food sector: Risk and regulation'. *Journal of Agrarian Change*, vol 15, no 4, pp. 541–548.
Weis, T. (2007) *The Global Food Economy: The Battle for the Future of Farming*, Black Point, Fernwood Publishing
Woertz, E. (2013) *Oil for Food: The Global Food Crisis and the Middle East*, Oxford University Press, Oxford
World Grain. (2014) 'Noble, COFCO complete agribusiness joint venture', available at: http://farmlandgrab.org/post/view/23993, accessed 15 April 2017
Zurayk, R., Chaaban, J. and Sabra, A. (2011) 'Ensuring that potential Gulf farmland investments in developing countries are pro-poor and sustainable', *Food Security*, vol 3, no 1, pp. 129–137

5 Opening the black boxes of finance-gone-farming

A global analysis of assetization

Stefan Ouma

Introduction

Finance has gone farming. Since the food price and financial crises of 2007/8, we have seen a rising interest of genuinely financial investors in all things agricultural. Their appetite for establishing a new 'alternative asset class' is part of a larger run on farmland/agriculture by agribusiness companies, high-net-worth individuals and states, who have poured large sums of capital into farmland or have purchased stakes in agribusiness companies in order to grow foods, animal feeds, agri-fuels and timber. These dynamics are frequently mentioned in the finance-focused 'land rush literature' (Buxton *et al.*, 2012; Daniel, 2012; Fairbairn, 2014; Salerno, 2014; Williams, 2014), but we are yet to develop a more grounded understanding of the operations behind the assetization of farmland and agricultural production.[1] To do so, we need to open five black boxes of the first generation of finance-gone-farming studies: the long history of the finance-farmland nexus; the flesh-and-blood institutions constituting the money management industry; the uneven geographies of capital placements into farmland; the assetization process; and, the limits to assetization. The notion of the black box, going back to the work of Bruno Latour, is particularly useful to unbundle a range of relations and practices lurking behind these analytical objects. In a more conventional sense, it refers to phenomena that evade the analytical gaze of researchers because they are physically not accessible, are not captured by dominant analytical paradigms, or their contents are simply taken for granted. More specifically, it refers to 'devices, practices, or organizations that are opaque to outsiders, often because their contents are regarded as "technical"' (MacKenzie, 2005, p. 555). As will become clear, such an approach equips us with the necessary tools to repoliticize the operations of assetization at the core of global finance.

To sustain my argument, I draw on several years of research, including participant observation of agricultural investment conferences, the analysis of industry intelligence and investment brochures, and interviews with key players in the agricultural investment chain (such as fund managers and market intermediaries). The approach adopted here is a global ethnography of assetization that closely follows a number of key sites of financial *praxis* in order to grasp how capital creates its very own form of value out of agriculture.

Finance goes farming: what we do (and don't) know

In 2009, *The Economist* had a striking headline: 'No matter how bad things get, people still have to eat' (*The Economist*, 2009). The headline epitomized the mood of the time. The global financial crisis ensuing from 2008 induced great uncertainties among investors. Due to declining or negative returns on mainstream assets, a fear of rising levels of inflation caused by counter-cyclical interventions, money printing and quantitative easing in 'core countries' such as the US, low returns on savings, and a rising general distrust in complex financial products, investors searched for new 'alternative asset classes'.[2] What was suddenly in demand was less 'financial engineering' and more 'real things' (Ouma, 2014).

Farmland/agriculture is such a 'real thing', and investment brochures and industry intelligence have been making a strong case for 'agriculture as an alternative asset class' by highlighting a uniform set of market fundamentals that would sustain it. These are: a growing world population; changing dietary preferences towards meat and protein in emerging markets such as China or Indonesia; a rising demand for agri-fuels (and carbon sinks) in the light of peak oil and climate change; the limited availability of agricultural land ('peak soil'); and, stagnant or decreasing productivity levels in core production regions and climate change-induced crop failures.

The financial industry anticipates that these factors will shape future demand–supply dynamics in the agro-commodities domain in crucial ways. In light of those dynamics, a standard narrative has evolved which emphasizes that investments in agricultural operations and the underlying farmland would guarantee stable returns on capital invested. As well, their 'value' is likely to appreciate when growing demand meets growing resource scarcity. Unlike gold, a favourite during times of financial crisis, farmland/agriculture stores *and* produces capital. Additionally, investments in farmland/agriculture are said to enhance portfolio diversification and efficiency, thereby increasing the robustness of portfolios with regard to external shocks (Hennings *et al.*, 2005). These promises, hand-in-hand with the relatively low complexity of farmland-based investment instruments and the tax-allowances granted on farmland investments in many countries, has made farmland/agriculture a space of 'other investment', rendered as exceptionally secure in a turbulent world.

Starting in 2005, but accelerating since 2008, there has been a sharp increase in farmland/agriculture-based investments in both 'mature' and 'emerging markets'. This surge in agri-investments has led to the proliferation of new investment vehicles, relations, and practices. Investor brochures/handbooks (Black Sea Farming, 2012; Colvin and Schober, 2012; Garner and Brittain, 2012), industry intelligence, and prestigious investor conferences, are domains where agriculture is currently being worked upon as an 'alternative asset class'. This rising interest in farmland by investors is also reflected in the evolution of the Global Farmland Index, launched by UK firm Savills in 2012 (see Figure 5.1).[3] It shows a significant upwards movement since 2002, even though several key regions have experienced (commodity) market-induced volatilities or even declines in value since 2012.

Exact numbers on the geographies, legal and organizational architecture, and scale of financial investments into farmland/agriculture, are difficult to come by. Global data banks such as the Land Matrix (www.landmatrix.org) do not adequately represent financial investments into farmland/agriculture. These are often channelled through far-flung networks cross-cutting several jurisdictions; are protected by non-disclosure agreements or hidden from the public because of the vehicles used (for example, private equity structures); are not listed publicly and are therefore exempt from legal requirements such as annual public reports (Tumusiime, 2012); or, cannot be separated from ordinary (so-called 'strategic') agribusiness investments. Furthermore, in many countries of the global South, recording is poor and even institutions of the same state may report different figures due to vested interests or lack of coordination (Cotula, 2013). Also, asset managers or companies looking for future investors may inflate numbers, making their investments larger than they actually are.

Nevertheless, some careful estimates are possible. Industry intelligence providers indicate that over the past decade approximately US$45 billion has been invested globally in farmland by institutional capital such as

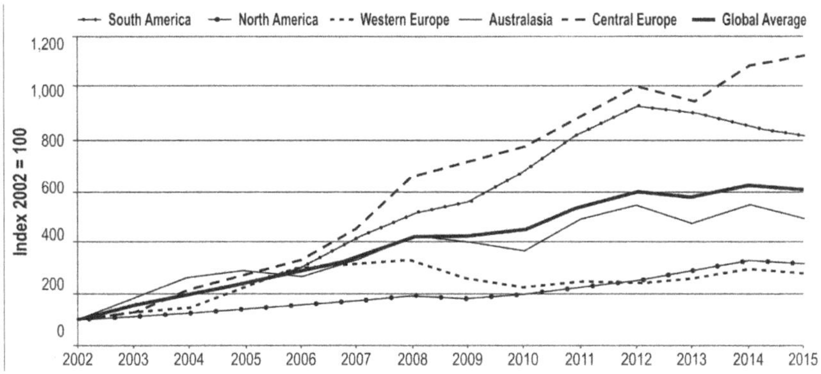

Figure 5.1 Development of the Global Farmland Index, 2002–2015.

Source: modified after Savills 2016, p. 2.

pension funds, insurance companies and endowments.[4] According to one prominent observer, institutional investment in farmland represents less than 0.5 per cent of the total value of farmland (US$1 trillion) globally (de Lapérouse, 2016, p. 1) and is still small in scale compared to the US$60 billion invested in timberland, where institutional ownership represents 20 per cent of the 'investible universe' (de Lapérouse, 2016, p. 2). Most institutional investors seem to only commit 1 per cent of their capital to direct investments in agriculture (Luyt *et al.*, 2013), which is still large in absolute terms given the amount of money these institutions manage.[5]

The long history of finance's coupling with agriculture

The existing literature often locates the financialization of agriculture within the period of neoliberalism, starting in the 1970s. However, a *longue durée* perspective suggests that finance has been co-producing agricultural landscapes in alliance with farmers and the state around the world for well over 100 years. The coupling between finance and agriculture/farming is less 'unnatural' than many critical accounts suggest. In the industry vernacular, it has been 'an integral part of the portfolio of wealthy individuals and families for centuries' (de Lapérouse, 2016, p. 4), making it the 'oldest asset class in the world'. In the US, finance and food production have been entangled since at least the late nineteenth century through commodity trading and farm insurance/financing schemes (Ouma, 2016; Koeninger, 2017). Even a deeper financial economization of farmland is not a thing of the recent past. The prominent book of Massey and Catalano (1978, p. 128) showed how institutional capital had transformed landownership in the UK, reporting that in 1973 institutions accounted for 28 per cent of all agricultural land sold (approximately 12,150 hectares) in Great Britain, and they were not the first ones to engage with this topic (Munton, 1977). In the US, with its particular institutional environment and agrarian structures, pension funds, university endowments, and wealthy families have been targeting farmland and timber more systematically since at least the late 1980s (Gunnoe and Gellert, 2011), with timber becoming a popular object of investment in the wake of the great farming crisis, which left many owner-operated farms bankrupt (Fairbairn, 2014).[6] Agricultural economist Peter Barry published an influential piece on farmland investments as early as 1980, drawing on ideas that had been circulating for at least two decades (Barry, 1980), and the first (unsuccessful) attempt to set up a farmland fund – to be sold through Merrill Lynch – occurred as early as 1977 (Koeninger, 2017). New Zealand agriculture – now heralded as one of the prime frontiers of the finance-driven land rush – coupled with global finance capital as early as the mid-1980s. For instance, superannuation funds from Australia had purchased 34 farming properties through the New Zealand Rural Property Trust by the late 1980s (Le Heron, 1991, p. 164).

These examples show that we must carefully examine how current phases of 'financialization' compare to earlier periods, as Christophers (2015a, pp. 23–24) rightly points out (see also Ouma, 2015).[7] However, just because finance has had a long (but by no means easy) entanglement with agriculture, does not mean that there is not something new about finance's run on all things agricultural. As Fairbairn (2015, p. 212) puts it:

> the expanded size and clout of institutional investors ..., the increasing prioritization of financial profits over productive profits among both financial and nonfinancial companies ... and the growing acceptance of financial investment as a morally legitimate activity ... can probably all help to explain why the historically recurring demand for land as a financial asset has taken a novel form over the last few years. These differences, though they may seem trivial from the eagle's eye view of history, have very real effects for the smallholder farmers that must now compete for land with multibillion dollar financial institutions.

But rather than mobilizing a historically blurry notion of financialization to capture these shifts (Ouma, 2016), I suggest that contemporary processes of 'financial economization' (Ouma and Bläser, 2015) in relation to farmland and other domains of the real economy should be referred to more accurately as *assetization*. Assetization entails the 'transformation of something into property that yields an income stream' (Birch, 2017, p. 468) as well as the creation of certain material, organizational, legal, and technological conditions so that future income can be effectively capitalized in a space of comparability.

The flesh-and-blood institutions behind finance-gone-farming

Abstract and disembodied narratives about Wall Street buying up farmland (a popular media heading) or the financialization of farmland (a popular academic heading) often obscure the internal heterogeneity of 'finance capital' and the culturally diverse 'flesh-and-blood institutions' that constitute the money management industry at large (Christophers, 2015b, p. 92). To get a better understanding of the operations behind farming's assetization (Ouma, 2016), we need to come to terms with the question of which actors are investing in farmland and agriculture, into which kinds of investment relations these are embedded, and which kinds of investment structures they employ to place capital into the countryside. Although some existing studies already give exploratory overviews on this (Buxton *et al.*, 2012; Fairbairn, 2013, 2014; Gunnoe, 2014; Salerno, 2014; Williams, 2014), I develop a more comprehensive heuristic in this section (see Figure 5.2).

When it comes to capital placements into farmland/agriculture, institutional investors – such as pension funds and insurance companies – are

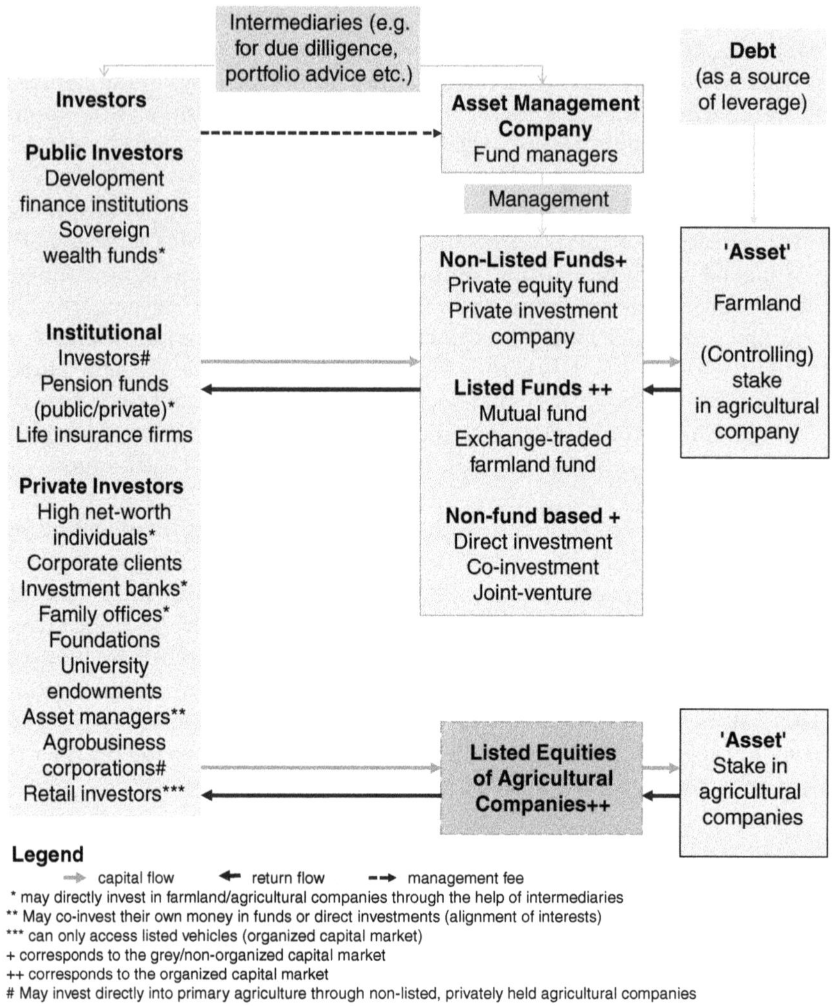

Figure 5.2 The flesh-and-blood institutions behind finance-gone-farming.

Source: revised after Ouma 2016.

the most important players. In addition, investment banks, university endowments, family offices, sovereign wealth funds, large agribusiness corporations, high net worth individuals (HNWIs), and, to a lesser extent, even retail investors have gained 'exposure' to farmland and agricultural production. Many of these players, such as family offices and HNWIs, have different 'risk appetites' than have pension funds or insurance companies. Many of these players also use a variety of investment vehicles (see Figure 5.2) such as private equity fund structures, joint-ventures or co-investment deals to channel capital into agriculture. These sophisticated investors, defined by their professional knowledge of financial markets, can be contrasted with retail investors, who are a product of the massification of financial investment among middle classes over the last decades (French *et al.*, 2011, p. 801). While these may already get exposure to farmland through mutual funds or through the acquisition of shares in listed agribusiness companies, they currently cannot get direct exposure to farmland due to the inaccessibility of more sophisticated investment vehicles to 'retail investors'. This may change in the near future due to the rise of exchange-traded farmland funds (ETFFs) such as Bonnefield in Canada, or Gladstone and the recently merged Farmland Partners and American Farmland in the US. As Fairbairn (2014) demonstrates, these investment structures represent a first attempt at securitizing farmland (Leyshon and Thrift, 2007) because they bundle income streams from the leasing out of several individual farm properties into a single vehicle that can be listed publicly (see also Stevenson, 2014).[8] This makes them tradable on stock exchanges and accessible to the masses, offering more liquidity than other investment vehicles.

Finally, development finance institutions are also playing an increasing role for agricultural investments, providing risk insurance and/or equity capital. These are also increasingly being called-upon by industry players to provide 'patient capital' (Palmer, 2010) in order to fund agricultural infrastructure in regions where it is not well developed (Brooks, 2016).

To understand the actions of investors in the 'marketplace', one has to understand how an object acquires financial value (Ortiz, 2013), or, better, financial worth which, in turn, requires an understanding of the investment rationalities and relations shaping the money management industry. The latter is led by institutional investors, acting as trustees for the original asset owners and usually delegating their funds to specialized asset managers (Clark and Monk, 2014). Their relationships are heavily contractualized and they contain as much public regulation (with regard, for example, to investor protection, information sharing, and division of labour) as they contain financial theory (MacKenzie, 2005; Ortiz, 2013). Both groups of players act according to the same models, which have been formalized since at least the 1980s, looking at a whole 'universe' of investment opportunities.

Against this backdrop, it is not surprising that many of the arguments for farmland/agriculture as an investment opportunity invoke the basics of modern portfolio management theory, which stipulates that 'diversification increases expected portfolio returns while reducing volatility' (Cumming *et al.*, 2013, p. 21). Institutional investors and delegated asset managers have to weigh future investment opportunities vis-à-vis existing ones in a whole investment 'universe' of investments (Ducastel and Anseeuw, 2017). These structural imperatives make fund managers 'face the constant challenge of maximizing returns on their investment portfolios, subject to risk and liquidity constraints' (Lins *et al.*, 1992, p. 544). While farmland/agriculture investments are not yet subjected to sophisticated mathematical portfolio construction and management techniques (Vormbusch, 2012), the underlying models still exert an influence on the performative constitution of an 'asset class in-the-making'.

Uneven geographies: where capital likes (and likes not) to go

Due to limits of space, I cannot engage with the intricacies and politics of knowledge production associated with the attempt to make finance's run on all things agricultural visible. It must be emphasized that many scholarly, NGO, and media, accounts have rendered it a 'total phenomena', seeing a structural coherence, global integration, and effective assetization where one is yet to emerge (see, for example, Buxton *et al.*, 2012; Fairbairn, 2014). Finance, with its global ambitions and inherently optimistic outlook, has equally contributed to this promethean narrative. When we zoom in, however, one arrives at a more complex and geographically variegated story. In the early land rush literature, it appeared as if investors were mainly pursuing large-scale land acquisitions in Africa, South America, and the Central and Eastern European (CEE) and Commonwealth of Independent States (CIS) regions. Africa, in particular, was frequently featured as the prime frontier region of the global land rush, including its finance-driven version (GRAIN, 2008). Today, a more differentiated picture is emerging – not only because scholars engage more critically with reported figures, but also because weighty institutional investors looking for large-scale agricultural investments seem to largely focus on North America, Brazil, Australia, New Zealand, and a few countries in Europe and Asia. Preqin (2016)[9] lists 180 fund managers with exposure to farmland, and 77 agriculture/farmland-focused funds closed since 2011, with 25 of them having a primary focus on North America and having raised US$5.7 billion, representing some 35 per cent of total capital raised (see Figure 5.3).[10]

Data regarding existing and pipelined investments in Africa need to be treated carefully. While Africa has experienced a surge in foreign investment in farmland given the *perceived* large-scale availability of cheap farmland (Cotula, 2013), and private equity funds are increasingly showing an interest in investing in African agribusiness (Silici and Locke, 2013), large institutional investments into African farmland/agriculture have been less widespread than more critical accounts tend to suggest. Even though there are some specialized asset managers who work for investors with a certain risk appetite or impact-oriented family offices or endowments, most (Western) institutional capital still shies away from investments on the continent due to potential reputational, regulatory, political, operational or market risks. Many asset managers are optimistic that this will change in the near future, partly due to the growing involvement of

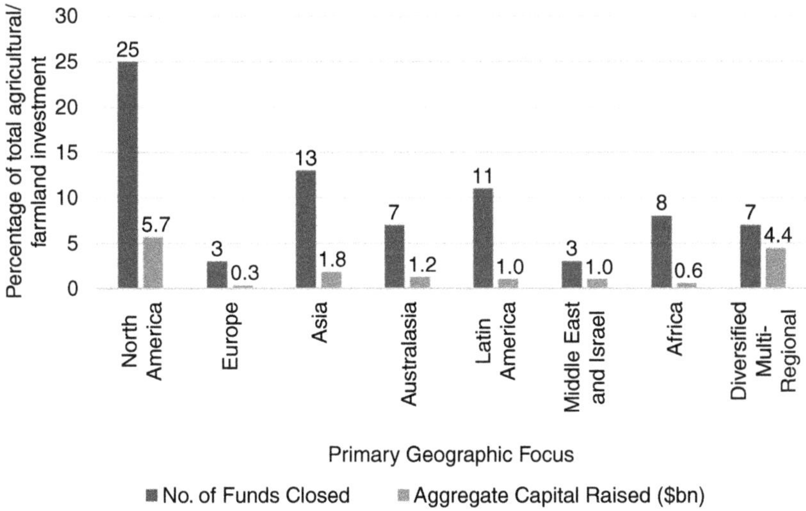

Figure 5.3 Unlisted agriculture/farmland-focused fundraising by primary geographic focus, 2011–2016 YTD (as at 20 July 2016).

Source: Preqin, 2016.

African institutional investors, but some admit that 'investment into agricultural land in Africa can be a minefield' (White, 2014). Indeed, none of the large institutional funding deals completed by 2015 had a focus on Africa (see Table 5.1).[11]

Overall, many investments still evolve on a bottom-up basis or as 'buddy transactions' – as one New Zealand asset manager put it in an interview in 2014 – not through liquid market structures or standardized procedures that could entice institutional capital. Exits have not occurred at a large scale or – if they do – are rarely talked about, indicating that some investments may have been less successful than envisioned. Furthermore, finance's engagement with farmland/agriculture has been a dynamic process, marked by moral and political struggles, as well as technical setbacks. For instance, the politics of NGOs has prompted investors to change their strategies in certain parts of the world or to refrain from certain 'geographies' altogether (White, 2014).

The assetization process

As outlined above, investors use a variety of structures/vehicles and intermediaries to assetize farmland/agriculture (Figure 5.2). The investment vehicles in use provide different 'exposures' to farmland/agriculture, which is a function of the depth of investment (direct/indirect), the

Table 5.1 The 10 largest unlisted agriculture/farmland-focused funds closed since 2016 (as at 20 July 2016)

Fund	*Firm*	*Headquarters*	*Fund size (mn)*	*Fund close date*	*Geographic focus*
TIAA-CREF Global Agriculture II	TIAA Asset Management	US	US$3,000	15 July	Global
TIAA-CREF Global Agriculture	TIAA Asset Management	US	US$2,000	12 May	Global
NCH Agribusiness Partners	NCH Capital	US	US$1,205	7 December	Central and East Europe
Paine & Partners Capital Fund III	Paine & Partners	US	US$1,204	7 April	Global
Altima One World Agriculture Development Fund	Altima Partners	UK	€756	8 November	Global
Mahaseel Agricultural Investment Fund	Kenana Agriculture	Sudan	US$1,000	12 November	MENA
Paine & Partners Capital Fund IV	Paine & Partners	US	US$893	14 December	Global
Macquarie Pastoral Fund	Macquarie Infrastructure and Real Assets (MIRA)	UK	AU$700	11 April	Australia
Black River Food Fund 2	Proterra Investment Partners	US	US$700	14 November	Emerging Markets
AMERRA Agri Fund II	AMERRA	US	US$535	13 January	North America, Latin America

Source: modified from Preqin, 2016, p. 3.

risk-return calculations of investors, the depth of property rights, the liquidity needs of capital invested, and the envisioned mode of exit. Different investment vehicles also result in different pathways of, and scopes for, assetization, express the specific 'portfolio needs' of investors, and cater to different investment cultures. Overall, liability issues are of considerable concern. As Ducastel and Anseeuw (2017, p. 201) argue, the 'source of capital ... weighs significantly on the investment policy'.

The socialities of assetization are complicated by the fact that many investors do not invest in agriculture on their own; rather, they have been approached by specialized asset managers who want to raise capital for their operations. Only a few large institutional investors such as the US retirement provider TIAA-CREF have the expertise to invest in farmland and agricultural production by themselves (TIAA-CREF, 2014). Some specialized asset managers use personal networks from previous activities in the financial sector to link up with clients, while others contract so-called 'placement agents', who are tasked with promoting an investment opportunity identified by an asset manager to potential investors. Many farmland investments are also strange hybrids – 'a combination of private equity, real estate and commodities', with private equity being perceived as a primary force (Schön, 2014).

Throughout the assetization process (see Figures 5.4, 5.5[12]), the relations between investors, asset managers, and investee farms are facilitated by a range of additional intermediaries. Besides placement agents, these include industry intelligence providers; consultants who provide due diligence services at the beginning of investment cycles; independent valuers who provide regular valuation services to establish the worth of invested assets; and, law firms providing advice on deal structuration as well as on contract corporate governance, remuneration, tax and investment law (Ouma, 2014).

While these intermediaries are often neglected in the more general literature on financialization (Pike and Pollard, 2010), they are crucial in dealing with problems of information asymmetry, information search, and coordination problems in financial markets (Bessy and Chauvin, 2013; Clark and Monk, 2014). They also play a central role when it comes to increasing 'structural coherence' (Tarim, 2012) of a still 'immature asset class' (de Lapérouse, 2016, p. 1).

The assetization of farmland/agriculture may take diverse forms. How this process unfolds depends on a range of factors. Investors and delegated asset managers calculate their decisions in a frame marked by four pillars: time, risks, returns, and access to 'resources'. Concerning the latter, investors have to take into account factors such as: availability, location and structure of land; soil, climatic, and macroeconomic conditions; property and investment regimes; the quality of regional infrastructure; and, distance to and size of commodity markets. The configuration of these pillars, and the interaction of investors with potential intermediaries

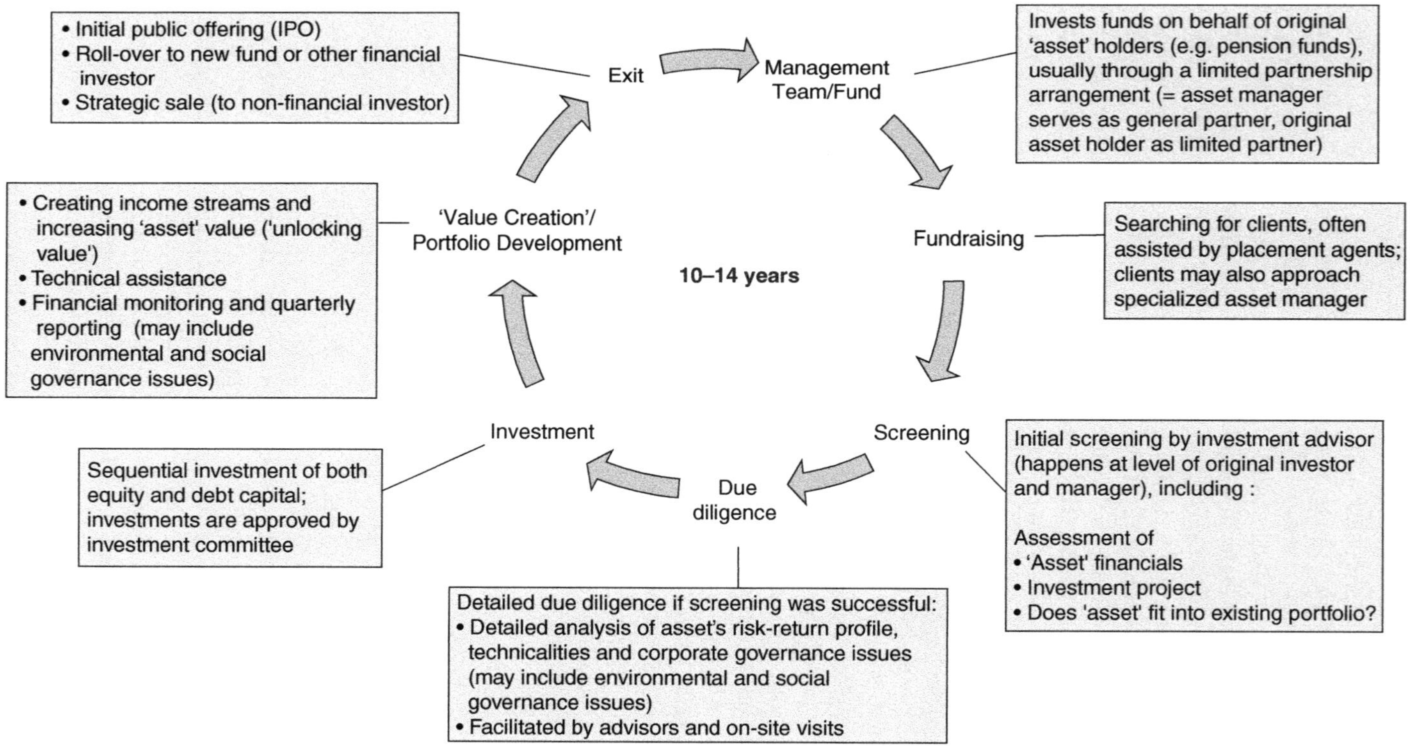

Figure 5.4 A stylized account of a private equity deal cycle in agri-finance.

Source: own research.

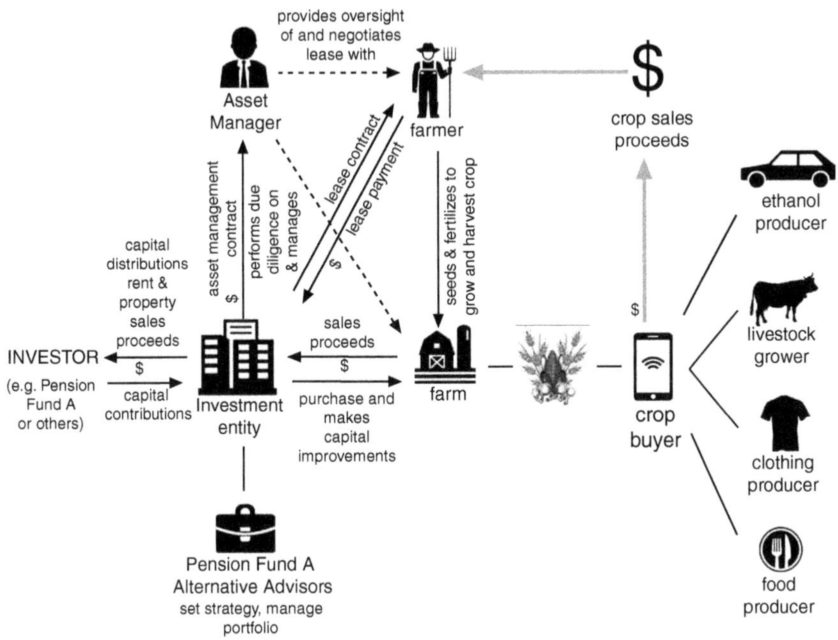

Figure 5.5 A stylized account of the architecture of farmland assetization (row crops).
Source: modified after TIAA-CREF (2014, p. 4).

such as asset management firms will result in different operational strategies, types of assets, investment geographies, and farm and firm structures (see Figure 5.6).

Although this represents a highly stylized account, it nevertheless captures some of the key aspects of farming's assetization. The mixture of these factors eventually conditions the strategy and depth of – along with the scale of and scope for – the assetization of farmland/agriculture. For instance, due to the availability of skilled farmland operators, many investors/asset managers in the US opt for an own-lease approach (Figure 5.6), where farmland is bought and leased out to farmers. This gives stable, but low, returns (Colvin and Schober, 2012). If investors/asset managers are willing to take higher risks, and have the expertise, they adopt an own-operate approach – which is also the case for regions with only a limited availability of skilled operators (e.g. Russia). In other regions, such as many parts of Africa where foreigners cannot own land on a freehold basis, investors/asset managers will actually lease land and operate it themselves. Particularly in Africa, this often involves considerable greenfield investments due to the limited availability of large-scale farms. In countries without freehold for foreigners, benefiting from 'asset price appreciation' – which rests on the premise of having secure and alienable property

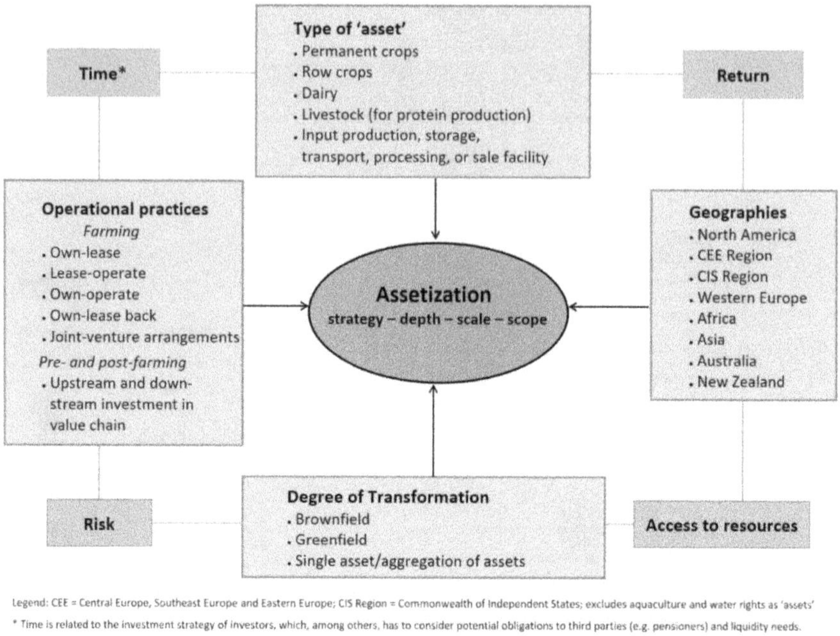

Figure 5.6 The assetization of farming: strategy, depth, scale, scope.

Source: own research.

rights over those assets – is not a straightforward affair. However, even if investors/asset managers cannot simply sell the land underlying the asset, they can still sell the lease right and the operations sustained by the land (see also Cotula, 2013). Thus, geography shapes finance's risk-return calculations. For instance, during an investment conference which I attended in 2013, industry experts reported *expected* risk adjusted-returns ranging from 8–12 per cent in North America (low-risk-low-return) to 20–30 per cent in Africa (high-risk-high-return).

The type of asset to be acquired is also an important factor in the investment calculus of asset managers. Crops may be sold in domestic, regional, and global markets – spaces shaped by different demand-supply dynamics and different currency and trade regimes. Some investors place capital across the whole agricultural value chain. Moreover, row crops such as maize and rice generally generate stable income and capital returns, while permanent crops such as fruit trees offer higher income, but also greater risk since they take longer to grow and lock-in considerable amounts of capital. Moreover, enrolling animals into financial circuits comes along with different challenges than doing so with crops, especially with regard to the valuation of assets. Valuation, such as through the discounted cash flow (DCF) model which is used to discount or actualize a future cash flow of an asset in present value, plays an important role in the world of global finance – organizing the future's attack on the present (Vogl, 2011, p. 12). While it is possible to build a dynamic DCF model for crops, 'so far the valuation of cattle remains on productivity per capita or per hectare, making it difficult to 'translate cattle production into financial language' (Ducastel and Anseeuw, 2017, p. 205). But even farmland, itself, due to its 'unique properties and unconventional accounting' is, according to Williams (2014, p. 663), 'not like most other investments' when it comes to valuation, and requires novel forms of metrics and standards.

In relation to exits, there is little 'track-record' since most of the funds established are in their fourth or fifth year. The standard narrative some years ago was that the assets would be rolled over into another fund, sold to other institutional investors, or made available to the public through an initial public offering (IPO). However, in some regions, such as many parts of Africa, regional stock exchanges are underdeveloped and IPO-oriented exits still pose a problem. A more likely strategy for such investors is a so-called trade sale, where the farmland/company is sold to a strategic buyer. The advantage to the seller rests on the fact that food processing companies, commodity traders and/or supermarkets are increasingly concerned about securing their supply chain, but want to avoid the risk of entering primary production directly, or, at least avoid entering it as pioneering investors.

The limits to assetization

Many critical observers attribute to finance a desire to assetize almost everything (Leyshon and Thrift, 2007; Birch, 2017). While we should not underestimate the power of financial elites and the problematic aspects of how finance extracts value from nature, economy, and society more generally, finance also often encounters limits in its venture to assetize the world. As should be clear by now, the run on farmland and agriculture by finance capital has been less spectacular than many accounts suggest. While we have been seeing rising interest by investors in farmland/agriculture since 2008, with institutional investments in farmland growing annually at 8–10 per cent over the past decade (de Lapérouse, 2016), institutional investors have allocated less capital to this 'alternative asset class' than many asset managers expected. This fact matters given the sheer amount of capital that institutional investors manage. Their financial power makes these 'supertankers', as one interviewee put it, market-makers in their own right. Without the engagement of large institutional investors there is no 'alternative asset class'. While many institutional investors are seeking an increased exposure to farmland and agriculture, a range of 'entry barriers' has deterred many others. There are seven main reasons for this.

First, there have been some spectacular failures by pioneering funds run by 'Wall Street guys', often targeting agrofuel investments. As one placement agent put it during an interview, their schemes, in combination with an often poorly incentivized fund and farm management, has put off many institutional investors and 'tilted the market in such ways that no one trusts each other anymore'. That one needs less financial engineering and more agricultural expertise has been mentioned in several of my interviews, with interviewees carefully distancing themselves from an 'investment banking mentality'.

Second, identifying and hiring asset managers who can combine the best of both worlds and who have a track-record seems to be a challenge due to the tender age of the 'asset class'. Many institutional investors cannot invest with first-time fund managers. However, in more recent times, structures have evolved where 'an investment management firm has created a joint venture with a farming operator, or an ex-farmer has joined forces with an ex-banker to establish a new investment firm' (Burwood-Taylor, 2014a) in order to create an institutional-grade investment structure.

Third, institutional investors have been scared away by high fee structures imported from other private equity domains. Critical industry observers argue that these fee structures neither reflect the quality of services asset managers are providing nor that agriculture is a long-term venture and may have to be run according to different performance criteria.

Fourth, many medium-sized institutional investors, who often do not have in-house expertise, are unsure how farmland and agriculture would fit within their portfolios – whether it is a real estate, private equity, or infrastructure 'play'. Here, the words of the representative of a European asset manager, whom I interviewed in 2014, are indicative:

> At the beginning, you always have to preach and explain. Agriculture is a system with … many equations and even more unknowns. This means that many decisions are made under conditions of uncertainty. You can never capture all factors 100 per cent. This is not industrial production. And I have the volatility, which is not only caused by prices, which everybody could deal with in a way or another, but also by physical yields. And these often multiply themselves. This means that yield volatility can be a multiple of the price volatility. And investors need to understand this. They need to know what they buy into, which is an average of 8 per cent return of capital (and) which may (produce) a robust return over 10 years … but can also range from minus 2 to plus 20, and investors need to understand this.

Fifth, the uncertainty looming over the 'asset class' is aggravated by the lack of structured, standardized, instruments – fairly similar to a situation experienced with real estate and infrastructure when they emerged as asset classes some 20 years ago (Ouma and Bläser, 2015). This situation poses a problem to the circulation of capital since 'financial markets can only be assumed to exist when there are routinized, systematic forms of trading, relatively stable settings, a minimal degree of standardization of financial securities, and established cognitive procedures for their evaluation' (Knorr Cetina and Preda, 2005, p. 6). While many observers believe that intrasectoral diversity is just a temporary phenomenon that disappears as the 'asset class' matures, many industry players are convinced that 'before the asset class can really develop and attract a broader range of investors, the industry will need to come up with some established structuring norms' (Burwood-Taylor, 2014b). Indeed, some asset managers have teamed up with specialized intermediaries to develop more structured investment approaches, which manage risks, returns, and financial flows in a more systematic way.

Sixth, many institutional investors have minimum investment thresholds, sometimes up to US$200 million per deal. This limits the range of farmland properties such investors can target. Globally, very few farming properties have this worth.

Seventh, as mentioned earlier, there has been reluctance to invest into certain geographies. While more structured investment approaches, as well as the adoption of responsible investment principles, are supposed mitigate political and reputational risks associated with certain geographies, it is clear that many Western institutional investors – aware of the potential of being hounded or attacked by critical NGOs – are avoiding investing in certain regions.

All-in-all, agriculture as an 'alternative asset' still suffers from 'structural incoherence' – a situation where analysts or market participants classify and interpret a financial object according to different heuristics (Zuckerman, 2004, p. 421).[13] It stems, in part, from the limited knowledge the finance industry has of agriculture (Williams, 2014) and its difficulty in translating the material world of farming into established frames of calculation and classification. This finding is at odds with the attempt of the financial industry and critical scholars alike to frame farmland as 'just another asset class'.

Conclusion: repoliticizing assetization

Preda (2013, p. 26) recently posed the question 'what is at stake in attempting to open the black box of finance?' Indeed, critical voices have argued that an 'opening the black box approach' – despite its valuable insights into micro-foundations of 'the social' or 'the economic' – may yet be blind to the big issues of financialized global capitalism. For some, the most disempowering outcome might be to open the black box only to 'find … it empty' (Winner, 1993) – emptied of power, culture, interests, inequality, class, race, gender, exploitation, surplus value, and so forth.

While this concern must be taken seriously, there is room for 'a micro-founded political economy of the investment chain' (Braun, 2016, p. 6). For Martin *et al.* (2008, p. 128):

> Reckoning finance into a practical activity discloses capital's own methods such that they might be both reappropriated and redeployed ... [I]t is an effort to specify what capital's movement does, both to itself and across a whole range of social sites and activities.

Such an effort is tightly associated with the repoliticization of modern finance's operations. This can happen along many lines, four of which are particularly important for this contribution. First, establishing how money creates more money can help us challenge the epistemic authority of financial elites who often claim to be the only knowledgeable subjects to speak about the operations of 'financial markets'. The ambition to understand the circulation of capital is fundamentally tied to the project of technical democracy (Callon *et al.*, 2009). Second, there is no such thing as an intrinsic value that just waits to be unlocked from the materiality of nature or economic life: instead, finance actively constructs *and* manages these values. This is a fundamentally normative activity. We need to open up these norms to a debate that is about much more than merely economics or technicalities. It is a fundamentally moral debate. Third, and tightly associated with how finance constructs value, is its problematic futurity. Key to the assetization of farmland and other economic domains is that future expectations about value exert operational pressures on the present – the present becomes a slave to the future. How this impacts on the management of farms, labour, property and nature must be a key concern of future research efforts. Finally, we need to recognize that the rise of the asset management industry has tied the reproduction of societies to the reproduction of various forms of capital (Vogl, 2011, p. 178). Indeed, besides organizations serving social elites, many pension funds investing in farmland are rather servicing the needs of the global 'mass affluent' (Seabrooke and Wigan, 2017, p. 4) – a direct result of the political drive to establish capital market-based pension systems around the world. Externalizing the costs of reproduction to new frontiers with the help of the asset management industry exacerbates existing inequalities. This happens through the further advancement of a variety of disjunctures: between value creation and wealth allocation (Seabrooke and Wigan, 2017, p. 2); between land/farm ownership and location; between surplus allocated to capital and surplus allocated to labour (Cochet, 2017); and, between different social groups with uneven access to such schemes of wealth generation and to the 'underlying assets'.

Notes

1 Note that 'agriculture as an asset class' comprises investments beyond farmland and primary production, such as soft commodities or investments along the value chain. I will exclude these from a detailed analysis, although I shall refer to them occasionally.

2 Traditional asset classes usually include stocks and government/corporate bonds. More 'mature' alternative asset classes include private equity (venture, buyout and mezzanine capital), hedge funds, commodities, infrastructure, and real estate investment trusts (REITS).

3 No data for Africa are available due to the opaque nature of regional land markets.

4 The term institutional investor does not have a common standard definition. One feature often mentioned is that these are 'not physical persons' but 'are organised as legal entities' (Celik and Isaksson, 2014, pp. 95–96). These can assume a wide variety of legal forms, from closed-end investment companies to private equity like limited partnerships to sovereign wealth funds. They 'may act independently or be part of a larger company group or conglomerate' (Celik and Isaksson, 2014, p. 96), such as mutual funds, which are often subsidiaries of banks or insurance companies. At times, the term is used synonymously with 'intermediary investors', i.e. an 'institution that manages and invests other people's money' (Celik and Isaksson, 2014, p. 96), but in some cases institutional investors may be the ultimate asset owners. They can be contrasted to retail investors, who are individuals and who can only access certain types of financial products.

5 Some older sources claimed that this would increase to a new optimum of 3–5 per cent over coming years (GRAIN, 2011).

6 As de Lapérouse (2016, p. 1) notes, this was driven by *public* regulation in 1974 following new rulings instituted in the US in 1974 by the Internal Revenue Service (IRS) and the passage of Employee Retirement Income Security Act (ERISA), in which qualified pension plans were mandated to diversify a certain portion of their investments beyond their traditional allocation to fixed income securities to alternative investment vehicles (including equities and commercial real estate in order to minimize the risk of large losses).

7 In light of this, one could even say that the 'financialization' of infrastructure – now an established 'alternative asset class' – is nothing new. The railways in the US could not have been built without what in the nineteenth century counted as international finance (Chandler, 1977, p. 90).

8 Thus, they basically mimic listed real estate investment trusts (REITs), which:

> enable investors to access the commercial real estate asset class indirectly through ownership of equity shares in a company whose assets consist primarily of commercial properties or mortgages and whose revenues derive primarily through commercial property leases or mortgage payments.
>
> (Baker and Filbeck, 2013, p. 9)

9 Obviously, not all financial investments in farmland use a fund structure, so these data need to be treated carefully.

10 Note, however, that:

> the pipeline for agriculture/farmland focused funds is strong, with 48 funds in the market, targeting US$12.8 billion in aggregate capital.... Sixteen of these primarily focus on North America, targeting US$3.5 billion in aggregate capital. Outside North America, the largest amount of capital is being sought for Africa, with five funds targeting US$2.6 billion.
>
> (Preqin, 2016, p. 4)

11 Nevertheless, investors from inside and outside Africa have shown an increasing interest in less risky brownfield projects (e.g. investing into existing production or processing companies) (Silici and Locke, 2013). Moreover, asset managers, banks and development finance institutions (DFIs) have worked out new structures to channel capital into African farmland and agriculture, promoting this as 'impact investing' or 'patient capital' (Brooks, 2016; Mawdsley, 2018).

12 For a slightly different set-up of a fund-based investment into South African agriculture, see Figure 1 in Ducastel and Anseeuw (2017).

13 It should be noted, here, that Zuckerman (2004) originally coined this term for his analysis of another 'asset class': stocks.

References

Baker, H. and Filbeck, G. (2013) 'Alternative investments: An overview', in H. Baker and G. Filbeck (eds) *Alternative Investments: Instruments, Performance, Benchmarks, and Strategies*, John Wiley and Sons, Hoboken

Barry, P. (1980) 'Capital asset pricing and farm real estate', *American Journal of Agricultural Economics*, vol 62, no 3, pp. 549–553

Bessy, C. and Chauvin, P.-M. (2013) 'The power of market intermediaries: From information to valuation processes', *Valuation Studies*, vol 1, no 1, pp. 83–117

Birch, K. (2017) 'Rethinking value in the bio-economy: Finance, assetization, and the management of value', *Science, Technology and Human Values*, vol 42, no 3, pp. 460–490

Black Sea Farming (2012) 'An investor's guide to agricultural opportunities in today's challenging economic climate', available at: http://blackseaagriculture.com/_guide_op/BSAs_Guide_to_Agriculture_Opportunities.pdf, accessed 26 March 2013

Braun, B. (2016) 'Gross, greed, and ETFs: The case for a micro-founded political economy of the investment', *Economic Sociology – The European Electronic Newsletter*, vol 21, no 3, pp. 6–13

Brooks, S. (2016) 'Inducing food insecurity: Financialisation and development in the post-2015 era', *Third World Quarterly*, vol 37, no 5, pp. 768–780

Burwood-Taylor, L. (2014a) 'Overcoming first fund fatigue', available at: www.agriinvestor.com/overcoming-first-fund-fatigue, accessed 08 July 2014

Burwood-Taylor, L. (2014b) 'Investment structure choice is good – for now', available at: www.agriinvestor.com/investment-structure-choice-good-now/, accessed 6 June 2014

Buxton, A., Campanale, M. and Cotula, L. (2012) 'Farms and funds: Investment funds in the global land rush', *IIED Briefing Papers*, available at: http://pubs.iied.org/17121IIED.html, accessed 20 March 2013

Callon, M., Lascoumes, P. and Barthe, Y. (2009) *Acting in an Uncertain World: An Essay on Technical Democracy*, MIT Press, Cambridge, Mass

Celik, S. and Isaksson, M. (2014) 'Institutional investors and ownership engagement', *Financial Market Trends*, vol 2, pp. 93–114

Chandler, A. (1977) *The Visible Hand: The Managerial Revolution in American Business*, Belknap Press, Cambridge, Mass

Christophers, B. (2015a) 'The limits to financialization', *Dialogues in Human Geography*, vol 5, no 2, pp. 183–200

Christophers, B. (2015b) 'Against (the idea of) financial markets', *Geoforum*, vol 66, pp. 85–93

Clark, G. and Monk, A. (2014) 'The geography of investment management contracts: The UK, Europe, and the global financial services industry', *Environment and Planning A*, vol 46, no 3, pp. 531–549

Cochet, H. (2017) 'Capital–labour separation and unequal value-added distribution: Repositioning land grabbing in the general movement of contemporary agricultural transformations', *Journal of Peasant Studies*, vol 1, no 191, pp. 1–22

Colvin, G. and Schober, T. (2012) *Investors' Guide to Farmland*, Createspace, New York

Cotula, L. (2013) *The Great African Land Grab? Agricultural Investments and the Global Food System*, Zed Books, London

Cumming, D., Hap, L. and Schweizer, D. (2013) 'The role of alternative investments in strategic asset allocation', in K. Baker and G. Filbeck (eds) *Alternative Investments: Instruments, Performance, Benchmarks, and Strategies*, John Wiley and Sons, Hoboken, pp. 19–36

Daniel, S. (2012) 'Situating private equity capital in the land grab debate', *Journal of Peasant Studies*, vol 39, no 3–4, pp. 703–729

de Lapérouse, P. (2016) *Agriculture: A New Asset Class Presents Opportunities for Institutional Investors*, Global AgInvesting, Boston

Ducastel, A. and Anseeuw, W. (2017) 'Agriculture as an asset class: Reshaping the South African farming sector', *Agriculture and Human Values*, vol 34, no 1, pp. 199–209

Fairbairn, M. (2013) 'Just another asset class? Neoliberalism, finance and the construction of farmland investment', in S. Wolf and A. Bonanno (eds) *The Neoliberal Regime in the Agri-Food Sector: Crisis, Resilience and Restructuring*, Earthscan, London, pp. 245–262

Fairbairn, M. (2014) '"Like gold with yield": Evolving intersections between farmland and finance', *Journal of Peasant Studies*, vol 41, no 5, pp. 777–795

Fairbairn, M. (2015) 'Reinventing the wheel? Or adding new air to old tires?' *Dialogues in Human Geography*, vol 5, no 2, pp. 210–213

French, S., Leyshon, A. and Wainwright, T. (2011) 'Financializing space, spacing financialization', *Progress in Human Geography*, vol 35, no 6, pp. 798–819

Garner, D. and Brittain, W. (2012) *Farmland as an Alternative Investment Asset Class. Fundamentals – Characteristics – Performance – Opportunities – Risks*, DGC Asset Management, Northampton

GRAIN. (2008) *Seized! The 2008 Land Grab for Food and Financial Security*, GRAIN Briefing, Barcelona

GRAIN. (2011) *Pension Funds: Key Players in the Global Farmland Grab*, GRAIN Briefing, Barcelona

Gunnoe, A. (2014) 'The political economy of institutional landownership: Neorentier society and the financialization of land', *Rural Sociology*, vol 79, no 3, pp. 478–504

Gunnoe, A. and Gellert, P. (2011) 'Financialization, shareholder value, and the transformation of timberland ownership in the US', *Critical Sociology*, vol 37, no 3, pp. 265–284

Hennings, E., Sherrick, B. and Barry, P. (2005) 'Portfolio diversification using farmland investments', *Selected Paper prepared for Presentation at the American Agricultural Economics Association Annual Meeting*, Providence, Rhode Island, 24–27 July

Knorr Cetina, K. and Preda, A. (2005) 'Introduction', in K. Knorr-Cetina and A. Preda (eds) *The Sociology of Financial Markets*, Oxford University Press, Oxford, New York, pp. 1–14

Koeninger, J. (2017) 'History of institutional farmland investment', available at: www.globalaginvesting.com/institutional-farmland/, accessed 1 June 2017

Le Heron, R. (1991) 'New Zealand agriculture and changes in the agriculture-finance relation during the 1980s', *Environment and Planning A*, vol 23, pp. 1653–1670

Leyshon, A. and Thrift, N. (2007) 'The capitalization of almost everything: The future of finance and capitalism', *Theory, Culture and Society*, vol 24, no 7–8, pp. 97–115

Lins, D., Sherrick, B. and Venigalla, A. (1992), 'Institutional portfolios: Diversification through farmland investment', *Real Estate Economics*, vol 20, no 4, pp. 549–571

Luyt, I., Santos, N. and Carita, A. (2013) *Emerging Investment Trends in Primary Agriculture: A Review of Equity Funds and other Foreign-led Investments in the CEE and CIS Region*, FAO Investment Centre, Rome

MacKenzie, D. (2005) 'Opening the black boxes of global finance', *Review of International Political Economy*, vol 12, no 4, pp. 555–576

Martin, R., Rafferty, M. and Bryan, D. (2008) 'Financialization, risk and labour', *Competition and Change*, vol 12, no 2, pp. 120–132

Massey, D. and Catalano, A. (1978) *Capital and Land: Landownership by Capital in Great Britain*, E. Arnold, London

Mawdsley, E. (2018) 'Development geography II: Financialization', *Progress in Human Geography*, vol 42, no 2, pp. 264–274

Munton, R. (1977) 'Financial institutions and their ownership of agricultural land', *Area*, vol 9, pp. 29–37

Ortiz, H. (2013) 'Financial value: Economic, moral, political, global', *HAU: Journal of Ethnographic Theory*, vol 3, no 1, pp. 64–79

Ouma, S. (2014) 'Situating global finance in the land rush debate: A critical review', *Geoforum*, vol 57, pp. 162–166

Ouma, S. (2015) 'Getting in between M and M' or: How farmland further debunks financialization', *Dialogues in Human Geography*, vol 5, no 2, pp. 225–228

Ouma, S. (2016) 'From financialization to operations of capital: Historicizing and disentangling the finance–farmland–nexus', *Geoforum*, vol 72, pp. 82–93

Ouma, S. and Bläser, K. (2015) 'Räume der Kalkulation, Kalkulation des Raumes', *Zeitschrift für Wirtschaftsgeographie*, vol 59, no 4, pp. 214–229

Palmer, K. (2010) 'Agricultural growth and poverty reduction in Africa. The case for patient capital', available at: www.evidenceondemand.info/agricultural-growth-and-poverty-reduction-in-africa-the-case-for-patient-capital, accessed 1 March 2016

Pike, A. and Pollard, J. (2010) 'Economic geographies of financialization', *Economic Geography*, vol 86, no 1, pp. 29–51

Preda, A. (2013) 'Pandora's box: Opening up finance to STS investigations', *Technoscienza*, vol 2, no 3, pp. 25–35

Preqin. (2016) *Preqin Special Report: Agriculture*, available at: www.preqin.com/docs/reports/Preqin-Special-Report-Agriculture-September-2016.pdf, accessed 1 March 2017

Salerno, T. (2014) 'Capitalising on the financialisation of agriculture: Cargill's private equity-driven land acquisition in the Philippines', *Third World Quarterly*, vol 35, no 9, pp. 1709–1727

Schön, D. (2014) 'Farm management, not real estate, is key to agri investing', available at: www.agriinvestor.com/?p=4952, accessed 01 July 2016

Seabrooke, L. and Wigan, D. (2017) 'The governance of global wealth chains', *Review of International Political Economy*, vol 24, no 1, pp. 1–29

Silici, L. and Locke, A. (2013) *Private Equity Investments and Agricultural Development in Africa: Opportunities and Challenges*, Future Agricultures, Brighton

Stevenson, A. (2014) 'Cash crops with dividends: Financiers transforming strawberries into securities', available at: http://dealbook.nytimes.com/2014/07/21/cash-crops-with-dividends-fi, accessed 21 June 2014

Tarim, E. (2012) 'Storytelling and structural incoherence in financial markets', *Journal of Interdisciplinary Economics*, vol 24, no 2, pp. 115–144

TIAA-CREF. (2014) *Responsible Investment in Farmland 2013: Report on Ethical Conduct and Responsible Stewardship of the Environment*, TIAA-CREF, New York

The Economist. (2009) 'Green shoots', available at; www.economist.com/node/13331189, accessed 18 September 2012

Tumusiime, E. (2012) 'Sector-wide analysis of US: Investment in large-scale land acquisitions', *Paper Presented at the International Academic Conference on Global Land Grabbing*, Ithaca, 17–19 October

Vogl, J. (2011) *Das Gespenst des Kapitals*, 2nd Edition, Diaphanes, Zürich

Vormbusch, U. (2012) 'Zahlenmenschen als Zahlenskeptiker. Daten und Modelle im Portfoliomanagement', in H. Kalthoff and U. Vormbusch (eds) *Soziologie der Finanzmärkte*, Transcript, Bielefeld, pp. 313–338

White, A. (2014) 'African farming is the new frontier for brave investors', available at: www.telegraph.co.uk/finance/economics/10901154/African-farming-is-the-new-frontier-for-brave-investors.html, accessed 26 June 2014

Williams, J. (2014) 'Feeding finance: A critical account of the shifting relationships between finance, food and farming', *Economy and Society*, vol 43, no 3, pp. 401–431

Winner, L. (1993) 'Social constructivism: Opening the black box and finding it empty', *Science as Culture*, vol 3, no 3, pp. 427–452

Zuckerman, E. (2004) 'Structural incoherence and stock market activity', *American Sociological Review*, vol 69, no 3, pp. 405–432

6 Farmland values

Media and public discourses around farmland investment in Canada and Australia

André Magnan

Introduction

Since the mid-2000s, Canada and Australia have become attractive targets for private and institutional investment in farmland. Both countries have highly industrialized, export-oriented, agricultural sectors and low farmland values by comparison to other countries in the global North (Savills, 2012). As a result of these and other factors, a range of domestic and foreign investors including pension funds, sovereign wealth funds, private equity companies, and wealthy individuals have acquired large tracts of farmland in each country (Magnan, 2015; Sippel, 2015; ATO, 2016; Desmarais *et al.*, 2017). Media coverage of these trends – which increased dramatically in the wake of the 'global food crisis' of 2008 – sparked public controversies over farmland investment and ownership. As large farmland deals were publicized, critics warned of 'selling the farm', 'losing our grip', and undermining family farmers, while others insisted on the benefits of finance's newfound interest in agriculture. In the wake of these controversies, governments in each country tightened their rules around farmland ownership. Cases such as these provide an opportunity to examine the politics of farmland across social contexts.

In this chapter, I trace the development of the 'farmland controversy' in each country through an analysis of media and public discourses between 2008 and 2014. In Canada, I focus on the case of Saskatchewan, the country's dominant agricultural producer and exporter.[1] I compare competing discourses on the benefits and risks of farmland investment; the role of the state in regulating investment flows and farmland ownership; the motivations of financial investors; and, the economic, social, and cultural values attached to farmland in each country. While there is an overlap in some of the discourses mobilized in each 'farmland controversy', cross-national differences in political economy, culture, and institutional contexts have led to debates with different parameters in each country. The cases reveal differences in how actors in each country view the relationship between farming, finance, and society, and as a consequence, how the financialization of agriculture is accommodated and contested.

Farmland values: financialization, discourse, and politics

As part of a larger literature on the financialization of agri-food (Burch and Lawrence, 2009, 2013; Clapp, 2012; Fairbairn, 2014b; Isakson, 2014; Williams, 2014; Field, 2016), there is growing interest in the financialization of farmland (Gunnoe, 2014; Ouma, 2014; Visser, 2015, 2017; Ducastel and Anseeuw, 2017). A number of studies have examined how financial actors of different stripes have acquired farmland in the global North, applying new business models and financial logics to farmland ownership and agricultural production (Fairbairn, 2014b; Mittal and Moore, 2014; Desmarais *et al.*, 2015; Magnan, 2015; Sippel, 2015; Sommerville and Magnan, 2015; Sippel *et al.*, 2017). These studies have documented the scale and scope of farmland investment, the motivations and business models of investors and, to a more limited extent, the tensions and debates around the financialization of farmland.

The financialization of farmland is a contested process, mediated by politics, discourse, and culture. Financial actors use a variety of discourses to promote farmland investment, to 'assemble' it as a financial asset, and to legitimize the financialization of agriculture (Fairbairn, 2014a; Li, 2014; Larder *et al.*, 2015; Visser, 2015). Yet, as several studies have noted, the process of turning farmland into a profitable investment is neither straightforward, nor inevitable (Magnan, 2015; Visser, 2015, 2017; Kuns *et al.*, 2016). Investment schemes sometimes collapse and investor interest ebbs and flows. In some ways, farmland is a less than an ideal financial asset – its ownership is highly fragmented, it is notoriously illiquid, and is subject to political risk. For these reasons, scholars have recognized that the financialization of farmland requires political, discursive, and cultural work (Fairbairn, 2014a; Li, 2014, 2015; Visser, 2017). As part of this discursive work, financial actors have drawn upon a range of legitimizing discourses to justify their activities. For instance, it is common for investors to connect financialization to rising global demand for food, suggesting that finance capital can provide the investments necessary to unlock the agricultural potential of different regions (Li, 2014; Larder *et al.*, 2015).

Tania Li (2014, 2015) has argued that land is inherently social and therefore political. Though it can be commodified to varying degrees, land is always more than a commodity – it embodies a series of material, cultural, and discursive qualities. Aside from its primary use is agricultural production, farmland has a number of other meanings and values (Desmarais *et al.*, 2017). Furthermore, because claims on land always involve the exclusion of some would-be users, 'political risk in land deals is endemic' (Li, 2015, p. 567). While Li's work has focused on 'land grabs' in the global South, her perspective can be extended to other social contexts. In the global North, farmland investors operate in well-established land markets in jurisdictions with strong government institutions and the rule of law. Yet, to the degree that investor purchases displace or exclude other buyers like family farmers or locals, they can be politically fraught. For farm families, farmland may be the source of economic security, a material link to one's ancestors, the site of one's residence, a connection to place, and a site of recreation. Changes to landownership patterns, whether real or perceived, can lead to strong reactions in rural communities, the farming sector, and society at large. As Sippel *et al.* (2017) have shown in their Australian field work, local community responses can range from 'acceptance' to 'accommodation' to 'unease'.

As scholarship on the financialization of agriculture matures, there is a need for more detailed analyses of farmland politics. My aim in this chapter is to provide a systematic comparison of debates around financialization in two industrialized countries. The chapter focuses on the arguments and discourses mobilized by actors such as investors, farmers, farm organizations, governments, and others, and how these construct controversies over farmland, including the process of re-regulation. Furthermore, I identify unstated assumptions that characterized the debates in each country. In doing so, the analysis can contribute to a more fully developed political debate around crucial questions such as: Who should own farmland and to what end? What is the value of farmland? How should the state regulate farmland ownership?

Context

In this section, I provide a brief overview of farmland financialization in Canada and Australia since the mid-2000s. In Saskatchewan, the financialization trend dates to the mid-2000s, when a small number of specialized farmland investment companies were launched (Sommerville and Magnan, 2015). These companies emerged in the wake of Saskatchewan's partial liberalization of farmland rules, in 2003, which extended ownership to Canadians outside the province for the first time since the 1970s. While the new rules opened new opportunities for Canadian investors, non-Canadians continued to be prohibited from purchasing Saskatchewan farmland. From about 2005 onward, investors began building substantial portfolios of Saskatchewan farmland, with investment activity peaking in 2012 (Magnan and Sunley, 2017). The investors have operated under a variety of structures including private equity funds, pension plans, private trusts, family-owned investment companies, and companies offering individually managed farmland portfolios (Magnan, 2015; Sommerville and Magnan, 2015). Desmarais *et al.* (2017) found that investors, including the entities reported in Magnan (2015) plus wealthy individuals and other lower-profile buyers, had acquired more than 338,000 hectares of land by 2014. Farmland prices increased dramatically – on average 16 per cent per year – between 2008 and 2015 (FCC, 2016a), the years of peak investment activity. Although it is clear that investors paid significantly more than other buyers for farmland (Magnan and Sunley, 2017), it is impossible to say to what extent investors contributed to farmland price inflation in the province.

Between 2008 and 2014, there was modest public and media interest in the activities of the *farmland* investors. When the Canadian Pension Plan entered the scene, however, controversy erupted. In late 2013, the plan's investment arm, the Canada Pension Plan Investment Board (CPPIB), announced it was purchasing the 115,000 acre portfolio of Assiniboia Capital Corp., a farmland investment firm (*Regina Leader-Post*, 13 December, 2013). The C$128 million farmland purchase was the largest commercial sale of farmland in recent memory – perhaps ever – in the province. The leader of the Progressive Conservative[2] party launched a campaign to oppose the deal, garnering substantial media coverage. As the debate intensified, the provincial government announced a review of farmland ownership rules in 2015. Based on the results of a public consultation, which revealed strong public opposition to foreign and institutional ownership of farmland (Government of Saskatchewan, 2015c), the government introduced tighter rules in late 2015. Specifically, the government prohibited pension funds and trusts of more than ten people from acquiring Saskatchewan farmland, increased penalties for non-compliance with ownership restrictions, and closed certain loopholes around the financing of farmland purchases (Government of Saskatchewan, 2015a).

In Australia, a wide range of foreign and domestic actors have been acquiring farmland and other agribusiness assets, contributing to the financialization of the agri-food sector (Lawrence and Campbell, 2014; Magnan, 2015; Sippel, 2015). In an earlier study, I documented a number of major players including pension funds, private investors, and specialized agricultural investment funds that had together purchased more than AU$2 billion of farmland in Australia by 2014 (Magnan, 2015). Since there are no laws prohibiting non-Australians from acquiring farmland in any State, many of these purchases were made by foreign investors, or firms backed by a combination of domestic and foreign capital. Domestic investors have had a more limited involvement, though there are some Australian firms acquiring farms. According to a government agency, the amount of farmland with some degree of foreign ownership increased by 4.7 million hectares between 2010 and 2013 (ABS, 2014). By 2016, the government reported that the proportion of agricultural land held by non-Australians had risen to 13.6 per cent (ATO, 2016). By size of farmland holdings, the UK, the US, the Netherlands, Singapore, and China, were the top five non-Australian farmland owners (ATO, 2016).

There has been a vigorous public debate around farmland ownership in Australia since 2010 (Larder *et al.*, 2015; Sippel, 2015). From 2011 to 2013, the Australian Senate held a widely publicized inquiry into foreign takeovers in the agribusiness sector. The testimony and reporting from the inquiry raised a number of concerns about foreign investors, including questions about the motivations of government-backed entities, threats to

Australian sovereignty, and tax avoidance (Senate of Australia, 2012, 2013). In 2012, controversy erupted over the sale of Cubbie Station, a nearly 1,000 km^2 sheep station, to Chinese investors (Grubel and Yueyang, 2012). This and other large-scale investments, especially the activities of Hassad, a subsidiary of the Qatari sovereign wealth fund (Sippel, 2015), raised the profile of the foreign investment trend. These issues simmered for several years, drawing increasing public scrutiny and political responses. As various political parties suggested better monitoring and regulating foreign investment in agriculture, farmland became an election issue in 2013. Ultimately, the new federal government – formed by a coalition of right-wing parties – introduced two important changes to the regulatory system, implemented in 2015. First, it reduced the threshold at which foreign farmland purchases must be examined by the Foreign Investment Review Board (FIRB), from more than AU$252 million to AU$15 million. Second, the government implemented a mandatory registration system for foreign owners of Australian farmland, the result of which is reported in a 'farmland register' (Government of Australia, 2015).

Methods

I collected news articles on Australia and Canada by two different methods. For Australia, I searched farmlandgrab.org, which systematically collates news articles on land deals. The website is maintained by the NGO GRAIN, which monitors online news services daily to add content to its repository (GRAIN, 2013). Using the tag 'Australia', I identified 171 relevant articles published in Australian news outlets between 2008 and 2015. While it is impossible to say if the selection of articles found on farmlandgrab.org captures all Australian news stories on farmland investment, the site provides a large and centralized repository of relevant stories. For Canada, I conducted an online search using the terms 'farmland' and 'Saskatchewan' in an agricultural newspaper (*The Western Producer*), a daily newspaper (*Regina Leader-Post*), and a national newspaper (*The Globe and Mail*). Articles were included if they reported on one of the following key themes: farmland purchases by investors; farmland ownership rules; and, debates around 'land grabbing' or farmland ownership. The search yielded 192 articles between 2008 and 2015. I believe the samples to be representative of the coverage of the 'farmland controversy' in each case. Figure 6.1 shows the distribution of articles over time, by case. I coded and analysed the news articles for key themes using NVivo. I also used NVivo's text search function to conduct word frequency analyses, which allowed me to compare the relative importance of certain key terms across the cases.

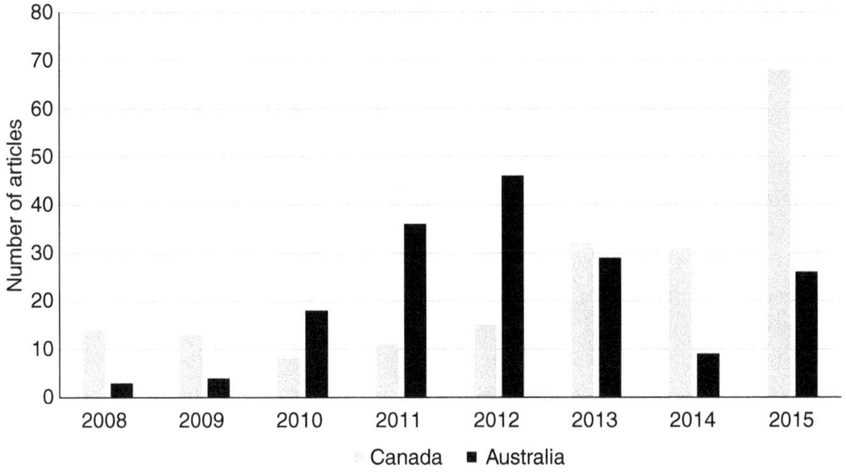

Figure 6.1 Number of news articles by year.

Public controversies over farmland investment in Canada and Australia

My analysis of farmland controversies in Canada and Australia is organized into five key themes. I compare across countries for each theme, and make connections among themes where appropriate.

Finance to the rescue

Various actors, especially investors, have made arguments articulating the necessity or benignity of farmland investment. An overarching theme, with some variation across the countries, is the idea that the agricultural sector needs outside capital in the form of private investment. Capital from outside the sector, it is argued, is needed to finance increased productivity, which in turn will allow farmers to meet growing demand for food, or to finance farm succession. In Australia, this argument was often accompanied by references to economic studies that have suggested that the country needs anywhere from AU$400 billion to AU$1 trillion to realize the opportunity to increase production and exports in the coming decades (McRobert, 2013; Marshall, 2015). Some articles convey a sense of urgency for the sector's need for external capital. The chair of PrimeAg, a publicly listed agriculture firm that was wound up in 2013 (Magnan, 2015), suggested that: 'What we desperately need is major unlisted funds to invest in water and rural land assets in Australia' (*The Land*, 2013). These types of arguments were sometimes used to justify foreign investment in agriculture specifically. An official with Corporate Agriculture Australia, a farming enterprise, explained:

> Foreign investment is a positive thing.... The level of equity in a lot of family-owned businesses has plummeted due to the variation of seasons in recent years and the issue really is the need for capital to come into agriculture.
>
> (Hinkley, 2011)

Commentators articulated a number of other potential benefits of foreign investment in farmland: supporting farmland prices, injecting new money into local economies, and creating jobs for farm managers.

In Canada, investors commonly presented similar arguments. In 2010, a representative of Bonnefield Financial, a farmland investor, suggested that investors provide agriculture with a 'badly needed' source of capital (Waldie and Leeder, 2010). The CPPIB published an opinion piece in the *Western Producer* justifying its farmland purchases as a means of helping Saskatchewan meet its targets for increasing agricultural production and exports (Leduc, 2015). Another firm, Skyline Agriculture, argued that its goal was to provide a form of 'flexible financing'[3] for farmers seeking to expand their operations (Playter, 2015). Investors and other commentators asserted that industrial agriculture is a capital intensive business, and that the industry would benefit by diversifying its sources of capital beyond traditional bank financing. In the Canadian case, no article explicitly called for foreign investment in Saskatchewan farmland, although some commentators suggested that the province's ownership rules were too strict.

The effort by some financial actors to portray farmland investment as a 'win–win' for farmers and finance is a notable theme in some of the news coverage. The CPPIB argued that its activities support the family farming model (Leduc, 2015). By purchasing land and leasing it to independent farmers, the CPPIB suggested that it helps retiring farmers exit the industry, allows young farmers to access land without having to buy it, and contributes to productivity improvements in the sector: 'We want to partner with local farmers to improve production techniques – and the livelihoods of those working in the sector' (Leduc, 2015). Meanwhile, Skyline Agriculture suggested that its 'flexible financing' model 'allows our clients [farmers] the option to purchase and farm an optimal number of (hectares)' and allows farmers to 'achieve their growth goals quickly, which is incentive to stay in Saskatchewan and farm full-time' (Playter, 2015). In Australia, public discourses justified farmland financialization in somewhat similar terms. One article touted the potential of a 'marriage of capital and farmland' to benefit both investors and farmers (Whitley, 2012). In another, external sources of capital are described as a 'lifeline for embattled farmers' in the dairy sector (*The Weekly Times*, 2013).

In both cases, there is a consistent narrative connecting the farmland investment trend to food scarcity and a 'food crisis'. Corporate officials and other commentators indicate that 'market fundamentals' including rising world populations, changing diets in emerging economies, and demand for biofuels, provide an impetus for investing in farmland. The global food situation was a stronger theme in the Australian articles, where 76 out of 169 articles included one of the terms 'food security', 'food demand', or 'food crisis'. By comparison, these terms appeared in only 14 of 205 articles in the Canadian sample of articles. In Australia, there has been considerable attention on Asia as source of demand for Australian exports. Commentators invoke the notion of Australia as a 'food bowl' for Asia in explaining the motivations of Chinese and other foreign investors (11 records from the Australian set used these terms). A typical example of this narrative is presented by a farmland investment manager: 'We are in close proximity to over three billion aspirational people with increasingly westernised, protein based diets' (Cawood, 2012). These findings are consistent with Larder *et al.*'s (2015, p. 593) observation that '"food security" is a crucial narrative that financial actors draw upon when engaging in agricultural production'.

Identifying investors: foreign and domestic

In the Australian news coverage, there is a preoccupation with foreign investors' country of origin. A word frequency analysis of the sample of articles showed that the word 'foreign' was the most common word, 'Chinese' was the eleventh most common, and 'China' the eighteenth most common.[4] The most frequently mentioned countries were China (123 articles), Qatar (28 articles), Canada (25 articles), and the US (23 articles). Some articles reported on lesser known sources of foreign investment such as Sweden, Denmark, Singapore, and others. There is far less coverage of domestic Australian firms acquiring farmland. This is despite the reality that some Australian farmland investment firms or joint Australian-foreign firms have made significant purchases (Magnan, 2015). When domestic investors are mentioned, it is often to lament their lack of interest of in the country's agricultural sector (*Reuters*, 2012; Wells, 2015) or to identify them as niche players (Marshall, 2012). As the CEO of the National Farmers' Federation put it in 2015, 'Australians aren't falling over themselves to invest in agriculture' (Wells, 2015). The announcement of an AU$150 million farmland purchase by an Australian pension fund, First State Super, in 2015, was reported in the context of a general lack of enthusiasm among the country's pension sector: 'There has been a growing outcry over the lack of investment by super funds in the sector, particularly by Agriculture Minister Barnaby Joyce, who has accused super funds of failing to understand the agricultural sector' (Cranston, 2015).

In Canada, news articles on the farmland trend also emphasize the fact that investors are outsiders of one form or another. Investors are characterized as 'foreign' or 'out-of-province' buyers, farmers and wealthy individuals from Alberta, and Chinese investors. In a word frequency analysis of the articles, the word 'foreign' appeared twenty-second on the list of most commonly used words, and 'Alberta', thirty-second. In 2012, controversy emerged over anecdotal reports that wealthy Chinese investors were funnelling off-shore money to their Chinese-Canadian relatives and contacts in order to circumvent the law. The media reported that the government had hired a special investigator to 'follow the money' for a small number of transactions where this scheme was suspected (Briere, 2012). In the end, the government was unable to find any instance of fraud, but admitted that it was limited in its ability to fully investigate the issue. Other commentators, particularly a Chinese Canadian realtor insisted that all of his Chinese clients interested in farmland were Canadian citizens or permanent residents (Briere, 2013).

Interestingly, the 'Canadian-ness' of entities like the CPPIB became a key debate in the farmland controversy in Canada. The CPPIB played on its identity as a Canadian investor to legitimize its investment activities. When the CPPIB deal was first announced, a spokesperson for the parties commented: 'It's a good deal for our investors and it's good deal for CPPIB.... Virtually every Canadian is getting the opportunity to invest in Saskatchewan farmland via their pension plan' (*Regina Leader-Post*, 2013). In an op-ed, the pension plan argued as follows:

> CPPIB is quintessentially Canadian. We work for the millions of Canadians who have contributed hard-earned dollars to the (Canada Pension Plan) and want to see that money put to work in investments like Saskatchewan farmland. Other than the traditional family farm, it is hard to imagine another owner that better represents both the interests of Saskatchewan residents and the province's own goals.
>
> (Leduc, 2015)

Yet, as explained below, this was not enough to shield the CPPIB from criticism from some farmers, agricultural policy groups, and politicians, who argued that the pension and other large entities presented unfair competition for farmers.

'Selling the farm' and 'Losing our grip': critiques of farmland financialization

What discourses have critics or sceptics of farmland investment mobilized to express their views? The critique is multi-faceted, but there are a number of common threads to the types of arguments used in each country. In Australia, the potential loss of sovereignty associated with increasing foreign farmland ownership is a key theme. Critics invoked the phrase 'selling the farm' when discussing the threat of foreign investment in farmland. Nick Xenophon Team Party leader, Senator Nick Xenophon, put it this way: 'We should be selling the milk not the cow, we should be selling the food not the farm' (Caldicott, 2013). The argument is that commercial agricultural exporting, not selling farmland to investors, is the appropriate way for Australia to contribute to the global food economy. For Senator Bill Heffernan, chair of the Senate inquiry into foreign investment in agriculture, ceding control over farmland ought to be considered a direct threat to sovereignty: 'Family farms are the mainstay of our sovereign production.... Once we sell big slabs of our sovereign assets to another government it won't care too much about what we think that land or water should be used for' (Marshall, 2011). As one of the leading Australian critics of foreign farmland deals, Senator Heffernan repeatedly warned of Australia losing control of its destiny.

The involvement of state-owned or state-backed entities is among the most controversial aspects of foreign investment in Australia agriculture (Sippel, 2015). Both China and Qatar have relied on state-owned businesses to secure farmland in Australia. The news record frequently cited concerns over the political and economic motivations of these actors. Of particular concern was the prospect of a state-owned company creating a 'closed supply chain' to export agricultural products directly to the investing country, by-passing normal commercial channels. Commenting on Chinese investment activity in Western Australia, a farmer remarked, 'I have no problem with the company wanting to invest in Australia but as far as it wanting to take WA produce out of the WA supply chain, I don't think that's right' (Hinkley, 2011). A similar concern was raised about Hassad Australia's investments in the State, with commentators speculating about the firm's desire to build its own grain-handling infrastructure in order to bypass Western Australia's dominant grain-handling company, CBH (Heard, 2012). This question was also highlighted in the report of the Senate inquiry into the National Interest Test (Senate of Australia, 2012, pp. 10–11).

The idea that large-scale investments by government-backed entities could undermine Australian food security was a related theme. In 2011, the President of the National Farmers' Federation warned that foreign investment in agriculture 'raises the question of transparency in the supply chain, potentially jeopardising competition at the farm gate and depressing the local market. At an extreme level, this could also lead to Australia's own food security goals being compromised' (Kelly, 2011). This argument was countered by others who observed that Australia is a net food exporter, and that it would require a massive transfer of resources to foreign interests to endanger the country's ability to supply its own population.

Critics of foreign investment in Australian farmland commonly cited the lack of transparency behind these activities. Politicians, leaders of farm organizations, and farmers expressed their frustration with the lack of information on the extent of foreign farmland purchases. A farm policy analyst, Mick Keogh, argued: '[W]here there is a black hole in terms of what is happening, it is very difficult for people to make a sensible assessment of the situation' (Bourke, 2013). Several stories pointed to the problem of 'creeping' acquisitions, an issue brought to light by the Senate inquiry into foreign investment into agriculture. Because the FIRB threshold for triggering a national interest review was set at AU$252 million prior to 2015, an investor could acquire a large portfolio by making a number of smaller purchases of farmland and evade regulatory scrutiny. Other news stories reported that investors were making purchases under secretive confidentiality agreements (Heard, 2012). These types of arguments were marshalled to make the case for a stronger monitoring and reporting mechanism, namely a 'register' of foreign farmland ownership.

In Canada, the critique of institutional investment in agriculture tended to revolve around the conditions under which different actors compete for farmland. Critics of the CPPIB-Assiniboia deal argued that the massive financial assets of an entity like the CPPIB created an uneven playing field in the farmland market (Pratt, 2014; Nickel, 2015). As Dan Patterson, former head of the Farmland Security Board argued, for young farmers, '(c)ompeting with pension funds and investment entities with access to hundreds of billions of dollars is a daunting and discouraging prospect' (Patterson, 2014). In a related narrative, some asserted that farmers should not have to 'compete with their own money' (that is, their accumulated pension savings) for farmland (Briere, 2015c; Mandryk, 2015).

Another critique related to the potential for institutional investment in farmland to alter the structure of the agricultural sector. The NFU published an influential report entitled, *Losing Our Grip*, in 2010, painting the farmland investment trend as a threat to family farmers (NFU, 2010). They suggested that, taken to its logical end, institutional farmland ownership could lead to neo-feudalism – a situation in which farmers come to depend on large, wealthy, landowners for their access to land. Dan Patterson, who became a vocal critic of the CPPIB deal, advanced a similar argument, suggesting 'Saskatchewan's rural economic and social structure would be radically and irrevocably degraded' by rampant institutional land purchases (Patterson, 2014). He argued that investors are motivated primarily by the profit motive, which in turn will determine their management of farmland. Over time, he suggests, investors will have an incentive to rent their land to very large farm operators such as corporate farm managers which in turn would undermine the viability of rural communities. Although this type of argument was less prominent in the Australian argument, a version of it was articulated by Bill Heffernan:

> It would be a great pity if Australia's farmers ended up as tenants and tenant farmers and unlike the prospect of capital growth to hand on to the family, only have the job to hand on and I guess there's a warning there.
>
> (Herbert, 2013)

Farmland is different

In both sets of articles, a variety of actors present arguments about the exceptionalism of agriculture and of farmland. Several Australian news stories characterize the debate around foreign investment in farming as 'emotive'. One article profiled a farming family dealing with financial difficulties, yet refusing to sell his land to foreign investors:

> The contentious issue of foreign investment hits a raw nerve for Mr Ward, who has been farming all his life. He was forced to put his Griffith farm on the market to fund a horticultural investment … that is rapidly going into debt. The venture was originally run in partnership with his … daughter … but she has had to make the tough decision to leave the farm in order to keep the business afloat. 'If you understood the passion and emotion, you would shake your head at the decline of the operation,' Mr Ward said.
>
> (Preece, 2012)

In some instances, the idea that farmland had become an 'emotive' issue was used by commentators who believed that opposition to foreign investment was misguided or irrational. Referring to the Australian government's decision to prohibit the sale of the S. Kidman and Co. cattle station, the country's largest farming aggregation, a commentator observed '(a)griculture in Australia is an emotive industry for everyone, whether they are involved in the industry or not' (McCarthy, 2015). In other instances, Australian actors also referred to farmland as a strategic resource or asset. Speaking of the sale of Cubbie Station to Chinese interests, Senator Nick Xenophon commented: 'It is a mongrel of a deal.… The conditions give a veneer of Australian involvement but result in Australia losing control of a strategic asset' (*Reuters*, 2013).

In the Canadian case, several actors consistently characterized farmland as unique or exceptional. A leading critic of the CPPIB deal, Rick Swenson, was among the first to express this idea. In 2014, he called the deal a 'precedent-setting sale of Saskatchewan's greatest resource' (Johnstone, 2014). Another article developed this theme at some length:

> I am sure farmland ownership has been a controversial issue since the first stones were piled around a field. For most prairie farmers, land ownership has been tied to survival of the family business.… As a result, farmers are often of mixed emotions when it comes to who should be allowed to own land and bid up its price.
>
> (Raine, 2015)

Another commentator drew a link between agriculture and the identity of Saskatchewan people: 'There's a professional side to this, a business side

... But it's also who we are, it's how we were bred and raised. It's what we're made up of, in many ways' (Briere, 2015a). When the Saskatchewan government decided to further restrict farmland ownership, in 2015, the Minister of Agriculture drew on similar themes: 'Our government understands that to many in the province, farmland is not just an asset.... It is a connection to our history and who we are as people' (Fleece, 2015). An editorial in a leading agricultural newspaper nicely captures this type of discourse:

> Farmland does deserve special attention compared to other businesses because much more is involved beyond just the land. Those who own it and work it form the basis of rural communities. They forge local governments, shop at local stores and organize the 4-H club, while spending weekends maintaining the local rink. There is a social fabric that must be both protected and nurtured at the same time.
>
> (*The Western Producer*, 2015)

Regulating farmland ownership

In both countries, governments introduced new regulations around farmland ownership in 2015. The news record provides an opportunity to examine discourses used to justify or challenge farmland ownership rules and the role of governments more broadly. In Australia, themes such as the 'national interest' and 'transparency' became central to the debate about regulating farmland ownership. As early as 2010, political actors began calling for regulatory changes including establishing a 'register' of farmland ownership and lowering the FIRB threshold for reviewing foreign farmland purchases. Stricter regulations were consistently championed by the Greens, who introduced the first legislative proposals to this end. In 2011, Senator Nick Xenophon argued: 'We are mugs compared to other countries that have very tight controls over foreign investment in prime agricultural land' (Williams, 2011). The right-wing Coalition initially opposed establishing a register, but expressed concern about foreign investment in 2010 (Pannett, 2010). By 2012, all parties, including the ruling Labor government favoured a register, but only the Coalition and Greens argued for lowering the FIRB threshold. A prominent Coalition politician, Joe Hockey, expressed his party's 2012 election platform on foreign farmland purchases as follows: 'There is a perception issue that there is unrestricted, unrestrained, uncontrolled foreign investment, particularly in agriculture. We may well flag changes to foreign investment rules' (Maley, 2012).

After the Coalition's election victory in 2013, it moved to implement the new rules. The government insisted that it welcomed foreign investment, but was responding to local concerns about the lack of transparency. In justifying the restrictions, Agriculture Minister Barnaby Joyce argued, 'We are entitled to have a proper oversight and in some instances a concern if people believe things are becoming excessive' (Rochfort, 2014). The decision to lower the FIRB threshold proved quite controversial as some farm organizations, business groups, and opposition politicians voiced their criticism. The National Farmers' Federation, for instance, worried that the new AU$15 million threshold 'could become a choke on foreign investment' in Australia (*The Australian*, 2015). Labor representative Joel Fitzgibbon argued that the new threshold 'sends the wrong signals to our investors in a very competitive environment' (*The Australian*, 2015). Groups such as the Business Council of Australia (Marshall, 2015) and real estate corporations (Uren, 2015) echoed the idea of sending the 'wrong signal'. In sum, the debate revolved around how to reconcile competing imperatives: providing appropriate scrutiny of foreign farmland purchases, while maintaining a welcoming investment climate.

In Saskatchewan, the government's[5] position on farmland ownership regulation evolved considerably over the course of the controversy. In 2012, the government raised concerns over the possibility that certain individual investors might be circumventing farmland rules by financing their purchases with off-shore money (see above). At the time, the Minister of Agriculture noted:

> We don't want Saskatchewan farmers to all be tenant farmers, farming the land for ownership that may be foreign-based. We're pleased to see that farm land values have responded to low interest rates, high commodity prices, and so on, but as far as international speculators affecting the price of land … we don't think that's a burden we necessarily want our farm families to bear.
>
> (Cited in Coolican, 2012)

At this time, however, the government did not consider revisiting the farmland investment rules. Even after the eruption of the political controversy around the CPPIB deal in 2014, the government insisted the farmland ownership rules did not need to be changed (Warick, 2014).

By early 2015, however, the government pivoted quite dramatically, announcing an official review of the legislation. The Minister of Agriculture justified the decision in these terms: 'Producers don't see that pension funds, with potentially billions of dollars to invest, are fair competition, and I get that' (cited in Pratt, 2015). In launching the review, the government emphasized the need to hear directly from farmers and other Saskatchewan residents about the issue. The results of the public consultation confirmed that a large majority of respondents opposed foreign (87 per cent) and institutional (75 per cent) investment in farmland (Government of Saskatchewan, 2015c). When it later tightened the rules, the government could therefore claim to be acting on the wishes of farmers and Saskatchewan residents: 'The people of Saskatchewan told us what they thought regarding farmland ownership and we listened' (Briere, 2015b).

Though clearly a politically popular stance, the government was nonetheless obliged to deal with a contradiction: how to reconcile its decision with its 'free market' philosophy? To that end, the Minister of Agriculture used arguments such as the following:

> We're very much open to investment in general terms. Farmland is a little bit exceptional – I know farmers feel, rightly or wrongly, that institutions and foreign investors that are well-heeled are not necessarily always fair competition.... Whether that's right or not, that's a strong feeling out in farm country ... (Agriculture is) a part of who we are and people take it pretty seriously. It's different here.
>
> (Cited in Nelson, 2015)

Elsewhere, the Minister argued that the new farmland rules would not deter investment in the province and insisted that the government welcomed the CPPIB's investment in other Saskatchewan assets (Johnstone, 2015a).

Some commentators strongly opposed any tightening of the already strict farmland ownership rules. The editorial board of *The Globe and Mail* and provincial media commentators argued that the province's ownership rules were already too restrictive, and that there was no reason to prohibit the CPPIB from owning farmland (Johnstone, 2015b; *The Globe and Mail*, 2015). An editorial in the *Regina Leader-Post* (15 April, 2015) argued that the government's actions amounted to 'pandering to a vocal base', namely farmers and rural residents, who opposed institutional investment in farmland. Understandably, the CPPIB expressed its dismay at the government's intention to tighten ownership rules (Leduc, 2015). The Saskatchewan Chamber of Commerce, a business lobby group, argued the new rules would harm the province's ability to attract investment (Fleece, 2015), a sentiment shared by the CEO of Bonnefield Financial, a firm with farmland investments across Canada: 'Capital is being scared away' as investors 'are aggressively putting their money elsewhere' (Skerritt, 2015).

Discussion

I have explored how various discourses have been used to frame financialization as a process that is benign or necessary, or alternatively, as a risky process that needs to be monitored and/or controlled. Actors who opposed the financialization of farmland, including some politicians, farm organizations, and other commentators, have drawn primarily on discourses such as secrecy, investor motivations, and the inequality of power/resources between investors and farmers. In some cases, critics of financialization have painted the process in somewhat hyperbolic terms. While it is legitimate to raise concerns about the trends, one must bear in mind the scale and scope of the phenomenon. In both countries, family farmers are still the dominant unit of production and a relatively small proportion of the agricultural land base is owned by investors or foreigners. On the other hand, proponents of investor schemes and financialization have also exaggerated the purported benefits to the sector. When boosters refer to a 'win–win' for farmers and finance, they ignore the fact that farmers almost universally prefer to own rather than lease farmland and prefer not to compete against investors for access to land. Investors are primarily interested in financial returns, and not the well-being of farmers or the agricultural sector. Nevertheless, investors seek to leverage the generally positive image of farm life by suggesting that their activities benefit family farmers.

The news coverage points to several 'blind spots' in the farmland debate. First, industry actors, commentators, and journalists generally failed to provide context to the agricultural sector's need for capital. Agricultural sectors in both Canada and Australia have been undergoing restructuring for several decades (Wiebe, 2012; Lawrence and Campbell, 2014). Guided by neoliberal ideology, governments have scrapped farmer-owned marketing structures (Magnan, 2016), withdrawn public investments in agriculture, and privatized agricultural research in favour of allowing 'market forces' to allocate resources in the sector. Globalizing trends, such as the liberalization of trade rules, have led to increased competition among farming sectors and greater volatility in farm commodity prices. Together, these conditions have led to low farm incomes, large-scale farmer exit, and rising debt. Between 2001 and 2015, farm debt in Canada rose by 126 per cent (FCC, 2016b) and average debt per farm increased by 96 per cent in Australia.[6]

Thus, if the agricultural sector is starved of capital, it has much to do with the effects of neoliberal restructuring over the last few decades. When governments or investors tout financial capital, whether domestic or foreign, as the solution to a lack of investment in the sector, they ignore alternative ways of addressing the problem such as public investment in agriculture, programs for beginning farmers, or alternative forms of land tenure and ownership. In this sense, the debates have been firmly rooted in a set of neoliberal framing assumptions. Furthermore, the notion that buying farmland is a form of productive investment can reasonably be questioned. One Australian critic, an elected official from Western Australia, expressed this sentiment: 'It's not 'investing', they're not opening mines or building dams or anything of that nature, they're just buying the asset that is there. So it's not investment, it's selling your country off' (*Yahoo News Australia*, 2015). Actors in both countries generally failed to distinguish between productive investments – those where capital is used to build new infrastructure, improve productivity, or create new jobs – versus simple land buys.

Second, the pre-occupation with certain types of investors (such as sovereign wealth funds and other foreign investors) tended to overshadow the activities of other players including wealthy individuals, private funds, and firms that bridged domestic and foreign capital. In Australia, the focus on sovereign-wealth funds and other government-backed actors sparked a debate about the motivations of investing countries. Firms like Hassad Australia combine commercial, food security, and financial motives in their investment activities (Sippel, 2015). Yet, no state-backed investor has engaged in the type of 'agro-security mercantilism' (McMichael, 2013) feared by critics. Furthermore, all purchases by government-backed entities in Australia were, throughout the period in question, subject to regulatory review through the FIRB. This is in contrast to private or institutional buyers, who faced no such automatic review. The coverage tended to overstate the risks presented by government-backed entities, often based on hypothetical situations rather than facts on the ground. In reality, firms like Hassad were engaged in the same types of activities as other farmland investors like overseas pension funds, wealthy individuals, and agribusiness corporations such as Cargill. Investments by American, Norwegian, and Canadian pension funds, or by firms like Laguna Bay Pastoral, an Australian firm managing foreign capital, raised less controversy than Chinese or Qatari investments. As Larder *et al.* (2015, p. 599) have shown, the Australian debate around financialization has tended to paint some investors as 'more foreign' than others. Furthermore, the domestic/foreign distinction is often a simplistic representation of the complex spatial and financial relations within and among investing firms (Fairbairn, 2015).

In the Canadian case, the debate revolved primarily around a particular investor, the CPPIB. The deal involved the transfer of a large portfolio of land from one investor to another, not from farmers to an investor. Assiniboia Capital Corp., a private trust, had been acquiring a large portfolio of land since the mid-2000s when it sold its holdings to the CPPIB in 2014. Although Assiniboia's activities were occasionally reported on in the news press, there was little controversy until the CPPIB entered the scene. The CPPIB is a well-recognized institution, which perhaps made it easier for non-farmers to relate to the controversy. Yet, there is no obvious reason why the CPPIB's purchase of land should be more problematic than that of other financial actors using the very same logic and business model. It is also worth noting that the single largest farmland owner in Saskatchewan is an individual investor from Alberta who reportedly owns more than 80,000 hectares.[7] The farmland ownership restrictions introduced by Saskatchewan in 2015 will not affect the activities of wealthy individuals who may wish to acquire huge tracts of land. The debate thus focused too heavily on the activities of one player, rather than on the broader phenomenon of investor purchases of farmland.

A notable difference across the cases is the way in which commentators understood the role of domestic institutional investors. In Australia, commentators decried the fact that domestic pension funds and other Australian investors were uninterested in agriculture. This argument implied that domestic investors could help solve the sector's need for capital, but was failing to do so because of misplaced motives or a lack of understanding of agriculture. In Canada, more was made of the distinction between farmer ownership and institutional ownership. In other words, the debate was framed around the question: 'Is it appropriate for large institutional investors to compete with farmers for land?' The difference is likely best explained by the history of land ownership regulation in each case. Whereas Australia has welcomed foreign ownership of farmland from its colonial origins, Saskatchewan has been much more restrictive. The parameters for the debate were therefore quite different. In Saskatchewan, there is a near consensus around prohibiting foreign ownership of farmland, and a strong majority also rejects institutional investment in farmland. In Australia, the debate is less concerned about institutions/ investors versus farmers, and more pre-occupied with Australian versus non-Australian ownership.

What do the regulatory responses of right-wing governments to each farmland controversy tell us about neoliberalism, in theory and practice? Both governments drew on the notion of agricultural exceptionalism, and more specifically, on the emotional, social, and historic values embedded in farmland to justify their decisions. The arguments are consistent with a form of right-wing populism in which governments must protect farmers and rural residents from outside interests. Nevertheless, each government continued to promote neoliberal ideals like investment, entrepreneurship, and a limited role for governments. Although the tightening of farmland ownership rules may have contradicted 'free market' principles, the changes were far from fatal to a broader neoliberal philosophy. In the Canadian case, the government restricted property rights for a very particular subset of people/ entities and arguably protected access to property (that is, farmland) for farmers. In Australia, the changes put into place greater scrutiny of foreign farmland purchases, but did not place new restrictions on foreign ownership. If the new Australian rules slow the pace of foreign investment, it will be as an indirect effect of imposing new reporting and registration requirements through the FIRB threshold and farmland 'register'.

Conclusion

In this chapter, I have explored how different actors have shaped political debates over the financialization of farmland. By systematically reviewing the news coverage of the farmland controversies, I have examined one of the important discursive arenas in which these debates took place. Actors and organizations with a range of agendas engaged publicly with these debates through news coverage. My analysis contributes to a better understanding of how both proponents and opponents of financialization draw upon values in justifying their positions. Whereas investors, political actors, and others who see promise in the financialization of agriculture draw upon values connected to addressing world food security and helping farmers, critics and sceptics by contrast drew upon values such as fairness and transparency. Actors across a range of views tended to adhere to a version of agricultural exceptionalism, which for some justified new government oversight over or restrictions of foreign/investor ownership. Despite the differences across cases, each demonstrates that political actors were sensitive to pressure from influential constituencies, including farmers and rural people.

Notes

1 In Canada, farmland ownership is regulated by the provinces. Saskatchewan accounts for about 40 per cent of all Canadian farmland, and in 2015, was responsible for 24.8 per cent of Canadian agricultural exports (Government of Saskatchewan, 2015b).
2 The party holds no seats in the provincial legislature, but is a rival of the governing right-wing Saskatchewan Party.
3 The firm has developed a complex business model that would allow farmers to retain ownership of their land, while exposing investors to income and profits from the farm operation and land appreciation. The company has yet to launch its operations because its model has been found to contravene Saskatchewan's strict farmland ownership rules. Skyline has repeatedly challenged these rules in court.
4 The word frequency analysis was based on exact matches, excluding words of three letters or less, and common grammatical elements like conjunctions and prepositions.
5 Through the whole period in question, the right-of-centre Saskatchewan Party has been in power.
6 Calculated from the Australian Department of Agriculture and Water Resources AGSurf service: http://apps.daff.gov.au/AGSURF/. It should be noted that the average debt per farm in Australia peaked in 2007, when it was 118 per cent higher than in 2001, and peaked again in 2010, at 116 per cent of the 2001 figure, before declining in subsequent years.
7 Desmarais *et al.*'s (2017) analysis pegged his landholdings at 65,000 hectares in 2014. Mr. Andjelic's personal website indicates that he now owns nearly 80,000 hectares (www.robertandjelic.com/).

References

ABS (Australian Bureau of Statistics). (2014) *Agricultural Land and Water Ownership*, available at: www.abs.gov.au/ausstats/abs@.nsf/Lookup/7127.0main+features 1Jun%202013, accessed 17 May 2017

ATO (Australian Taxation Office). (2016) *Register of Foreign Ownership of Agricultural Land*, available at: http://firb.gov.au/about/publication/register-of-foreign-ownership-of-agricultural-land/, accessed 16 May 2017

Bourke, E. (2013) 'Clash of concerns over foreign ownership of Australian agriculture', 1 March, *Australia Broadcasting Corporation*, available at: http://farmlandgrab.org/post/view/21719-clash-of-concerns-over-foreign-ownership-of-australian-agriculture, accessed 6 June 2017

Briere, K. (2012) 'Sask. appoints farmland investigator', *The Western Producer*, 14 December, available at: www.producer.com/2012/12/sask-appoints-farmland-investigator/, accessed 6 June 2017

Briere, K. (2013) 'Chinese land buy fears calmed', *The Western Producer*, 1 March, available at: www.producer.com/2013/03/chinese-land-buy-fears-calmed/, accessed 6 June 2017

Briere, K. (2015a) 'Lawyer says land seen as investment, lifestyle', *The Western Producer*, 4 June, available at: www.producer.com/2015/06/lawyer-says-land-seen-as-investment-lifestyle/, accessed 6 June 2017

Briere, K. (2015b) 'Lender appeals Sask. land ownership rules', *The Western Producer*, 29 October, available at: www.producer.com/2015/10/lender-appeals-sask-land-ownership-rules/, accessed 6 June 2017

Briere, K. (2015c) 'Producer input on land ownership urged', *The Western Producer*, 25 June, available at: www.producer.com/2015/06/producer-input-on-land-ownership-urged/, accessed 6 June 2017

Burch, D. and Lawrence, G. (2009) 'Towards a third food regime: behind the transformation', *Agriculture and Human Values*, vol 26, no 4, pp. 267–279

Burch, D. and Lawrence, G. (2013) 'Financialization in agri-food supply chains: private equity and the transformation of the retail sector', *Agriculture and Human Values*, vol 30, no 2, pp. 247–258

Caldicott, P. (2013) 'Foreign investors grab for SA farmland', *7 News Adelaide*, 9 August, available at: http://farmlandgrab.org/post/view/22416-foreign-investors-grab-for-sa-farmland, accessed 6 June 2017

Cawood, M. (2012) 'Land the perfect investment for foreigners', *Stock & Land*, 26 January, available at: http://farmlandgrab.org/post/view/19949-land-the-perfect-investment-for-foreigners, accessed 6 June 2017

Clapp, J. (2012) *Food*, Polity Press, Malden, MA

Coolican, L. (2012) 'Sask. farm land in demand', *Regina Leader-Post*, 9 December, available at: http://search.proquest.com/docview/1240799417/A9A46996F1EF440BPQ/34?accountid=13480, accessed 6 June 2017

Cranston, M. (2015) 'Super fund's $150m farm buy', *Farm Weekly*, 21 August, available at: http://farmlandgrab.org/post/view/25235-super-fund-s-150m-farm-buy, accessed 6 June 2017

Desmarais, A., Qualman, D., Magnan, A. and Wiebe, N. (2015) 'Land grabbing and land concentration: Mapping changing patterns of farmland ownership in three rural municipalities in Saskatchewan, Canada', *Journal of Canadian Food Studies*, vol 2, no 1, pp. 16–47

Desmarais, A., Qualman, D., Magnan, A. and Wiebe, N. (2017) 'Investor ownership or social investment? Changing farmland ownership in Saskatchewan, Canada', *Agriculture and Human Values*, vol 34, no 1, pp. 149–166

Ducastel, A. and Anseeuw, W. (2017) 'Agriculture as an asset class: Reshaping the South African farming sector', *Agriculture and Human Values*, vol 34, no 1, pp. 199–209

Fairbairn, M. (2014a) 'Just another asset class? Neoliberalism, finance and the construction of farmland investment', in S. Wolf and A. Bonanno (eds) *The Neoliberal Regime in the Agri-Food Sector: Crisis, Resilience and Restructuring*, Routledge, New York, pp. 245–262

Fairbairn, M. (2014b) 'Like gold with yield': evolving intersections between farmland and finance', *Journal of Peasant Studies*, vol 41, no 5, pp. 777–795

Fairbairn, M. (2015) 'Foreignization, financialization and land grab regulation', *Journal of Agrarian Change*, vol 15, pp. 581–591

FCC (Farm Credit Canada). (2016a) *Historic Farmland Values: 1985–2015*, available at: www.fcc-fac.ca/fcc/about-fcc/corporate-profile/reports/farmland-values/farmland-values-historic-2015.pdf, accessed 13 May 2017

FCC (Farm Credit Canada). (2016b) *Outlook for Farm Assets and Debt 2016 -17*, 7 September, available at: www.fcc-fac.ca/fcc/about-fcc/corporate-profile/reports/farm-assets-report/outlook-for-farm-assets-and-debt-2016-17.pdf, accessed 6 June 2017

Field, S. (2016) 'The financialization of food and the 2008–2011 food price spikes', *Environment and Planning A*, vol 48, no 11, pp. 2272–2290

Fleece, T. (2015) 'Mixed reaction to farmland ownership changes', *Regina Leader-Post*, 22 October, available at: http://leaderpost.com/business/agriculture/mixed-reaction-to-farmland-ownership-changes, accessed 6 June 2017

Government of Australia. (2015) 'Government tightens rules on foreign purchases of agricultural land', 11 February, available at: http://jbh.ministers.treasury.gov.au/media-release/005-2015/, accessed 17 May 2017

Government of Saskatchewan. (2015a) 'New laws for farmland ownership proclaimed', 21 December, available at: www.saskatchewan.ca/government/news-and-media/2015/december/21/farmland-ownership-laws, accessed 17 May 2017

Government of Saskatchewan. (2015b) *Saskatchewan Agricultural Exports 2015*, available at: http://publications.gov.sk.ca/documents/20/83881-Saskatchewan%20Agriculture%20Exports%202015.pdf, accessed 16 May 2017

Government of Saskatchewan. (2015c). *Saskatchewan Farmland Ownership: Summary of Results from Public Consultations*, available at: www.saskatchewan.ca/government/public-consultations/past-consultations/farmland-ownership-consultation/results, accessed 16 May 2017

GRAIN. (2013) 'Collating and dispersing: GRAIN's strategies and methods', *Journal of Peasant Studies*, vol 40, no 3, pp. 531–536

Grubel, J. and Yueyang, M.L. (2012) 'Australia approves China farm purchase, sparks investment concern', *Reuters*, 3 September, available at: http://farmlandgrab.org/post/view/20964-australia-approves-china-farm-purchase-sparks-investment-concern, accessed 22 June 2017

Gunnoe, A. (2014) 'The political economy of institutional landownership: Neorentier society and the financialization of land', *Rural Sociology*, vol 79, no 4, pp. 478–504

Heard, G. (2012) 'Hassad wraps up purchase', *The Land*, 3 May, available at: http://farmlandgrab.org/post/view/20436-hassad-wraps-up-purchase, accessed 6 June 2017

Herbert, B. (2013) 'Chinese purchase of WA farmland divides opinions', *Australia Broadcasting Corporation*, 23 July, available at: http://farmlandgrab.org/post/view/22342-chinese-purchase-of-wa-farmland-divides-opinions, accessed 6 June 2017

Hinkley, B. (2011) 'Chinese company push for Western Australia farmland', *Farm Weekly*, 12 August, available at: http://farmlandgrab.org/post/view/19067-chinese-company-push-for-western-australia-farmland, accessed 6 June 2017

Isakson, R. (2014) 'Food and finance: The financial transformation of agro-food supply chains', *Journal of Peasant Studies*, vol 45, no 1, pp. 749–775

Johnstone, B. (2014) 'Farmland deal 'wrong', Swenson; PCs want sale to be reviewed', *Regina Leader-Post*, 22 July, available at: http://search.proquest.com/docview/1547435529/EFC437E071714C14PQ/5?accountid=13480, accessed 6 June 2017

Johnstone, B. (2015a) 'Farmland ownership changes won't deter investors', *Regina Leader-Post*, 21 December, available at: http://leaderpost.com/business/agriculture/farmland-ownership-changes-wont-deter-investors, accessed 6 June 2017

Johnstone, B. (2015b) 'If the law ain't broke, don't (a)mend it', *Regina Leader-Post*, 21 February, available at: http://search.proquest.com/docview/1657161901/A9A46996F1EF440BPQ/28?accountid=13480, accessed 6 June 2017

Kelly, J. (2011) 'Food security the fear as foreign investors rush in', *The Australian*, 4 July, available at: http://farmlandgrab.org/post/view/18880-food-security-the-fear-as-foreign-investors-rush-in, accessed 6 June 2017

Kuns, B., Visser, O. and Wastfelt, A. (2016) 'The stock market and the steppe: The challenges faced by stock-market financed, Nordic farming ventures in Russia and Ukraine', *Journal of Rural Studies*, vol 45, pp. 199–217

Larder, N., Sippel, S. and Lawrence, G. (2015) 'Finance capital, food security narratives and Australian agricultural land', *Journal of Agrarian Change*, vol 15, no 4, pp. 592–603

Lawrence, G. and Campbell, H. (2014) 'Neoliberalism in the antipodes: Understanding the influence and limits of the neoliberal political project', in S. Wolf and A. Bonanno (eds) *The Neoliberal Regime in the Agri-Food Sector: Crisis, Resilience, and Restructuring*, Routledge, New York

Leduc, M. (2015) 'Sask. farmland a good fit for CPP', *Regina Leader-Post*, 17 February, available at: http://search.proquest.com/docview/1655647512/EFC437E071714C14PQ/20?accountid=13480, accessed 6 June 2017

Li, T. (2014) 'What is land? Assembling a resource for global investment', *Transactions of the Institute of British Geographers*, vol 39, pp. 589–602

Li, T. (2015) 'Transnational farmland investment: A risky business', *Journal of Agrarian Change*, vol 15, no 4, pp. 560–568

McCarthy, M. (2015) 'Foreign investors could be deterred by extra scrutiny of farm and agribusiness sales', *Australia Broadcasting Corporation (ABC)*, 24 November, available at: http://mobile.abc.net.au/news/2015-11-25/changes-to-firb-threshold-could-deter-investors/6971658?pfm=sm&section=qld, accessed 6 June 2017

McMichael, P. (2013) 'Land grabbing as security mercantilism in international relations', *Globalizations*, vol 10, no. 1, pp. 47–64

McRobert, K. (2013) 'Don't sell our farms: Greens', *Farm Weekly*, 17 August, available at: http://farmlandgrab.org/post/view/22436-don-t-sell-our-farms-greens, accessed 6 June 2017

Magnan, A. (2015) 'The financialization of agri-food in Canada and Australia: Corporate farmland and farm ownership in the grains and oilseed sector', *Journal of Rural Studies*, vol 41, pp. 1–12

Magnan, A. (2016) *When Wheat was King: The Rise and Fall of the Canada-UK Grain Trade*, UBC Press, Vancouver

Magnan, A. and Sunley, S. (2017) 'Farmland investment and financialization in Saskatchewan, 2003–2014: An empirical analysis of farmland transactions', *Journal of Rural Studies*, vol 49, pp. 92–103

Maley, J. (2012) 'Coalition to target foreign farm buyers', *Sydney Morning Herald*, 6 January, available at: http://farmlandgrab.org/post/view/19852-coalition-to-target-foreign-farm-buyers, accessed 6 June 2017

Mandryk, M. (2015) 'Right supports nanny state, too', *Regina Leader-Post*, 9 October, available at: http://leaderpost.com/opinion/columnists/mandryk-right-supports-nanny-state-too?__lsa=e6d0-79ad, accessed 6 June 2017

Marshall, A. (2011) 'Arab farm investment push', *Stock & Land*, 10 February, available at: http://farmlandgrab.org/post/view/18142-arab-farm-investment-push, accessed 6 June 2017

Marshall, A. (2012) 'Buy a farm, says investment guru', *Stock and Land*, 10 June, available at: http://farmlandgrab.org/post/view/20622-buy-a-farm-says-investment-guru, accessed 6 June 2017

Marshall, A. (2015) 'Don't frustrate foreign buyers: BCA', *Farm Weekly*, 7 December, available at: www.farmweekly.com.au/news/agriculture/agribusiness/general-news/dont-frustrate-foreign-buyers-bca/2749114.aspx?storypage=0, accessed 6 June 2017

Mittal, A. and Moore, M. (2014) *Down on the Farm: Wall Street, America's New Farmer*. Oakland: Oakland Institute, available at: www.oaklandinstitute.org/sites/oakland institute.org/files/OI_Report_Down_on_the_Farm.pdf, accessed 16 May 2017

Nelson, J. (2015) 'Saskatchewan farmers split on allowing pension funds to buy land', *The Globe and Mail*, 19 April, available at: http://license.icopyright.net/user/viewFreeUse.act?fuid=MjE3NzU4MTU%3D, accessed 6 June 2017

NFU. (National Farmers Union). (2010) *Losing Our Grip*, available at: www.nfu.ca/sites/www.nfu.ca/files/06-07-losing_grip.pdf, accessed 6 June 2017

Nickel, R. (2015) 'Exclusive: Canadian province eyes tougher rules on investor farm buys', *Reuters*, 26 January, available at: www.reuters.com/article/2015/01/26/us-canada-investment-farmland-exclusive-idUSKBN0KZ2AN20150126, accessed 6 June 2017

Ouma, S. (2014) 'Situating global finance in the land rush debate: A critical review', *Geoforum*, vol 57, pp. 162–166

Pannett, R. (2010) 'Australia opposition may tweak foreign investment laws', *The Australian*, 2 August, available at: http://farmlandgrab.org/post/view/14655-australia-opposition-may-tweak-foreign-investment-laws, accessed 6 June 2017

Patterson, D. (2014) 'Government must address land investment deals', *The Western Producer*, 20 November, available at: www.producer.com/2014/11/govt-must-address-land-investment-deals/, accessed 6 June 2017

Playter, T. (2015) 'The case for flexible land financing', *The Western Producer*, 6 August, available at: www.producer.com/2015/08/the-case-for-flexible-land-financing/, accessed 6 June 2017

Pratt, S. (2014) 'Farm owner blames ownership rules', *The Western Producer*, 27 November, available at: www.producer.com/2014/11/farm-owner-blames-ownership-rules/, accessed 6 June 2017

Pratt, S. (2015) 'Sask. reviews farmland ownership', *The Western Producer*, 1 January, available at: www.producer.com/2015/01/sask-reviews-farmland-ownership/, accessed 6 June 2017

Preece, L. (2012) 'Don't sell off our farms', *Stock & Land*, 2 February, available at: http://farmlandgrab.org/post/view/19983-don-t-sell-off-our-farms, accessed 6 June 2017

Raine, M. (2015) 'Farmland ownership major concern', *The Western Producer*, 15 October, available at: www.producer.com/2015/10/farmland-ownershipmajor-concern/, accessed 6 June 2017

Regina Leader-Post. (2013) 'CPPIB buys Assiniboia Farmland for $128 million', 13 December, available at: http://search.proquest.com/docview/1468076167/EFC437E071714C14PQ/7?accountid=13480, accessed 6 June 2017

Regina Leader-Post. (2015) 'Farmland review puzzling', 15 April, available at: http://search.proquest.com/docview/1673701199/A9A46996F1EF440BPQ/6?accountid=13480, accessed 6 June 2017

Reuters. (2012) 'Analysis: Foreign investors in Australian farms; rash or prescient?' 19 November, available at: http://farmlandgrab.org/post/view/21325-analysis-foreign-investors-in-australian-farms-rash-or-prescient, accessed 6 June 2017

Reuters. (2013) 'Australia approves China farm purchase, sparks investment concern', 3 September, available at: http://farmlandgrab.org/post/view/20964-australia-approves-china-farm-purchase-sparks-investment-concern, accessed 6 June 2017

Rochfort, S. (2014) 'Tough new rules as foreigners gouge huge chunk of Australia', *Daily Telegraph*, 12 November, available at: http://farmlandgrab.org/post/view/24217-tough-new-rules-as-foreigners-gouge-huge-chunk-of-australia, accessed 6 June 2017

Savills. (2012) *Focus 2012: International Farmland*, available at: http://pdf.euro.savills.co.uk/global-research/international-farmland-focus.pdf, accessed 16 May 2017

Senate of Australia. (2012) *Examination of the Foreign Investment Review Board National Interest Test*, Interim Report, Rural and Regional Affairs and Transport Reference Committee, available at: www.aph.gov.au/Parliamentary_Business/Committees/Senate/Rural_and_Regional_Affairs_and_Transport/Completed_inquiries/2012-13/firb2011/interimreport/index, accessed 17 May 2017

Senate of Australia. (2013) *Examination of the Foreign Investment Review Board National Interest Test*, Final Report, Rural and Regional Affairs and Transport Reference Committee, available at: www.aph.gov.au/Parliamentary_Business/Committees/Senate/Rural_and_Regional_Affairs_and_Transport/Completed_inquiries/2012-13/firb2011/report/index, accessed 17 May 2017

Sippel, S. (2015) 'Food security or commercial business? Gulf State investments in Australian agriculture,' *Journal of Peasant Studies*, vol 42, no 5, pp. 981–1001

Sippel, S.R., Larder, N. and Lawrence, G. (2017) 'Grounding the financialization of farmland: Perspectives on financial actors as new land owners in rural Australia,' *Agriculture and Human Values*, vol 34, no 2, pp. 251–265

Skerritt, J. (2015) 'Canada farmland boom cools as Saskatchewan mulls investor limits', *Bloomberg Business*, 7 May, available at: www.bloomberg.com/news/articles/2015-05-06/canada-farmland-boom-cools-as-saskatchewan-mulls-investor-limits, accessed 6 June 2017

Sommerville, M. and Magnan, A. (2015) '"Pinstripes on the prairies": examining the financialization of farming systems in the Canadian prairie provinces', *Journal of Peasant Studies*, vol 42, no 1, pp. 119–144

The Australian. (2015) 'Foreign investment fears misplaced, forum told', 15 April, available at: http://farmlandgrab.org/post/view/24780-foreign-investment-fears-misplaced-forum-told, accessed 6 June 2017

The Globe and Mail. (2015) 'Saskatchewan's farmland ownership rules are too restrictive', 22 April, available at: http://license.icopyright.net/user/viewFreeUse.act?fuid=MjE3NzU5MTY%3D, accessed 6 June 2017

The Land. (2013) 'TIAA-CREF could fund more PrimeAg sales', 18 February, available at: http://farmlandgrab.org/post/view/21662-tiaa-cref-could-fund-more-primeag-sales, accessed 6 June 2017

The Weekly Times. (2013) 'Foreign investors key for dairy: MG', 26 August, available at: http://farmlandgrab.org/post/view/22462-foreign-investors-key-for-dairy-mg, accessed 6 June 2017

The Western Producer. (2015) 'Rewards of land investment must be balanced with risks', 5 March, available at: www.producer.com/2015/03/rewards-of-land-investment-must-be-balanced-with-risks-2/, accessed 6 June 2017

Uren, D. (2015) 'Foreign farm buys above $15m will need FIRB approval', *The Australian*, 11 February, available at: http://farmlandgrab.org/post/view/24515-foreign-farm-buys-above-15m-will-need-firb-approval, accessed 6 June 2017

Visser, O. (2015) 'Finance and the global land rush: Understanding the growing role of investment funds in land deals and large-scale farming', *Canadian Food Studies*, vol 2, no 2, pp. 278–286

Visser, O. (2017) 'Running out of farmland? Investment discourses, unstable land values and the sluggishness of asset making', *Agriculture and Human Values*, vol 34, pp. 185–198

Waldie, P. and Leeder, J. (2010) 'Pinstripes on the prairie', *The Globe and Mail*, 24 November, available at: http://search.proquest.com.libproxy.uregina.ca:2048/docview/808449949/E6B4D48F125D4DA1PQ/4?accountid=13480, accessed 6 June 2017

Warick, J. (2014) 'No changes to farmland ownership rules', *Regina Leader-Post*, 1 December, available at: http://search.proquest.com/docview/1629513118/A9A46996F1EF440BPQ/5?accountid=13480, accessed 6 June 2017

Wells, L. (2015) 'Does Australia's farm sector need foreign investment?' *BBC News*, 1 March, available at: http://farmlandgrab.org/post/view/24598-does-australia-s-farm-sector-need-foreign-investment, accessed 6 June 2017

Whitley, A. (2012) 'Australia selling the farm as 20% price slump lures influx of global funds', *Bloomberg*, 30 January, available at: http://farmlandgrab.org/post/view/19968-australia-selling-the-farm-as-20-price-slump-lures-influx-of-global-funds, accessed 6 June 2017

Wiebe, N. (2012) 'Crisis in the food system: The farm crisis', in M. Koç, J. Sumner, and A. Winson (eds) *Critical Perspectives in Food Studies*, Oxford, Don Mills, Ontario, pp. 155–170

Williams, J. (2014) 'Feeding finance: A critical account of the shifting relationship between finance, food and farming', *Economy and Society*, vol 43, no 3, pp. 401–431

Williams, R. (2011) 'Who's really buying the farm?', *The Age*, 6 March, available at: http://farmlandgrab.org/post/view/18261-who-s-really-buying-the-farm, accessed 6 June 2017

Yahoo News Australia. (2015) 'Foreign investors on hunt for Aussie farms', 5 July, available at: http://farmlandgrab.org/post/view/25103-foreign-investors-on-hunt-for-aussie-farms, accessed 6 June 2017

Part II

Financialization of the food industry

7 Riding on waves of crises

Finance, food, and political tumult in Egypt

Marion W. Dixon

Introduction

In 2010, wheat and other food prices had begun to rise globally after falling from an historic high in the beginning of 2008. By the first months of 2011, global food prices reached the highest level since recording began in 1990 (FAO, 2012) and food inflation in Egypt reached over 20 per cent (Harrigan, 2014, p. 107). This time the crisis coincided precisely with a mass popular uprising first in Tunisia and then in Egypt and throughout the region. In Egypt, immediately after President Hosni Mubarak was forced to step down during the 25 January 2011 mass uprising, many of those who were most closely associated with the former President were arrested or threatened with arrest. Some of the members of Mubarak's 'old guard' fled the country or attempted to flee. Mubarak's sons Gamal and Alaa were jailed after being charged with a host of crimes, including embezzlement of public funds for personal benefit and, in 2012, insider trading, along with Hassan Heikal and Yasser El Mallawany – two executives from one of the country's largest investment banks, EFG-Hermes.

Much of the revolutionary activism, public commentary, and scholarly analysis has focused on the illicit ways in which wealth and privilege have accumulated among the few during the Mubarak regime. Mubarak is often referred to as having cronies and the economic system that was built during the neoliberal period as one of crony capitalism (Armbrust, 2011; Ismail, 2011; Joya, 2011). Although the dominant classes clearly gained wealth through illicit means, they have been able to gain class cohesion primarily through a transnational infrastructure of doing business legally and legitimately. In the absence of significant limits to their wealth and privilege through institutional mechanisms (e.g. the tax system and property law) leading up to the 2011 uprising, finance and family businesses, in competition and cooperation with transnational corporations (TNCs) and multinational corporations (MNCs),[1] were able to rapidly grow and consolidate the formal economy.

The focus of this chapter is on one sector of the formal economy – the agri-food industry – and the processes of financialization that led to its exponential growth and consolidation, from the early 1990s until the 2011 uprising. The agri-food industry is the integrated sector of the country's corporate agri-food system (Dixon, 2013, 2014). This system has three main markets: an agro-export market (of fresh and processed fruits, vegetables, and ornamentals like flowers); food service (franchises like McDonalds); and, food retail (including supermarkets and hypermarkets). Upstream includes corporate seed varieties (high-yielding and GMO), chemical fertilizers, pesticides, and other agricultural inputs. However, corporate food value chains – from land to labour to markets – crisscross in multiple ways with the informal, non-integrated, agriculture and food sector, blurring clear distinctions between the formal and the informal. The system grew from the 1970s by funnelling resources – land, labour, irrigation water, state agricultural revenue, and so forth – from staple food production for subsistence and national consumption to corporate food production for hypermarkets and supermarkets in Egypt and abroad (Dixon, 2013, 2014, 2015, 2017).

I demonstrate that the growth of corporate food from the 1990s, through the 2007–8 crisis, and up to the 2011 uprising, contributed to structural vulnerabilities in Egypt to global food price volatility. These vulnerabilities came in two forms: chronic national indebtedness and class-based vulnerabilities to food price hikes. Corporate food is largely for the global consumer class (in Egypt and abroad) and is dependent on imports (of energy, food ingredients and machinery). As food prices stayed relatively high (and were predicted to continue to rise) in the period between 2008 and 2011, the corporate agri-food system continued to grow and import dependency was maintained. Thus, with the onset of the next crisis in 2011, the Egyptian government had few options to mitigate class-based vulnerabilities to global food price fluctuations – not just among the poor, but also among the middle classes who became food insecure as a result of the price increases.

The growth of corporate food further contributed to political vulnerability – the legitimacy crisis of the Mubarak regime – as financial firms and family businesses accumulated capital rapidly and rather ostentatiously during this period and, in part, through investments in agriculture and food. These very activities to secure their wealth and privilege, in turn, engendered popular resentment and anger. That is why, in addition to their close association with Gamal Mubarak, the political class of finance and family business became a target during the 2011 uprising. In this analysis, the financialization of agriculture and food heralds two reinforcing processes. The first is the reshaping of agriculture and food for the benefit of the global consumer class (a class project). The second process is the re-constitution of the class power of finance, which, following Bourdieu (1986), has an enduring tension: as the dominant classes secure official, direct means to secure wealth and privilege and attempt to cement their position within the social structure, doing so engenders a popular (and subversive) critique of their entitlement to govern and to accumulate wealth.

In this chapter I first situate agri-food financialization in Egypt in three broad literatures: on the 2011 popular uprisings or 'an Arab Spring' (Dixon, 2011); on US-led financialization and US hegemony; and on financialization in Egypt and the rest of the Middle East and North Africa (MENA) region. In the following two sections, I provide an analytical framework that combines US-led financialization with financialization in Egypt and the rest of the region to trace two waves of financialization in agriculture and food in Egypt. These two waves are partially outcomes of US-centred economic downturns and political dilemmas (potential crises) that were avoided, in Krippner's (2011) terms, by instigating war and occupation and by prompting the opening of markets in Egypt and throughout the indebted world. The first wave, from the early 1990s to the early 2000s, followed a US-centred economic downturn and sovereign debt crisis in Egypt which led to comprehensive structural adjustments. The second wave, from 2003–4 to 2011, followed a US-centred economic downturn and a string of political scandals in the US, and led to far-reaching banking reforms from the pro-business Prime Minister Ahmed Nazif in 2004. In the last section, links between the growth of corporate food and chronic national indebtedness and class-based vulnerabilities to global food price volatility are drawn.

Political revolution – what does agri-food system change have to do with it?

A young, small-scale, vegetable vender's suicide in Tunisia in December 2010 sparked a revolutionary process in the region. Yet, there has been little research devoted to concrete structural changes that profoundly shaped the production and distribution of food and the lives of those gaining their livelihoods from agriculture and food. Of the vast literature on the Arab Spring, there are only a handful of works that attempt to relate the uprisings that swept the region to agri-food system change (Zurayk, 2011; Bush and Ayeb, 2012; Hanieh, 2013; Harrigan, 2014). As well, the literature that attempts to establish a direct relationship between the food price hikes of 2007–8 and 2011 and the revolutionary (and counter-revolutionary) processes that ensued tends to explain the food crises in terms of a growing gap between demand for food (and particularly energy-intensive foods like meat and dairy) and food production (UNDP, 2011; Harrigan, 2014). In order to establish a direct connection between the crises and the uprisings, however, the questions of who controls agriculture and food, and who gets to eat, need to be addressed.

In establishing the extent of financialization in agriculture and food in Egypt, I follow Krippner (2005) in examining long-term structural changes – in this case, the Egyptian economy. However, rather than adopt an 'accumulation-centred' perspective on where profits are generated, as Krippner (2005, pp. 175–176) advocates, I analyse changes in the organization of economic activity (primarily through mergers and acquisitions), the substance of those activities (the production of industrial, durable, agri-foods), along with the political character of class (re-)formation, that is, the class power of finance capital. Further, I heed Krippner's (2011, p. 4) warning not to reify the distinction between finance and production, or between the speculative and the real economy, by focusing on the mutually reinforcing processes of sectoral financialization, or the growing role of the financial sector (as opposed to production and trade) for capital accumulation, and non-sectoral financialization, that is, the growing reliance on portfolio income among non-financial firms (Krippner, 2005, p. 182). Thus, this analysis captures the co-production of finance and commodity production and trade (see also Fairbairn, 2014).

The growing role of finance in the Egyptian agricultural and food sectors, and the formal economy more generally, is placed within a hegemonic world order, in which US-centred financialization since the 1970s has heralded shifts in capital circuits worldwide (including shifts in the forms of capitalist economic enterprise and in class formation) (Arrighi, 1994; Konings, 2011). This framing of financialization is not merely theoretical, but rather, following Carroll (2010), Epstein (2005), and others, I attempt to establish the mechanisms through which US-centred financialization led to structural shifts in a peripheral, lower-income, country like Egypt. Studies on financialization in Egypt tend to focus exclusively on events and processes within the country. While financialization in Egypt is tied tightly to processes of financialization in the MENA region, the world-historical perspective that I offer is one of the region becoming more integrated into a US-centred world economy at particular conjunctures.

In order to understand the relationship between financialization and political revolt, I address the class power of finance capital (see Epstein, 2005). In Bourdieu's (1986) terms, their power is exercised through their ability to convert various capitals (particularly economic, social and cultural) into profits and future profits, and to use these capitals for accumulating even greater advantage through institutional mechanisms.[2] Class power is represented in the (re-)organization of capitalist enterprise around kinship. In contradistinction to the corporate model, which is based on the separation of ownership from management, 'there is an indelible relation between the accumulation of corporate capital and the production and reproduction of great fortunes' (Carroll, 2010, p. 133). And, these great fortunes are largely based on patriarchal kinship social units – or what Zeitlin and Ratcliff (1988) call the kinecon group (see Credit Suisse, 2011 on family business groups in South and East Asia; Zeitlin and Ratcliff, 1988 on Chile; and, Janelli, 1993 on South Korea). In this formulation, finance capital as a political class is made up of the executives and owners (shareholders) of financial firms and of family business groups – and the networks or interlocks among them through family, education and economic interests. While this class is connected tightly to the centres of finance globally, and may accumulate capital primarily through global capital circuits (for example, FDI and mergers and acquisitions across borders), they are indelibly tied to a country or region politically, socially, culturally. In Carroll's (2010, p. 35) words, there is a certain disjuncture between class formation as a sociocultural process and the economic process of capital accumulation. Therein lies a tension in the constitution of their class power. Through socio-cultural practices and economic re-organization they have attempted to cement their class position (Montgomerie, 2008; Carroll, 2010) but, in doing so, they have become a focus of subversive critiques that bring to light the arbitrariness of their entitlement (Bourdieu, 1986).

Cementing their class position through the growth and consolidation of the agri-food sector, in particular, has been fraught. This is because of the double entendre of what McMichael (2000) calls 'the power of food': given food's material and symbolic functions that link nature, human survival, livelihoods, culture, and health, food becomes subject to greater corporate control, while also fostering resistance to growing corporate power. Although there are studies that implicate agri-food financialization in the explosion of trading on the commodity futures markets (Crotty, 2005; Clapp, 2014), the literature on the financialization of agriculture and food largely misses this power (see Isakson, 2014 for a comprehensive literature review). It is through food that finance capital cements class power, while simultaneously and paradoxically undermining its class position.

Riding on crises: wave I, 1991–2003

The story of financialization in Egypt is usually presented in the following manner. A 2003 banking law was pushed through Parliament after Gamal Mubarak, son of then President Hosni Mubarak, rose within the ruling National Democratic Party (NDP) and became head of the party's Policies Secretariat in 2002, paving the way for private banks, especially foreign banks and investment banks, to increase their influence (Roll, 2010; Iyer, 2011). Gamal Mubarak had an investment career in London at the Bank of America and then, in 1996, formed an investment banking firm which, in turn, formed EFG-Hermes private equity (MacFarquhar *et al.*, 2011). EFG-Hermes is one of the largest investment banks in the Middle East today, and Egyptian Financial Group (EFG), established in 1984, had been a 'key player in creating, analyzing and managing the privatization program for the Government of Egypt' (Ismail, 2009, p. 74). Following the passage of the 2003 banking law, which Gamal Mubarak and other executives of financial firms and banks had backed, came the administration of Prime Minister Ahmed Nazif in 2004. His ministers were overwhelmingly the patriarchs of the country's most powerful family businesses.

The rise to political power of a pro-finance coalition within the ruling party and within the Cabinet was important to processes of financialization in Egypt – and to the political climate during the decade leading up to the 2011 uprising. However, this narrative does not sufficiently address the questions 'why, then?' and 'where did the financing come from?'. I address these questions by tracing the first wave of financialization to the country's sovereign debt crisis of 1990 and the passage of the Economic Reform and Structural Adjustment Programme (ERSAP) in 1991, which included the re-opening of the stock market, the privatization of more public banks, and the floating of exchange rates (Berthélemy and Bentahar, 2004; Roll, 2010). ERSAP also paved the way for the (re-)entry of TNCs and MNCs and the country joining the World Trade Organization (WTO) in 1995 (along with the bi- and multi-lateral trade agreements that followed). This wave does not mark the birth of finance but rather signals a moment in which there was a significant shift toward finance. The US-centred economic downturn of the late 1980s to the early 1990s, precipitated in part by speculation on currency markets and an increase in oil prices and the Gulf War, had triggered this moment of crisis (see Mitchell [2002] for the case of Egypt). The result was a decade-long period of recurring banking, currency, and debt crises in Southern indebted countries and nearly two decades of relative economic stability in the Northern core (see Figure 7.1). In Egypt this Northern boom/Southern crisis led to record levels of economic growth as Egyptian family business groups, TNCs, and MNCs bought public enterprises and rapidly consolidated the formal economy, including the agriculture and food sector.

During the Northern boom/Southern crisis, the trans-Atlantic powers liberalized their financial sectors. The launch of the euro and the liberalization of the European financial system encouraged financial trade within

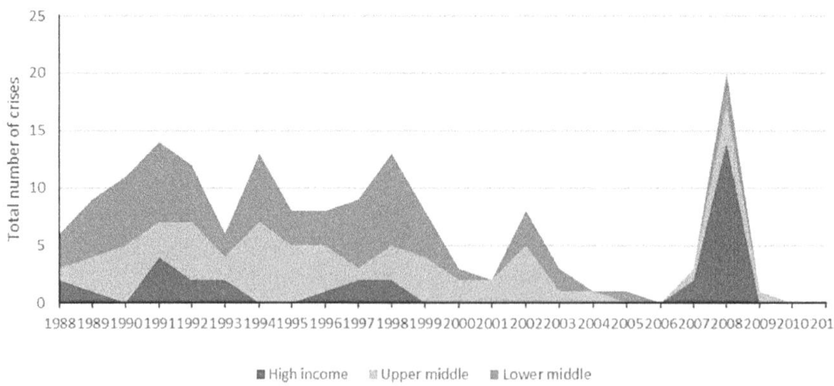

Figure 7.1 Frequency of banking, currency, and debt crises, 1988–2011.

Source: compiled from Broner *et al.*, 2012.[3]

the Eurozone and across the Atlantic (Lane, 2012). By the end of the 1990s there were significant financial liberalization measures in the US – including the repeal of the Glass-Steagall Act and the Commodity Futures Modernization Act (CFMA). Trans-Atlantic financialization meant not just that transnational and local banks could lend more (and that many banks grew exponentially), but that there was growth in new financial markets (securities and derivatives) (Clapp, 2014). It was European banks that were active in purchasing asset-based securities in these markets, which likely accelerated the growth of the asset-based securities markets in the US (Lane, 2012). Concomitant to the securitization of the trans-Atlantic financial markets was the proliferation of private equity, hedge funds, and other financial firms (Clapp, 2014). As the next section shows, these combined policy changes made possible the second wave of financialization in Egypt (and throughout the MENA region).

In Egypt the first wave, from roughly 1991 to 2003, led to the corporatization of family business groups[4] and to non-sectoral financialization. By the late 1990s, family business groups began to set up their own investment firms. For example, the El Mansour family (the two heads, Mohamed and Yasseen, made the Forbes billionaire list in the 2000s), joined their cousins in the El Maghraby family in 1996 to form El Mansour and El Maghraby Investment and Development (MMID). In 2001, MMID became an important shareholder in EFG-Hermes, the investment firm connected to the Mubaraks (Table 7.1). In 2004, Rashid Mohamed Rashid, who at the time was the head of Unilever Mashreq (a Middle East Foods and the Home and Personal Care Division of the Unilever Group International), also bought into the investment bank. That same year Ahmed Nazif became Prime Minister, and Rashid Mohamed Rashid, Mohamed Mansour, and Ahmed Maghraby all became ministers in the newly formed Nazif administration.

In agriculture and food, a corporate agri-food system grew rapidly: a handful of TNCs and MNCs re-entered the country and existing agri-food companies began to expand their market shares. Family business groups diversified into and within agribusiness either vertically or horizontally, in competition as well as in cooperation with TNCs and MNCs. For example, family business groups bought into state-held food companies. Under the Public Enterprise Law of 1991, about 13 publicly owned agribusinesses were privatized. Family business groups also began to buy into fast food franchises. A handful of agribusiness executives and representatives that I interviewed in 2011 had captured a significant share of the poultry industry before moving into the agro-export market and/or food processing.

Also, after Egypt joined the WTO in 1995 and signed bi- and multilateral trade agreements, the country became a trading hub of agri-food corporations. Corporations began: importing food (and other) inputs into the country, or buying crops produced in the country; packaging/processing them; and, then selling them to super/hypermarkets in the

Table 7.1 Sample of Egyptian financial firms with agri-food investments (1990s to 2011)

Name	*History/Background*	*Interlocks*	*Agri-food holdings/Activities*	*Geographic reach*
Actis	Formed in 2004; Spinout of Commonwealth Development Corporation (British corporation established in 1948 to invest in former colonies).	Large shareholder in Commercial International Bank (by 2009).	One of the first to enter agri-foods – in 2003 – with the acquisition of Rashidi El-Mizan company. Most activity was between 2003 and 2008.	International (former colonies)
Beltone	Founded in 2002 with regional finance.	Co-founders worked on Wall Street before forming Hermes Financial, which merged with Egyptian Financial Group to former EFG-Hermes.	Not particularly active in Egypt, but in 2009 in partnership with the Kenana Sugar Company in Sudan established a $1billion agriculture investment fund.	Subsidiaries in Egypt, Lebanon, and China
Citadel Capital (now Qalaa Holdings)	Co-founded in 2004 by two former EFG-Hermes executives.	One co-founder received his MBA from Harvard Business. The other is the brother of the former CEO of EFG-Hermes.	With the largest agri-food holdings. Through holding company, Gazour, was very active in mergers and acqusitions in agri-foods from 2007 to 2009.	Regional (North Africa, East Africa, Middle East); 15 countries (in 2011)

Concord International	An SEC registered investment management company founded in 1988.	Co-founder has MBA from Harvard Business.	In 2005 acquired a majority stake in Bisco Misr, the leading public company of baked goods and confectionary.	Based in New York City and Cairo
EFG-Hermes	Egyptian Financial Group (EFG) and Hermes Financial merged in 1996.	Gamal Mubarak, son of the former President and former London-based investment banker, co-founded private equity arm of EFG-Hermes.	Was active in the agri-foods sector between 2007 and 2009.	International (Europe, Middle East, North Africa)
El Mansour and El Maghraby Investment and Development (MMID)	Formed in 1996 by El-Mansour and El-Maghraby families (as investment arm of the two family businesses).	Important shareholder in EFG-Hermes (by 2001). Mohamed Mansour and Ahmed Maghraby become ministers under Ahmed Nazif administration (in 2004).	Its main agri-food holdings have included two of the country's largest high-end supermarket chains.	Regional/MENA

Source: compiled by author from Dixon, 2013, 2014.

Middle East, the European Union, and/or North America. For example, the EU–Egypt Association Agreement of 2004 stipulates that Egyptian industrial products enter the EU with no tariffs or quotas and that Egypt liberalizes trade and tariff barriers over 12 years to allow the entry of European products (Dixon, 2013).

Riding on crises: wave II, 2004–2011

During field research in Egypt between 2008 and 2012, I attempted multiple times – and unsuccessfully – to meet with, and interview, representatives of financial firms in Cairo. The elusiveness of financial firms reflects a growing tension in the constitution of finance capital's class power in the intervening years between the 2007–8 crises and the 2011 uprising, the period of my field research. A class of finance capitalists, which re-emerged post-2003, became important to the reconstitution of the dominant classes. Through an intertwining of kinship, sociocultural institutions and practices (through education and speaking English), and corporate structures (such as family businesses), the dominant classes cemented their position within the Egyptian social structure. At the same time, this engendered popular critiques and dissent. This political class then responded by attempting to conceal their activities and wealth.

The economic boom from roughly 2003–4 to 2007–8 was short-lived and in many ways far surpassed the economic boom of the 1990s in its reach and depth (see Figure 7.1 on the steep decline in crises globally between these years). Because the proliferation of cheap debt financing characterized this upswing, it is often referred to as a securitization boom (Lane, 2012; Clapp, 2014). For example, globally private equity alone grew threefold between 2004 and 2008, from US$141.7 billion to US$445 billion in 2008 (Ismail, 2009, p. 13). In the MENA region, financialization – as measured in terms of the level of stock market capitalization, the level of competition in banking, and so on (El Safti, 2007) – was uneven. The extent of financialization in the region prior to 2003 was low compared to East and Southeast Asia and Latin America (Kaminsky and Schmukler, 2003; Creane *et al.*, 2004, p. 13).

What spawned this global economic boom? Merely one year after record global FDI levels in 2000 (UNCTAD, 2001, p. 1), the economic bubble (of the 1990s) burst as the US became embroiled in large-scale corporate accounting fraud scandals and the ensuing dot-com crash (as well as the 11 September 2001 terrorist attacks). The US-centred economic downturn was short-lived: gross capital flows in the US increased markedly between 2002 and 2004 (Broner *et al.*, 2012) as the US attacked and occupied Afghanistan and then Iraq. Furthermore, the US government's response was weak in terms of placing controls on financial market expansion through stricter accounting measures, and the like.[5] During the economic upswing, capital flows were increasingly directed to so-called emerging markets, especially the east-Asian region as it recovered from the 1997–98 sovereign debt crisis (UNCTAD, 2004; Lane, 2012).

Countries in the MENA region responded to the increase in global capital flows by passing potent financial liberalization measures. In Egypt, banking reform was passed in 2003, and in the Gulf Cooperation Council a series of financial liberalization measures were passed in 2004 (UNCTAD, 2005, p. 282). While a Wall Street-led campaign against financial regulation had pointed to this growth in capital inflows outside 'the Triad' – the US/ Canada, the European Union) and Japan – as evidence that regulations were scaring away investment in the US (Stoltenberg *et al.*, 2011), growing investments in the MENA region (and other regions) is said to have helped fuel the economic boom in the US and EU (see Lane, 2012).

The US-centred economic downturn and political scandals of 2001–2 spawned an economic downturn in Egypt, and the 2003 Unified Banking Law was justified as a means to stabilize prices in the context of growing inflation (Iyer, 2011). The 2003 United Banking Law departed significantly from previous finance reforms because it expanded the Central Bank's authority and authorized the selling of public banks (via joint mergers) and the mergers of private banks. These policy measures led to significant foreign capital penetration: transnational banks (such as Barclays, Credit Agricole, Societe Generale) and regional banks (including Greek bank Piraeus, Lebanese bank BLOM, Abu Dhabi Islamic Bank, and the National Bank of Kuwait) played a key role in consolidation of the banking sector. Also, select private Egyptian banks bought existing (private/public) banks – and these banks were largely investment banks, like MMID (Global Investment House, 2008). In terms of market capitalization, the capitalization to GDP ratio rose significantly: from US$28 billion in 2002 (29 per cent of GDP) to US$82 billion in 2010 (40 per cent of GDP) (Diwan and Chekir, 2014, p. 188). Accompanying the exponential increase in the value of newly formed publicly traded corporations (now on the stock exchange) was a sharp increase in FDI from 2003–4, and FDI, with the largest sources from the US and UAE, continued to go upward only dipping slightly after 2007–8 (Kapadia, 2011, p. 47).

The 2003 law also precipitated the rapid growth of private equity funds, which rose from US$533 million in 2004 to US$6.4 billion in 2008 (Ismail, 2009, pp. 6, 61). Most private equity funds in Egypt are created with private co-investors – such as sovereign wealth funds, institutional funds (such as pension funds), and wealthy Gulf individuals with the firm's managers, who generally commit 10–20 per cent of the funds, and institutional investors like the World Bank's International Finance Corporation (IFC). In fact, the IFC and other institutional lenders have been major guarantors of private equity activity in so-called emerging markets like Egypt (Daniel, 2012, p. 722).

The 2003 Banking Law did not just usher in a process of sectoral financialization (to marry non-sectoral financialization), it also led to the re-emergence of a political class of finance. EFG-Hermes was at the centre of this class. Many of the heads of private equity firms worked for EFG-Hermes (Table 7.1). Further, executives of financial firms (including banks) rose to prominence within the dominant classes and reconstituted them in the post-2003 period. Their place within the private sector think tank Global Trade Matters (GTM) reflects their rise to prominence. GTM was a leading Cairo-based think tank for 'economic policy and reform', organizing conferences, seminars, and award ceremonies among the 'who's who' of the private sector. In 2010, GTM hosted the 'Egyptian Business Oscars', with all the familiar actors within finance were there: Best CEO Under 40 was Hisham El Khazindar, CEO of Citadel Capital; Best Company Overall was Commercial International Bank; and, Best Investment Bank was CIB's CI Capital Holding. Finance, as a political class, secured its position within the dominant classes through the intertwining of kinship, sociocultural institutions, and corporate structures. As a member of the GTM network said, 'We are one big family' (interview, Cairo, 13 August 2010). And, this 'family' consolidated the formal economy and gained political power as the formal economy grew from the 1990s. Wealth creation in agriculture and food, in particular, contributed to the Mubarak regime's legitimacy crisis by determining who owns what and who has access to nutritious and adequate food.

From roughly 2003 to 2011 the corporate agri-food system underwent a second wave of growth and consolidation, as financial firms (especially private equity) began to invest in agriculture and food – and the formal economy, more generally. Financial firms' mergers and acquisitions in the agri-food industry followed from the 2003 banking law and grew exponentially after the 2007–8 crisis (and until 2011), when food prices rose and were expected to continue to rise in the long term (Dixon, 2013, 2014). The primary activity of private equity is acquiring large companies (including formerly state-owned ones) in order to make them more competitive by increasing their size and/or their geographical reach and, then, re-selling them to corporations or other financial firms. The private equity firm Citadel Capital (now an investment bank, Qalaa Holdings) created the largest holding company within the agri-food industry, Gozour, during this period, 'with the purpose of creating a vertically integrated regional agriculture and multi-category consumer foods conglomerate with primary lines of business' (Ismail, 2009, p. 178 and see Citadel Capital, 2010). Until 2015, when Citadel Capital changed to an investment bank and began divesting from Gozour, the holding company was made up of three integrated parts: agri-foods and dairy (Gozour Agri), fast-moving consumer goods (Gozour Foods), and intermediate industries such as sugar and corn processing (Gozour Intermediate) (see Figure 7.2). For example, Gozour first acquired (and then sold) Dina Farms, the largest private farm with nearly 10,000 *feddans*,[6] which supplies Enjoy's milk – Enjoy being one of the largest producer of packaged milk, juices, and yogurt. Gozour

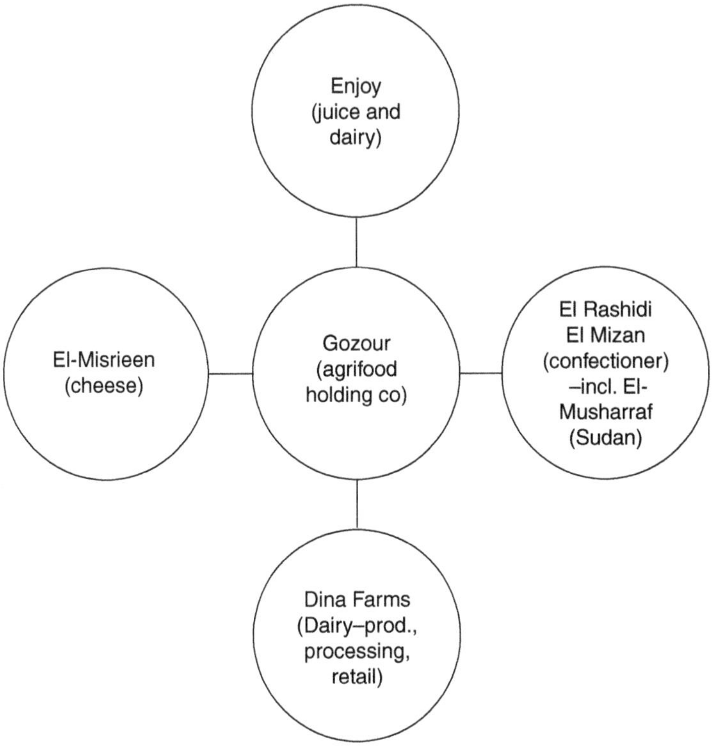

El Rashidi El Mizan (halawa and tahini co): Timeline

1889	*2000*	*2002*	*2007*	*2009*	*2015*
Family business founded	Acquired by Best Foods/ Unilever	Acquired by Actis private equity	Acquired by Citadel Capital	Acquires Sudan's El-Musharraf co.	Acquired by Olayan Financing Co. (Saudi)

Figure 7.2 Citadel Capital's Agri-food Holdings, 2014.

Source: compiled from Dixon, 2013, 2014.

combined the packaging facilities of Mom's Food and Enjoy, and Enjoy was in line to supply Al Misrieen, a cheese and juice company, with fruit juices. Citadel Capital had plans, not only to integrate its companies, but to expand them. The history of Rashidi El Mizan, the leading sesame producer in the country, is illustrative: Gozour had been expanding Rashidi El Mizan since buying the company from private equity firm Actis in 2008. Rashidi El Mizan had since gained a near monopoly of the sesame product market, with 60 per cent of the halawa market and 80 per cent of the tahina market. Then, Rashidi El Mizan acquired a majority stake in al Musharraf, one of the largest producers of biscuits and sweets in Sudan.

During these waves of corporatization and financialization, the corporate agri-food system grew and consolidated. In 1995 annual real investment in the food industry was at roughly LE 400 million and by 2008 rose to about LE 1,900 million – the greatest annual real investment in Egypt for 2008 (Abdel-Latif and Schmitz, 2011). Between 1992 and 2002 food processing alone grew from 12.6 per cent to 71.1 per cent of total manufactured value added (IMC, 2005). Most of the largest food processors have been the largest agro-exporters and, in 2004, the agro-export market was already heavily concentrated in the hands of the largest national and multi/transnational food corporations, including Faragalla and Juhayna (two of the largest food processors), Farm Frites (a regional MNC that partnered with Americana), and Cadbury and Nestle (two of the world's largest food TNCs). In 2010 the top 6.5 per cent of exporting companies made up 41 per cent of the total value of agricultural exports (EDF, 2011). In food processing, by 2005 ten corporations were dominating the production of frozen vegetables and seven were dominating the processing and packing of dairy products (IMC, 2005). In the poultry industry in 2006 broiler production was heavily concentrated in the hands of five poultry corporations (Dixon, 2015).

Chronic national indebtedness and class-based vulnerabilities to global food crises

While it is difficult to measure the extent to which the corporate agri-food system as a whole contributed to import dependency, the industry relied heavily on the top imports of 2007 – machinery, petroleum, wheat, and corn. In fact, five corporate poultry breeders are the near exclusive consumers of imported corn (Dixon, 2013, 2015). Most of the other major food import commodities are essential ingredients of corporate food (edible oil, sugar, red meat, powdered milk) (FAO STAT, n.d.). Leading up to the crisis, the industry was contributing disproportionately to the import bill by transforming food ingredients in multiple ways into largely durable commodities that often travelled long distances. Importantly, corporate food is energy-intensive (in terms of transportation, refrigeration, on-farm mechanization, and chemical agricultural inputs) and import-dependent.

The agro-export market of non-traditional, high value, foods and ornamentals – a cornerstone of the corporate agri-food system – was supposed to counteract declining terms of trade. In interviews in Cairo in 2010 and 2011, development practitioners and government consultants and officials largely expressed pride in the rapid growth of agro-exports. In one interview a development manager of a multilateral organization boasted that horticulture exports increased six-fold during the 2000s (interview, Cairo, 9 August 2011). There was also a way in which agro-export growth was equated with Egypt's food security. For example, when asked to assess the state of the country's food security, a development consultant replied that food security would improve if exports continued to grow (interview, Cairo, 3 June 2010). And, while overall export revenues did grow, import costs rose even faster. Between 1995 and 2007 the value of exports to imports fell from 62 per cent to 38 per cent.[7] In the inter-crisis period, as prices rose the trade deficit was only partially addressed – the value of exports to imports rose from 38 per cent to roughly 46 per cent by 2010.[8] Therefore, the country remained a price-taker in the world economy.

Moreover, the unfavourable terms of trade contributed to class-based vulnerabilities during the global food crises of 2007–8 and 2011. The growth of corporate food is often attributed to growing consumer demand as the middle classes use their disposable income to buy global consumer goods (World Bank, n.d.; USDA, 2015). As well, the consumer market for corporate food lies along a fairly wide spectrum – from 'low end' (kiosks) to 'middle end' (discount supermarkets) to 'high end' (European supermarkets) (Dixon, 2013). In the agri-food industry – and the global food economy of which it is apart – industrial, processed, foods represent a space of dietary convergence across class. The wealthy and working classes alike consume 'soft drinks, packaged snacks and so-called convenience foods that are full of fats, sweeteners, artificial flavours and colouring' (Weis, 2007, p. 15). A comparison of Egypt's National Nutrition Institute cross sectional survey data conducted in 1981 and then again in 1998 indicates that in the intervening years: animal protein intake rose sharply; 'meals away from home' more than doubled; and, the disparity in energy intake between the wealthy and poor grew (NNI, 2000; see also Galal, 2002). While these surveys do not reveal the type or level of processing of the food consumed, the tremendous growth in food processing alone beginning in the 1990s attests to a level of dietary convergence across class. The wide spectrum of the consumer market has certainly enabled more population groups, especially youth, to buy corporate food – whether it be a soda at a kiosk or a burger at McDonalds.

The breadth of the market spectrum of corporate food should not obscure the fact that the spectrum is heavily skewed toward the upper end for the global consumer classes both at home and abroad. While the consumer market of corporate food grew, a significant part of the population was spending a sizable percentage (35–55 per cent) of their income on food (Harrigan, 2014, pp. 110–111). An estimate is that over half a million people fell below the poverty line as a result of the crises – an increase of around 4.5 per cent (Harrigan, 2014, p. 97). Given the close relationship between poverty and food insecurity, the government measures to mitigate the 2007–8 crisis – from increasing expenditures on food subsidies, improving public sector wages, and instituting a temporary ban on rice imports – could only do so much to shield these groups. Furthermore, the continued growth of corporate food and the maintenance of import dependency in the inter-crisis period undermined the ability of the government to recover from the 2007–8 crisis and, thus, to be in a better position to insulate vulnerable population groups from the shocks of 2011.

Conclusions

Financialization in the US since the 1970s has led to uneven processes of financialization around the world. In the case of Egypt, I argue that the two waves of financialization since the early 1990s rode on (potential) crises. During the first crisis, the US had attempted to address economic and political dilemmas by encouraging (and enforcing) the opening of new markets abroad, and the inflow of finance capital from the US, in particular, and the Triad, more generally, made financialization in Egypt possible. The passage of the comprehensive structural adjustment package (ERSAP) in 1991 and membership in the WTO in 1995 (and the subsequent bi- and multi-lateral trade agreements) led to the rapid growth and consolidation of the formal economy. Transnational corporations and multinational corporations (re-)entered the country, and in competition and partnership with them, Egyptian family business groups corporatized and (vertically and/or horizontally) integrated their holdings, in part through the development of investment arms. These processes of non-sectoral financialization, in which non-financial firms come to rely increasingly on portfolio income (from interest payments, dividends, capital gains on investments, and the like), were part-and-parcel of the making of a corporate agri-food system. Family business groups and TNCs/MNCs bought into agriculture and food – from fast food franchises and super/hypermarkets to public sector food processing. Corporate food grew exponentially during this first wave until the early 2000s.

During the second wave of financialization from 2003–4 to 2007–8, during the so-called securitization boom, corporations and family business groups came to rely increasingly on portfolio income. As well, international and regional financial institutions offered institutional support to Egyptian financial firms (especially private equity). But the level of financialization pre-2003 in Egypt (and the greater MENA region) was lower than in other Southern regions (namely, in southeast Asia and Latin America). It can be inferred then that the deepening of financialization post-2003 was significant in the region compared to other semi-peripheral regions. Furthermore, during this second, deeper, wave of financialization, finance as a political class greatly increased its influence within the Egyptian social structure, in part because of the unique role of Gamal Mubarak. Through the intertwining of kinship, sociocultural institutions, and corporate structures, finance reconstituted the dominant classes, which subsequently attempted to cement their economic position.

These processes of sectoral financialization expanded exponentially the agro-export market of fresh and processed fruits and vegetables for the super/hypermarkets of the Middle East and Europe. This expansion was particularly pronounced in the inter-crisis period, between 2007–8 and 2011, in response to the global commodity price hike. Yet, imports of food ingredients, energy, and other inputs for the corporate agri-food system continued to grow. Chronic national indebtedness, rather than being chipped away at as neoliberal proponents predicted, was maintained. These unfavourable terms of trade exacerbated class-based vulnerabilities to food price volatility, both in 2007–8 and 2011. The government faced difficulties insulating those who were (marginally) food insecure from the price shocks, all while finance and family business cemented, seemingly successfully, their class position. However, herein lies the tension of the class power of finance: finance and kinship-based economic units as ways to accumulate wealth and privilege are the most direct, with few disruptions in the transmittal of forms of capital (Bourdieu, 1986), and at the same time, defy most clearly any notions of meritocracy. As financial firms and family businesses attempted to cement their position, the popular critique of their entitlement grew in concert.

Notes

1 The main distinguishing characteristic among TNCs and MNCs here is their reach. Transnational corporations operate globally and multinational corporations operate regionally. In Egypt, most MNCs are from southern Europe (e.g. Greek, Italian), the Middle East (e.g. Lebanon, Saudi Arabia, Kuwait), and (more recently) China.

2 For Bourdieu (1986, p. 243) economic capital is immediately and directly convertible into money and may be institutionalized in the form of property rights; cultural capital may be institutionalized in the form of educational qualifications; social capital may be institutionalized in the form of titles; and, both cultural capital and social capital may be convertible into economic capital in certain conditions.

3 According to Broner (2002), there are 38 high-income countries (examples are Australia, Austria, Canada, Denmark, Finland, France, Germany, Greece, Italy, Japan, Netherlands, New Zealand, Norway, Spain, Sweden, United Kingdom, and United States); there are 26 upper-middle-income countries (including Argentina, Brazil, Chile, Costa Rica, Croatia, Iran, Libya, Malaysia, Mexico, Russian Federation, South Africa, Turkey, Venezuela); and, there are 38 lower-middle-income (which include Albania, Algeria, Angola, Bolivia, China, Ecuador, Egypt, India, Indonesia, Jamaica, Morocco, Namibia, Nicaragua, Thailand, Tunisia, Ukraine, and Vietnam).

4 Corporatization refers to companies being bought and sold through a legal framework of mergers and acquisitions; separating (to some extent) ownership and management with the establishment of a Board of Directors; instituting legally-binding regulatory and accounting procedures; and (possibly) becoming a publicly traded company on the stock exchange through a set of reviews, value assessments, and so on.

5 It should be noted that re-regulation, in the form of new accounting measures with the Sarbones Oxley Act (SOX) in 2002, applied only to publicly traded companies.

6 A *feddan* is roughly equivalent to an acre.

7 Given its energy intensity, the agri-food industry is tied tightly to the oil industry. Crude oil prices rose steadily during the first decade of the twenty-first century (a price hike that precipitated the 2007–8 crisis), and the value of crude oil exports from Egypt rose accordingly. The oil export value was at approximately US$2.2 billion in 1995, and began to increase exponentially in 2003, reaching US$7.4 billion in 2007. Exports of crude oil decreased while exports of higher-grade oil rose during this period. Yet, the value of finished oil imports rose more quickly than that of exports: Oil imports to exports rose from 39 per cent in 1995 to 55 per cent in 2007.

8 Total import export values for both periods – 1995–2007 and 2007–11 – were calculated from the datasets of the Observatory of Economic Complexity (http://atlas.media.mit.edu/en/). And the oil export and import values were taken from EconStats (www.econstats.com/weo/V025.htm, www.econstats.com/weo/V026.htm).

References

Abdel-latif, A. and Schmitz, H. (2011) 'The politics of investment and growth in Egypt: Experimenting with a new approach', *Development Policy Review*, vol 29, no 4, pp. 433–458

Armbrust, W. (2011) 'The revolution against neoliberalism', *Jadaliyya*, 23 February, available at: www.jadaliyya.com/pages/index/717/the-revolution-against-neoliberalism-, accessed 10 March 2011

Arrighi, G. (1994) *The Long Twentieth Century: Money, Power and the Origins of Our Times*, Verso, London and New York

Berthélemy, J. and Bentahar, N. (2004) 'Financial reforms and financial development in Arab countries', Paper presented at the International Conference on Institutions and Development Performance, Cairo, available at: www.cedej-eg.org/IMG/pdf/BenTaharBerthelemy.pdf, accessed 22 January 2016

Bourdieu, P. (1986) 'The forms of capital', in J. Richardson (ed.) *Handbook of Theory and Research for the Sociology of Education*, Greenwood Press, New York

Broner, F., Didier, T., Erce, A. and Schmukler, S. (2012) 'Gross capital flows: Dynamics and crises', *Journal of Monetary Economics*, vol 60, no 1, pp. 113–133

Bush, R. and Ayeb, H. (eds) (2012) *Marginality and Exclusion in Egypt*, Zed Books, New York

Carroll, W. (2010) *The Making of a Transnational Capitalist Class: Corporate Power in the 21st Century*, Zed Books, London and New York

Citadel Capital. (2010) *Annual Report 2010*, available at: www.citadelcapital.com/downloads/Citadel-Capital-Annual-Report-2010.pdf, accessed 2 July 2012

Clapp, J. (2014) 'Financialization, distance and global food politics', *Journal of Peasant Studies* vol 41, no 5, pp. 797–814

Creane, S., Goyal, R., Mobarak, A. and Sab, R. (2004) 'Evaluating financial sector development in the Middle East and North Africa: New methodology and some new results', *Topics in Middle Eastern and North African Economies*, vol 6, available at: http://ecommons.luc.edu/cgi/viewcontent.cgi?article=1052&context=meea, accessed 31 January 2016

Credit Suisse. (2011) *Asian Family Business Report 2011*, available at: https://mobile.credit-suisse.com/index.cfm?fuseaction=mOpenArticle&coid=293542&aoid=322410&lang=EN, accessed 5 July 2012

Crotty, J. (2005) 'The neoliberal paradox: The impact of destructive product market competition and 'modern' financial markets on nonfinancial corporation performance in the neoliberal era', in G. Epstein (ed.) *Financialization and the World Economy*, Edward Elgar, Cheltenham, UK, pp. 77–110

Daniel. S. (2012) 'Situating private equity capital in the land grab debate', *Journal of Peasant Studies*, vol 39, nos 3–4, pp. 703–729

Diwan, I. and Chekir, H. (2014) 'Crony capitalism in Egypt', *Journal of Globalization and Development*, vol 5, no 2, pp. 177–211

Dixon, M. (2011) 'An Arab Spring', *Review of African Political Economy*, vol 38, no 128, pp. 309–316

Dixon, M. (2013) 'The making of the corporate agri-food system in Egypt', PhD Thesis, Cornell University, Ithaca, New York

Dixon, M. (2014) 'The land grab, finance capital, and food regime restructuring: the case of Egypt', *Review of African Political Economy*, vol 41, no 140, pp. 232–248

Dixon, M. (2015) 'Biosecurity and the multiplication of crises in the Egyptian agri-food industry', *Geoforum*, vol 61, pp. 90–100

Dixon, M. (2017) 'Plastics and agriculture in the desert frontier', *Comparative Studies of South Asia, Africa, and the Middle East*, vol 31, no 1, available at: https://marionwdixon.com/refereed/, accessed 28 April 2017

EDF. (2011) *Agricultural Crops Program of the Export Development Fund*, Export Development Fund, Cairo

El Safti, A. (2007) 'Financial sector reforms in the Arab countries', *AMF Economic Studies*, no. 8, Arab Monetary Fund, Abu Dhabi, available at: www.amf.org.ae/content/financial-sector-reforms-arab-countries, accessed 22 January 2016

Epstein, G. (ed.) (2005) *Financialization and the World Economy*, Edward Elgar, Cheltenham, UK

Fairbairn, M. (2014) '"Like gold with yield": Evolving intersections between farmland and finance', *Journal of Peasant Studies*, vol 41, no 5, pp. 777–795

FAO. (2012) *FAO Food Price Index. World Food Situation*, available at: www.fao.org/worldfoodsituation/wfs-home/foodpricesindex/en/, accessed 30 January 2012

FAO STAT. (n.d.) *Country Profile: Egypt – Top Imports – 2011*, available at: http://faostat.fao.org/desktopdefault.aspx?pageid=342&lang=en&country=59, accessed 10 August 2012

Galal, O. (2002) 'The nutrition transition in Egypt: Obesity, undernutrition and the food consumption context', *Public Health Nutrition*, vol 5, no 1A, pp. 141–148

Global Investment House. (2008) *Egypt Banking Sector*, available at: www.globalinv.net/research/egypt-banking-092008.pdf, accessed 26 January 2016

Hanieh, A. (2013) *Lineages of Revolt: Issues of Contemporary Capitalism in the Middle East*, Haymarket Books, Chicago

Harrigan, J. (2014) *The Political Economy of Arab Food Sovereignty*, Palgrave Macmillan, New York

IMC. (2005) *Egyptian Processed Food Sector Review: Final Report*, Industrial Modernisation Centre, Cairo

Isakson, S. R. (2014) 'Food and finance: the financial transformation of agro-food supply chains', *Journal of Peasant Studies*, vol 41, no 5, pp. 749–775

Ismail, A. (2009) *Private Equity and Venture Capital in Emerging Markets: A Case Study of Egypt and the MENA Region*, PhD Thesis, Massachusetts Institute of Technology, Massachusetts

Ismail, S. (2011) 'A private estate called Egypt', *Guardian*, 6 February, available at: www.theguardian.com/commentisfree/2011/feb/06/private-estate-egypt-mubarak-cronies, accessed 1 March 2011

Iyer, D. (2011) 'Building the capacity to regulate: Central bank reform in Egypt, 2003–2009', available at: https://successfulsocieties.princeton.edu/sites/successfulsocieties/files/Policy_Note_ID156.pdf, accessed 30 January 2016

Janelli, R. (1993) *Making Capitalism: The Social and Cultural Construction of a South Korean Conglomerate*, Stanford University Press, Stanford, CA

Joya, A. (2011) 'The Egyptian revolution: Crisis of neoliberalism and the potential for democratic politics', *Review of African Political Economy*, vol 38, no 129, pp. 367–386

Kapadia, S. (2011) *Egypt's Financial Liberalisation: Why Didn't It Do What It Said It Would On the Box?* MA Thesis, International Institute of Social Studies, available at: https://thesis.eur.nl/pub/10688/, accessed 30 January 2016

Kaminsky, G. and Schmukler, S. (2003) *Short-Run Pain, Long-Run Gain: The Effects of Financial Liberalization*, available at: www.imf.org/external/pubs/ft/wp/2003/wp0334.pdf, accessed 31 January 2016

Konings, M. (2011) *The Development of American Finance*, Cambridge University Press, New York

Krippner, G. (2005) 'The financialization of the American economy', *Socio-Economic Review*, vol 3, pp. 173–208

Krippner, G. (2011) *Capitalizing on Crisis: The Political Origins of the Rise of Finance*, Harvard University Press, Cambridge, MA

Lane, P. (2012) 'Financial globalisation and the crisis', Paper presented at the 11th BIS Annual Conference on 'The Future of Financial Globalisation', Lucerne, 21–22 June, available at: www.lse.ac.uk/researchAndExpertise/units/growth Commission/documents/pdf/contributions/lseGC_lane_FinGlob.pdf, accessed 16 February 2017

MacFarquhar, N., Rohde, D. and Roston, A. (2011) 'Mubarak family riches attract new focus', *New York Times*, 12 February, available at: www.nytimes.com/2011/02/13/world/middleeast/13wealth.html?_r=1&pagewanted=all, accessed 29 June 2012

McMichael, P. (2000) 'The power of food', *Agriculture and Human Values* vol 17, pp. 21–33

Mitchell, T. (2002) *Rule of Experts: Egypt, Techno-Politics, Modernity*, University of California Press, Berkeley, CA

Montgomerie, J. (2008) 'Bridging the critical divide: Global finance, financialisation and contemporary capitalism', *Contemporary Politics*, vol 14, no 3, pp. 233–252

NNI. (2000) *Food Consumption Pattern and Nutrients Intake Among Different Population Groups. To update Food Consumption Pattern Survey conducted in 1981*, National Nutrition Institute, Cairo

Roll, S. (2010) '"Finance matters!" The influence of financial sector reforms on the development of the entrepreneurial elite in Egypt', *Mediterranean Politics*, vol 15, no 3, pp. 349–370

Stoltenberg, C., George, B., Lacey, K. and Cuthbert, M. (2011) 'The past decade of regulatory change in the US and EU capital market regimes: An evolution from national interests toward international harmonization with emerging G-20 leadership', *Berkeley Journal of International Law*, no 577, available at: scholarship.law.berkeley.edu/bjil/vol. 29/iss2/4, accessed 31 January 2016

UNCTAD. (2001) *World Investment Report: FDI Policies for Development: National and International Perspectives*, United Nations Conference on Trade and Development, New York and Geneva

UNCTAD. (2004) *World Investment Report: FDI Policies for Development: National and International Perspectives*, United Nations Conference on Trade and Development, New York and Geneva

UNCTAD. (2005) *World Investment Report: FDI Policies for Development: National and International Perspectives*, United Nations Conference on Trade and Development, New York and Geneva

UNDP. (2011) *Food Security Challenges in the Arab States/MENA Region in the Context of Climate Change: The Role of the Regional Directors Team*, Regional UNDP, Arab States/Middle East and North Africa Position Paper: Nexus of Climate Change and Food Security, UNDP, New York

USDA. (2015) *Egypt: Food Processing Ingredients, Annual*, USDA GAIN Report, 9 November, available at: http://gain.fas.usda.gov/Recent%20GAIN%20Publications/Food%20Processing%20Ingredients_Cairo_Egypt_11-9-2015.pdf, accessed 14 February 2016

Weis, T. (2007) *The Global Food Economy: The Battle for the Future of Farming*, Zed Books, London and New York

World Bank. (n.d.) 'GNI per capita, Atlas method (current US$)', available at: http://data.worldbank.org/indicator/NY.GNP.PCAP.CD?locations=ZQ, accessed 1 February 2016

Zeitlin, M. and Ratcliff, R. (1988) *Landlords and Capitalists: The Dominant Class of Chile*, Princeton University Press, Princeton, N.J

Zurayk, R. (2011) *Food, Farming, and Freedom: Sowing the Arab Spring*, Just World Books, Charlottesville, Virginia

8 Financialization in Japanese agri-food regimes

Uncovering the role of *sogo-shosha* in global soy investment

Midori Hiraga

Introduction

This chapter argues that Japanese *sogo-shosha* (so-called 'general trading companies') have been some of the most significant early actors in the financialization of food systems, both in Japan and the world. Since the birth of modern food systems more than a century ago, Japanese *zaibatsu* (so-called 'financial conglomerates') and *sogo-shosha* have been operating as indivisible institutions combining industrial-mercantile-finance capital, making an essential contribution to both the nation-state project and the capitalist development of Japan. Their influence remains highly significant today, and what was originally a pivotal role in Japanese national economic development has now expanded greatly across the global food system, especially in Asia. Building on these strong, albeit unusually configured, economic institutions Japan operates as an influential investor in the global finance sector. Japan's Government Pension Investment Fund (GPIF) is regarded as the world largest institutional investor. With its asset size of 135 trillion yen (US$1.2 trillion) at the end of fiscal year 2015, it pours capital into most major funds and corporations in Japan, while also investing about 40 per cent of its finances into foreign bonds and equities (GPIF, 2015). This percentage is likely to increase as a result of further financial deregulation. In addition to the pension fund, Japanese banks are among the world's largest, and its *sogo-shosha* are actively investing globally.

One of their targets is food and agriculture, and financial investors tend to follow the profit-making trends in that sector. Some Japanese bank groups and *sogo-shosha* have already been profiled for their investments in Southeast Asia's commodity sectors (RAN, 2017). Japanese *sogo-shosha* have been active overseas, especially in Asia. Itochu, together with the CP Group, invested a combined total of about 1,200 billion yen in CITIC, China's state-owned company and one of its largest conglomerates, in 2015. With this strategic business alliance and capital participation with CITIC and CP Group, Itochu has become one of the most influential actors in China's food system as well as in the wider Asian region (Itochu

2016, 2017). Marubeni, another Japanese *sogo-shosha*, claims it handles about 20 per cent of soybean imports to China as Japan's largest grain trader (Marubeni, 2012). Both Itochu and Marubeni, together with other Japanese *sogo-shosha* like Mitsui & Co. and Mitsubishi Corporation, have similar strategies in the global food system – actively investing in food industries in Asia (especially in China) to increase the demand for ingredients and at the same time investing in production and supply facilities in North and South America to increase procurement capacity.

Most crucially, although they are described as traders, *sogo-shosha* today have become more 'general investment companies' in a wide range of industries, and they work closely with other financial institutions, including ex-*zaibatsu* corporate groups (more details in later sections). In other words, they retain and are building on the hybrid industrial-mercantile-financial institutional character that has typified economic institutions in Japan since the very beginning of capitalist development. Thus, given rising concerns about the financialization of the global food system, it is vitally important that we have analyses that reveal the otherwise hidden ways that *sogo-shosha* are operating as agents of financialization, contributing to Japan's huge financial capital flows and its significant investments throughout the agri-food system – and in particular in Asia. Furthermore, while financialization has been observed as a global phenomenon, particularly since the mid-1980s, such global trends are, as is revealed in the Japanese case, mediated and facilitated by economic institutions that have very specific historical and regional configurations.

The financialization of food and agriculture

Financialization of the global food system has attracted attention in English academia since the 2007–8 food price crisis. It is argued that capital itself has made land and food a major sector for investment, speculation, and technical change, with financial investment having increased in farmland and agricultural commodity markets (Friedmann, 2016). Detailed analysis of the exact roles and influences that financial actors are playing is difficult to assess because their involvement is obscured by 'distancing' in the global food system: the number and types of actors involved is becoming difficult to track as sophisticated agricultural commodity derivatives evolve (Clapp, 2014).

Rather than quantifying the extent of current financial capital, some scholars have sought to position recent financialization of the food system within a wider theorization of the changing dynamics of capitalism, including in the context of food regimes. The food regime is an analytical framework to study the role of food as a driver of both capitalist development and its associated political structures (Friedmann and McMichael, 1989; McMichael, 2013a, 2013b). It positions the 'first food regime' under UK hegemony as the regime in which basic food commodities from the settler-states entered world trade and provided cheap food for capitalist development in the core imperial/industrializing countries. Subsequently, during the second food regime, this time under US hegemony, agri-food complexes developed, linking multiple nation-states in global trade and aid arrangements, and helped to proliferate industrial agriculture and manufactured foods at a global scale. Out of this regime, a series of political transformations saw the next 'corporate food regime' emerging from the post-Fordist era and economic stagnation of the 1970s (Friedmann and McMichael, 1989; McMichael, 2005). Financialization of the food system is a logical extension of the elaborating global corporate food regime: following state deregulation of finance sectors under neoliberalization, finance capital is increasingly present in a variety of ways, in speculation as well as productive investment in food and agriculture, impacting the operation of agri-food supply chains from production right through to the retail sector (Burch and Lawrence, 2009; McMichael, 2012). The importance of financial motives, financial markets, financial actors and financial institutions had been growing in many economies (Epstein, 2005). The US economy, which leads global finance, has undergone significant financialization in the last several decades. That is, increasingly loosely regulated financial activities have become more and more dominant in the US economy, and the US policy regime has also shifted away from regulation and towards promotion of the financialization of the economy (Krippner, 2011).

In this context, it is highly instructive to empirically analyse how Japanese financial capital and financial actors have been contributing to the financialization of the global food system. The lack of transparency and the distance in the financialization of economic institutions and supply chains, however, stand as obstacles to fully understanding Japan's contribution to financialization. For example, there are significant difficulties in distinguishing whether investments are for pure business operation or to speculate for financial gains (Clapp, 2014). This challenge is exacerbated by the fact that Japan's financial system in general contrasts strongly with that in the US – being more based on a main-banking system rather than equity investments – and the academic research on financialization of Japan's economy has only just begun, not to mention research on financialization of the agri-food system.[1] To help meet this analytic challenge, this chapter aims to provide insight by focusing on Japan's *sogo-shosha* as the active actors in the financialization of agri-food systems, both in Japan and around the world, based, in particular, on their historical development trajectory in the capitalist development of Japan. Japan's *zaibatsu* and their *sogo-shosha* have long been dominant institutions in food systems in Japan and surrounding nations. They have developed a unique intertwining structure of finance and agri-food sectors since the birth of the nation-state of Japan. In fact, it might be possible to argue that Japan's modern food system has been financialized by these actors from the very beginning

– with financial conglomerates and their group banks working together with their traders, who invested and speculated in addition to trading. They have played significant roles in the capitalist development of Japan, and as a part of its project, they have developed the modern food system in Japan. Officially, *zaibatsu* were dismantled when Japan was defeated in World War II, but the structure of ex-*zaibatsu* and *sogo-shosha* can be found distinctively even today. For example, Mitsubishi Corporation, a *sogo-shosha* which is dominant in the current food system, is one of the leading components of the Mitsubishi conglomerate together with the Bank of Tokyo-Mitsubishi UFJ. Likewise, Mitsui & Co. (*sogo-shosha*) is prominent in the Mitsui conglomerate with Sumitomo Mitsui Banking Corporation (Toyo Keizai, 2016). Marubeni and Itochu are called non-*zaibatsu sogo-shosha* as they originated from textile traders about a century ago, but they have developed into 'general' traders and active financial investors as they work closely with other banking groups. These *sogo-shosha* have been involved in the proliferation of the industrial diet in Japan, and are now significant actors in the financialization of food and agriculture on a global scale.

This chapter first investigates the development of *zaibatsu* and *sogo-shosha* to reveal how they established their foundations in food and agriculture as a part of the nation-state project of Japan. Then, it focuses on the outbound investment and corporate strategy of Japan's *sogo-shosha* in the global agri-food system, especially in Asia. I focus on soybeans as one of today's 'flex crops' which – in food, feed, and fuel markets – is receiving increased financial investment (Alonso-Fradejas *et al.*, 2015; Oliveira and Schneider, 2015). Although the soybean is usually regarded as essential for food security, its commodification as a raw material for the industrial market was undertaken by *zaibatsu* and *sogo-shosha*. Before arriving at this analysis of the contemporary politics of soy complexes, it is essential to review the longer institutional relationships that were essential to the stabilization of food regime relationships throughout Japan's modern history.

Using the food regime framework to map the development trajectory of Japan's soybean and vegetable oil complex

The food regime framework originally focused upon relations between Europe and the so-called 'settler states' (Friedmann and McMichael, 1989), with little mention of Asian countries including Japan. Some literature discusses Japan as one of the recipients of US grains in the Second food regime (Friedmann, 1993). Indeed, Japan was one of the major destinations for US grains in the Second food regime. After World War II, a defeated Japan was incorporated into the US-centred intensive meat (soy-corn-animal) complex and the durable foods complex (processed foods), with heavy involvement of the US government. With the reliance upon the US firmly established, Japan today still remains a major importer of wheat, soybeans, and corn (maize). *Sogo-shosha* handle most of these imports as trading companies, as well as investing in large-scale feed and ingredient companies, livestock, food processing, and fast-food industries both domestically and globally.

General trading companies are said to be unique in Japan. They are called 'general', because they trade almost anything – from mineral water to communication satellites,[2] and from procuring raw materials (up-stream) to retailing products (down-stream) (Mitsubishi Corporation, 2013). In fact, 'trading' may no longer fully represent their characteristics because they are not only trading or supporting Japanese manufacturers to expand their international trading. Instead, *sogo-shosha* are equally, or possibly more, active in the financial investment and development of new projects. As such, it may be more accurate to call them 'investment companies' (JFTC, 2014).

For *sogo-shosha*, food is one of a wide range of sectors in which they invest – including energy, metal and non-metal resources, machinery, chemicals, development infrastructure, textiles, information technology, and finance. Their association reports that the *sogo-shoshas*' food investments contributed to about 20 per cent of their total profit in 2009 (JFTC, 2009). For the Japanese food industry, however, *sogo-shosha* play an extremely dominant and influential role. *Sogo-shosha* hold shares in many companies in a wide range of food industries, including: wholesale distributers; large-scale food companies, especially those based on global commodities like sugar, meat (feed), and vegetable oils; retail (supermarkets and convenient stores); and, eating-out (Table 8.1). Just as the Western-based ABCD companies operate in farm and manufacturing processes while also participating actively in the financialization of food and agriculture (Murphy *et al.*, 2012), Japanese *sogo-shosha* are involved in a suite of up-stream and down-stream processes. For example, Mitsubishi Corporation holds shares in KFC Japan as well as companies in chicken processing, the livestock sector, feed manufacturing, and the grain import sector (Mitsubishi Corporation, 2013).

The development of these dominant Japanese *sogo-shosha* was not only a response to the demand of consumers, but was also part of the political process of Japan's modernization over a century ago. *Sogo-shosha* worked for the government and played a significant role in Japan's industrial and capitalist development. They have shifted their roles and functions over time and have become more interested in global expansion.[3] An elaboration of the food regime framework for Japan reveals this nation-building role, the hybrid character of the institutions, and the subsequent 'global shift' in focus.

Table 8.1 Examples of *sogo-shosha* and their stakes in major food companies

Mitsubishi Corporation Wholesale distributer: Mitsubishi Shokuhin 62 per cent Sugar: Dai-Nippon Meiji Sugar 100 per cent Frozen fish: Toyo Reizo 81 per cent Meat: Yonekyu 62per cent, Ito Ham 24 per cent Cornstarch: Nihon Shokuhin Kako 59 per cent Fastfood: Japan KFC 65 per cent Convenience store: Lawson 32 per cent Supermarket: Life Corporation 19 per cent	*Mitsui & Co.* Wholesale distributer: Mitsui Foods 100 per cent Brazil grain: Multigrain AG (Brazil) 100 per cent Sugar: Mitsui Sugar 32 per cent
Itochu Wholesale distributer: Itochu Shokuhin 50 per cent Wholesale distributer: Nippon Access 93 per cent Fruits: Dole (Asia fruit section) 100 per cent Meat: Prima Meat Packers 39 per cent Veg oil: Fuji Oil 25 per cent Convenience store: Family Mart 35 per cent	*Marubeni* Meat: Wellfam Foods 100 per cent Feed: Marubeni Nisshin Feed 60 pe cent Sweets: Yamaboshiya 95 per cent US grain: Gavilon 100 per cent Sugar: Toyo Sugar Refining 39 per cent Veg oil: Nisshin OilliO Group 15 per cent Supermarket: Tobu Store 31 per cent

Source: Toyo Keizai, 2016; summarized and translated by the author.

Imperial Japan (1880s–1930s): capitalist development and the First Food Regime[4]

When Japan opened its economy in the middle of the nineteenth century and the modern state was established by the Meiji Restoration in 1868, the Japanese government sought to resist being 'colonized'. The state of Japan limited the activities of foreign traders only to designated ports (starting with Yokohama), discouraged foreign investment, and prohibited foreign traders from entering Japan's domestic market. The state also quickly modernized Japan's own financial institutions from the 1870s by establishing national banks, the central Bank of Japan, and special banks to handle foreign trade. Financial capital that had existed since the feudal era, including Mitsui, Mitsubishi, and Sumitomo, established modern banking institutions in this era. The government also sought to obtain foreign currency by building Japan's own trading companies. Many of today's active *sogo-shosha* were established with the support of these policies. Mitsui & Co. was founded in 1876, Mitsubishi Corporation in 1918, and Sumitomo Corporation in 1919. The origin of Marubeni and Itochu can be traced back to the 1850s, when their common founder began trading. Suzuki Shoten, which once used to be Japan's largest *zaibatsu*, was founded in 1874 (it is one of the origins of today's Sojitz).

Many of the *sogo-shosha* had a close relationship with the government, especially in their early days. Thus, they were called *sei-sho* (government traders) before they came to be called '*sogo-shosha*' in a later era. With strong government support for the active development of *zaibatsu* and *sogo-shosha*, Japan managed to conduct its industrial and capitalist development unlike other colonized Asian countries. Beginning in the 1880s, Japan's consulate office, special banks, and traders like Mitsui & Co. expanded into Asian countries. As one of its imperial, political, and economic activities in Asia, Japan began the international trade of Manchurian soybean products (bean, meal, and oil), not as food but mainly in order to supply soy meal to fertilize Japanese farmland to produce more export commodities. The international trade in soybeans started from Manchuria, today's Northeast part of China, around the beginning of the twentieth century. Although most of the English literature on soybeans focuses on the US and Brazil, Manchuria had become a world centre of soybean production and export by the 1930s. Local and European traders were active as well, but the emerging nation-state of Japan and its *sogo-shosha* contributed to the transformation of soybean from a traditional Asian food into a globally traded commodity mainly to be crushed into meal and oil for industries.

Mitsui & Co. is regarded as the pioneer of trading Manchurian soybean products. It helped establish the trade of Manchurian soy meal to Japan by the 1890s, and exported the first soy oil to Liverpool in the UK in 1908. Japan's imports of Manchurian soy meal grew rapidly especially after the Sino-Japanese War in 1894–5. A trade pattern was established to export soy meal to Japan, soy oil to Europe and the US, then later, soybeans to Japan and to Europe, as the trade data show (see Figure 8.1).

Zaibatsu were active in implementing the state policy of expansion into Asia, promoting industry and international trade. Okura *zaibatsu* – looking for a business opportunity to expand to Manchuria following this policy – collaborated with a fertilizer broker and together founded the Japan–China Bean Meal Manufacturing Company in 1907, which was the origin of Nisshin OilliO Group, still one of today's two dominant edible oil companies in Japan. The new company first built its main factory in Dalian and began crushing Manchurian soybeans for the fertilizer industry in Japan. As their name indicates, their main product was soy meal for fertilizer, with soy oil as a by-product. The origin of the other dominant oil company today, J-Oil Mills Group, was established as Honen Oil in 1922 (Honen Oil, 1963). The company had started with the investment of Suzuki Shoten, the largest *zaibatsu* in the early twentieth century, to succeed the factory and technology developed by the South Manchurian Railway. These companies started with their main factories in Manchuria.

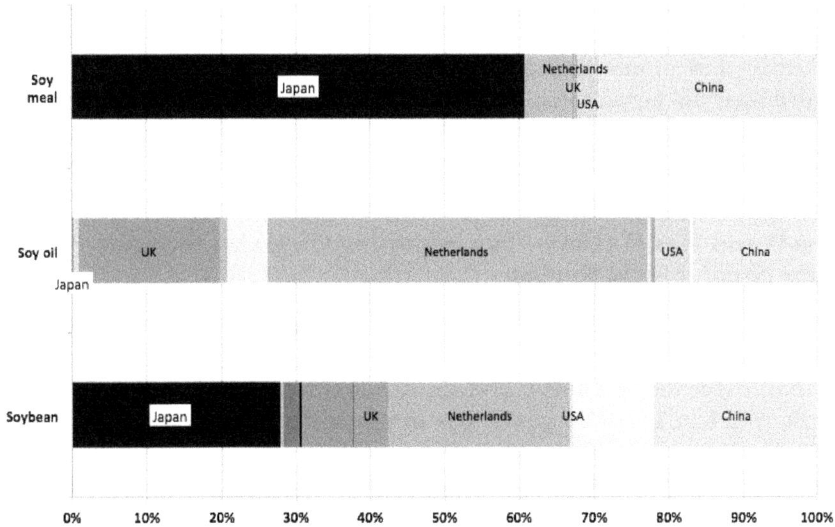

Figure 8.1 Soy meal, soy oil and soybean exports from three key ports in Manchuria, including destinations, 1930.

Source: compiled by author from North China trade returns, 1930.

Then, Japanese capital, especially Suzuki Shoten, also began to build soybean crushing factories located in port areas of Japan to crush imported Manchurian soybeans. Most of these 'sea factories' (large-scale oil crushing and refinery plants based at sea ports) were large-scale with leading-edge technology for oil pressing or solvent extraction, which required huge capital investment. Some documents record that the main purpose of these Japanese companies in Manchuria was, in fact, to speculate in the Manchuria soybean market, and they regarded their soybean crushing factories as the buffer institution to handle extra soybeans (Mitsubishi Corporation, 1986; Sugiyama, 1957).

In summary, Japan promoted outbound expansion and international trade in order to avoid colonization and to accumulate capital among Japanese hands to urgently progress industrial and capitalist development. In this nation-state-building project, the government collaborated with the existing financial capital of *zaibatsu*, and strongly supported 'government traders', the origins of today's *sogo-shosha*. Many of these government traders were leading components of *zaibatsu*, together with *zaibatsu* banks. For example, Mitsui *zaibatsu* contained Mitsui & Co. (trading company) and Mitsui Bank, as well as hundreds of companies in major industries. These hybrid institutions combining industrial-mercantile-financial capital worked closely together in the same *zaibatsu* group. They played significant roles in implementing the state policy of expanding across Asia and promoting international trade, and at that time, they found opportunities in Manchurian soybeans. In this way, while soybeans were transformed from a traditional Asian bean into a raw ingredient for fertilizer and oil industries and an internationally traded commodity, Manchurian soybeans contributed to Japan's nation-state project and capitalist development, which I argue to be a configuration of the First Food Regime in an Asian context (Hiraga, 2015).

Post-World War II (1945–70s): importing US grains to Japan in the Second Food Regime

As Japan entered various wars in the 1930s and 1940s – expanding its battlefields to China, Southeast Asia and the Pacific – the export of Manchurian soybeans to Europe and the US eventually ceased. Following its wartime defeat in 1945, Japan was occupied by the US and entered a period of severe food crisis. *Zaibatsu* and major *sogo-shosha* were dismantled, and international trade as well as foreign currency exchange came under the direct control of the government, thus the US. Japan lost imports of rice, sugar, soybean, oil crops, salt, and other important materials that the economy had become dependent upon, and which came from its Asian territories like Taiwan, Korea, Manchuria and China, and Southeast Asia. Japan's domestic agriculture was greatly damaged by war. Starvation in Japan was severe, and it opened the way for the US to supply grain, especially wheat and soybeans, as food aid (Friedmann, 1982).

In the first few years of turmoil after World War II, food aid was provided by the Licensed Agencies for Relief in Asia (LARA) and the US Government Appropriation for Relief in Occupied Areas (GARIOA). Some of the aid was used to provide school lunches for starving Japanese children, mainly with bread and powdered milk (Cwiertka, 2006). International trade resumed, mainly importing food and cotton from the US, under the strict control of the government. Several public corporations (*ko-dan*) were established to manage the food supply and trade, while the fragmented parts of ex-*sogo-shosha* and newly established traders gathered under these public corporations. In this restricted trading environment, non-*zaibatsu* traders who used to specialize in textile trading, like Itochu and Marubeni, expanded their business to handle food and other available commodities while working closely with the trade-controlling authorities.

Vegetable oil companies that had developed in the pre-war era – based on Manchurian soybeans – like Nisshin Oil and Honen Oil, lost their major production facilities and assets overseas, and their facilities in Japan were partially damaged by war. Still, these oil companies quickly re-structured themselves in a few years, and the government allocation of imported soybeans gave advantages to these existing large-scale companies. Under government control, for example, nearly half of the imported soybean quota in 1953 was allocated to three companies

– Honen Oil (20.4 per cent), Ajinomoto (14.1 per cent), and Nisshin Oil (10.5 per cent) – while thousands of oil crushing factories existed in Japan (MAFF, 1956). Such controlled allocation of imported soybeans favoured the larger oil companies and secured markets for their products (Nakajima, 1967).

After the US lost its European wheat market as Europe recovered from World War II, the US turned to food aid through PL-480, beginning in 1954, to dispose of surplus agricultural commodities. US soybeans, as well as wheat, were major commodities exported under this policy (Friedmann and McMichael, 1989). The trade of oil crops, including soybean, started to accelerate in the middle of 1950s as Japan entered the era of high economic growth. This trade was liberalized gradually, starting with soybeans in 1961. Vegetable oil supplies in the Japanese domestic market increased rapidly, mainly based on imported soybeans from the US, crushed at the 'sea factories'. Meanwhile, the number of small-scale oilseed crushing factories, so-called 'mountain factories', which crushed locally produced rapeseed for domestic consumption, declined rapidly in the 1960s.[5]

About the same time, the so-called 'westernization' of the Japanese diet was promoted by both the US and Japanese governments, as well as Japanese companies. The US Grains Council was established in 1960 with a task to develop the US grain market in Japan. In the following year, the Japanese Feed and Grains Council was established and became a counterpart of the US Grains Council to promote meat and dairy products in Japan. They promoted modern industrial livestock production with grain feed mainly imported from the US (Suzuki, 2003). The 'Nutrition Improvement Movement' promoted the Western diet – including cooking with oils (the so-called 'frying pan' movement of the time encouraged Japanese housewives to use more oil – mainly soybean oil – for cooking by using frying pans). There was a rapid increase and prevalence of vegetable oils (over animal fats) in Japan after World War II (Figure 8.2).

Sogo-shosha regained their strength as the Japanese economy and society recovered, and international trade grew. In addition to that, *sogo-shosha* increased their significance in Japan's food sector by controlling feed-to-livestock and from ingredients to food processing industries, as well as through the 'food kombinat' project in the 1960s and via incoming foreign investment into fast foods in the 1970s.

The Korean War of 1950–3 and the Cold War led to changes in occupation policy in Japan, allowing the once-dismantled *sogo-shosha* to reassemble. Mitsubishi Corporation incorporated again in 1954, and Mitsui & Co. in 1959. Trade was gradually liberalized, and *sogo-shosha* resumed as the major players of trade and investment as well as the organizers of business and industry development. *Sogo-shosha* and other Japanese companies re-formed more loose financial and business groups around the major banks instead of officially connected *zaibatsu* or holding companies. *Sogo-shosha* also became influential actors as Japan's livestock industries

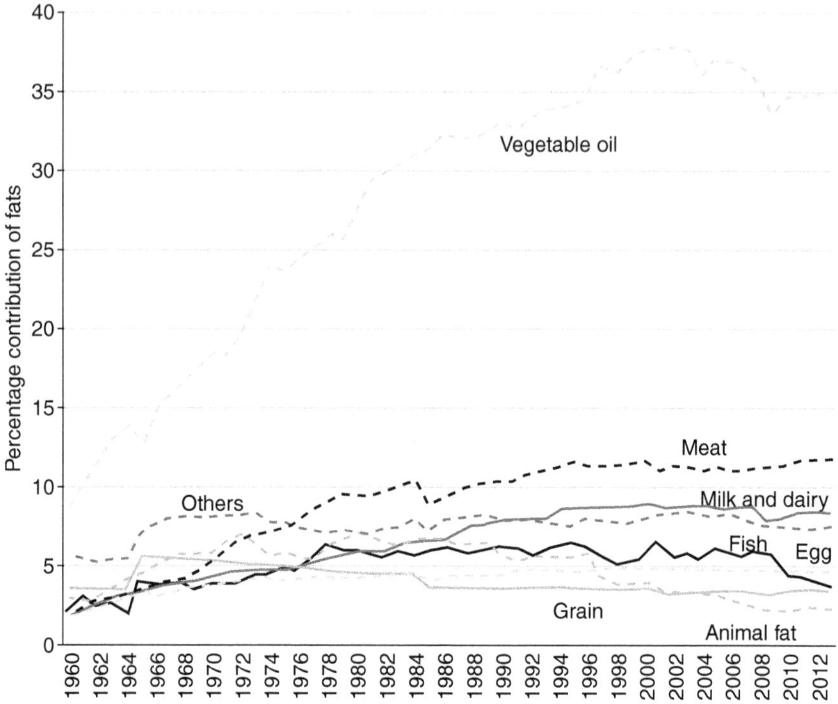

Figure 8.2 Foods that supplied fats to the Japanese 1960–2013.

Source: MAFF food balance data.

developed rapidly in the 1960s based on imported grains, mainly soybean and corn (maize) from the US. Not only operating as the traders to import these grains, *sogo-shosha* also invested in the feed industries, as well as broiler and other livestock operations, and contributed to structuring livestock integration in Japan and increased the supply of animal products (milk, eggs, and meat) in the Japanese market (Integration Research Group, 1971).

On the procurement side of the integration, *sogo-shosha* collaborated with the state's national planning to develop the 'food kombinat'. Since the 1960s the Japanese government promoted the 'food kombinat'[6] project in order to modernize Japan's food industries. New technologies, new forms of capital investment, and changes in the workplace were introduced in an attempt to make Japanese firms competitive against foreign large-scale food companies. On the distribution side of this grand project of integration, *sogo-shosha* supported foreign fast-food chains entering the Japanese market. Foreign direct investment in Japan's restaurant sector was gradually deregulated and – by the beginning of the 1970s – there was a rush of American-based fast-food into Japan (Kentucky Fried Chicken arrived in 1970, McDonalds in 1971, Mister Donut and Dunkin' Donuts in 1971, and Dairy Queen in 1972) (Kozuka, 1999). Often the *sogo-shosha* worked as the partner of American fast-food in Japan. In this way, *sogo-shosha* structured the feed-to-livestock integration both as direct participants and as providers of financial capital: from the import of grains, processing the grain at the food kombinat on the port, supplying feed and operating livestock, as well as processing meat and proliferating fast-food restaurants.

Other food companies developed as well in post-World War II Japan. As the supply increased for basic ingredients like wheat, vegetable oil, and sugar based on imported grains, the Japanese food industry also began to sell processed foods using these ingredients. Oil-containing products like salad dressings came into the market at the end of the 1950s, meat sauce for pasta in 1959, and ready-made curry roux in 1960. Other products proliferating at this time were: margarine, ice cream, mayonnaise and other processed foods. Japanese housewives learned how to prepare new cuisine like salads, spaghetti, and curries with the products supplied by these Japanese food companies, rather than the traditional way of preparing dishes from scratch (Iwamura, 2010). Instant noodles, a deep-fried processed food, were invented by the Japanese entrepreneur Momofuku Ando. He founded the Nissin Food Company, which began selling instant noodles in 1958 and the cup-type instant noodle in 1971. One of his aims of inventing instant noodle was to industrialize noodle manufacturing into a large-scale operation in order to absorb large supplies of wheat flour, which the traditional small-scale noodle manufacturing could not cope with. Mitsubishi Corporation soon became the distributor of instant noodles as well as the supplier of wheat flour to Nissin Food (Ando, 2002).

In summary, although *zaibatsu* and major *sogo-shosha* were once officially dismantled, they soon re-assembled into more loose financial and corporate groups that returned to their prior highly strategic role in national economic development, and contributed to Japan's rapid economic growth in the post-World War II era. By the 1980s, Japan had established the following structure: *sogo-shosha* imported basic grains to food kombinat located at ports and large-scale food companies (often related with *sogo-shosha*) processed the grains into basic ingredients for food companies and feed for the livestock sector. *Sogo-shosha* played a significant role in receiving food aid from the US, then incorporating Japan into the US-centred intensive meat (soy-corn-animal) complex and the durable foods complex (processed food) in the Second Food Regime. Although they were called trading companies, they were active not only in trade, but also investing in a wide range of industries along the food value chains.

Sogo-shosha *in the Corporate Food Regime (1980s–current)*

Japan had experienced significant food dependency through its food trade with the US in the Second Food Regime. Beginning in the 1980s, along with the changing financial condition and the roles of *sogo-shosha*, Japan established a more multilateral approach to the food trade as it developed the 'Japan-centered East Asian import complex' in the corporate food regime (McMichael, 2000). Japan significantly shifted its policies for trade, agriculture, and financial institutions, especially from the 1980s. For example, there has been significant restructuring in political and economic institutions since the Plaza Accord of 1985,[7] when there was an international intervention in the currency market to devalue the US dollar against the yen in an attempt to discourage the exporting of Japanese goods. Japanese companies began to seek new production bases overseas and Japan's outbound foreign direct investment massively increased from US$3,612 million in 1983 to US$48,024 million in 1990 (JETRO, 2015), with the food and agriculture sectors well represented as targets for investment. In 1986, the Committee on Economic Structural Adjustment for International Cooperation in Japan submitted a policy recommendation, the so-called Maekawa report, to the Japanese government, recommending that the government promote outbound foreign direct investment, and to increase imports of agricultural products (including processed food) that could be imported more cheaply than produced domestically (with the exception of staple agricultural products) (Maekawa Report, 1986). Around this time, China began opening its economy to encourage incoming foreign investment, which coincided with Asian countries like South Korea, Hong Kong, and Taiwan exploring options for overseas lower-cost manufacturing locations. These policy changes, and the following 'bubble economy' inside Japan, encouraged Japanese capital and companies to invest overseas, including in real estate, financial sectors and food and agriculture sectors.

On the one hand, financial operations became more profit-making than productive activities, and '*Zai*-tech' (financial manipulation technique) became a buzzword of the bubble economy in the latter half of the 1980s. Many non-financial companies and individuals became enthusiastic about financial investment rather than their original business operations (Steven, 1990; Kikkawa, 2006). With the bursting of the bubble, the stagnation of the 'lost decade', and the Japanese 'big bang' from the 1990s to 2000s, a significant restructuring of Japan's financial institutions and corporate strategies occurred.

On the other hand, many more traditionally configured manufacturing companies (including food processors) moved overseas, especially to China and other Asian countries. For example, the instant noodle industry was among those which began heavily investing in China, with Japanese and Taiwanese instant noodle companies having an active presence since the 1980s. As a consequence, China's instant noodle industry developed from being inconsequential to becoming the world's largest, producing some 8.28 million tonnes in 2011. China Data Online show that over half (54.4 per cent) of the gross industrial output value of China's convenience food (mostly instant noodle) market was produced by foreign funded enterprises.

Japanese *sogo-shosha*'s changing roles and global strategies

In the past, *sogo-shosha* regularly claimed that they 'fed Japan' by procuring food from overseas. Their annual reports in recent years, however, proclaim that they are now feeding not only Japan, but Asian and other emerging nations in the world. For example, Marubeni's Food Materials Division aims to provide reliable food supplies and improved diets in developing nations, saying: 'Specifically, we are focusing on the food products business, including livestock feed, oil pressing, and milling' (Marubeni, 2013, p. 20). Marubeni has been increasing its presence in the global grain trade. It situates itself as the top Japanese grain trading company, with a trading volume of some 67 million tonnes in the year ended March 2016. Marubeni bought Gavilon (except its energy business) for approximately US$2.7 billion in 2013 to strengthen its procurement capability of corn (maize) and soybeans grown in North America. It owns Helena Chemical Company, the second largest retailer of agricultural chemicals, fertilizers, seeds, and other agricultural material in the US. Marubeni also holds 210,000 hectares of afforestation sites in four projects in three countries. Its water activities impact upon nine million people in Central and South America, China, the Middle East, Australia, and Europe (Marubeni, 2016).

Mitsui & Co. plans to expand the manufacturing capacity of feed additives with Novus in order to respond to the growing demand for food throughout the world (Mitsui & Co., 2016). Similarly, the Mitsubishi Corporation acquired a global salmon aquaculture company Cermaq ASA to ensure stable procurement of animal protein '(as) demand for food continues to grow worldwide' (Mitsubishi Corporation, 2016, p. 89). Itochu acquired the Asian fresh produce and worldwide packaged foods business of Dole Food Company, Inc., of the US in 2013. The company aims to further develop a global market with its Dole brand and to expand joint business activities with CITIC and the CP Group. Itochu now promotes the motto of 'Generating Earnings in China and Asia' (Itochu, 2016, p. 3) and aims to become 'Asia's largest integrator of agricultural products' (Itochu, 2017, p. 84).

The global strategy maps in annual reports of Marubeni, Itochu, Mitsui & Co., and Mitsubishi Corporation all shows a similar picture: an integrated project involving investment in North and South America to increase the supply of grains like wheat, corn, soybeans, and feedstock, then parallel investment in food industries including food processing, restaurants and retailing (especially convenience stores) in China and Asian countries. Thus *sogo-shosha* have become a major vector for investment and integration of key supply chains at a global scale.

While China attracts attention for its rapid increase in soybean imports since 1996 – becoming by far the largest soybean importing and crushing country in the world – Japan, in parallel, reduced its raw soybean imports. Japan, instead, has been increasing imports of soy meal, especially from China and India. Japan imported about half of China's total soy meal exports between 2012 and 2015, thus becoming the largest soy meal exporting destination for China (JETRO, 2015). While Japan decreased its imports of soybeans, Japan increased imports of meat and meat-containing processed food. Although more empirical analysis is required to show exactly how much of China's soybean imports eventually are consumed by the Japanese, these trends suggest the structural shift is happening around the soybean-animal complex in Asia, and Japan's *sogo-shosha* have been central contributors to this process.

Japanese **sogo-shosha** *and soy: China and Brazil*

Marubeni claims that it handles about 20 per cent of soybean imports to China, about ten million tonnes out of China's total soybean imports of 55 million tonnes (Marubeni, 2012). Mitsui & Co. also says that it handles about four million tonnes of soybeans in this intermediary trade, mostly from Brazil to China, with one-tenth, about 400,000–500,000 tonnes, going to Japan (Monthly Oils, 2015). Itochu also works closely with China's COFCO and Ting Hsin, the largest instant noodle producer in China, and many other companies (Itochu, 2015, 2016). Other Japanese *sogo-shosha* also invest in food industries in China and Asian countries, as well as helping other Japanese food companies with outbound investment. Their reasons is that the Japanese market has matured thus forcing them to target growth in wider Asian markets.

As they actively expand into the food system in Asian countries – especially China – Japanese *sogo-shosha* have also been actively investing in Brazil's soybean supply. Mitsui & Co. acquired Multigrain AG as a 100 per cent subsidiary and has been strengthening logistics capability by making an alliance with Vale S.A., Brazil's largest logistics operator. Mitsubishi Corporation holds large shares in Los Grobo Ceagro do Brasil S.A., and Agrex do Brasil, and has been building the value chain from agricultural materials, seeds, and grain transport in Goias State. Sumitomo Corporation, Marubeni, Itochu, Sojitz, and Toyota Tsusho also have been investing in Brazil in recent years (Hall, 2015; AGRI in Asia, 2016; Sano, 2016).

In fact, *sogo-shosha* are not alone in investing in Brazil, and this is not the first time that Japan's public and private players have actively invested in Brazil to increase the soybean supply. Japan invested about 68 billion yen (28 billion yen of which is Official Development Assistance) between 1979 and 2001 to develop Brazil's Cerrados into a world-scale production site for soybeans. This Japanese–Brazilian Cooperation Program for Cerrado Development (PRODECER) is regarded as a trigger that made Brazil a world soybean producer and it is praised as a 'success' story, especially for Japan's International Cooperation Agency (JICA), which served as an administrative agency for Japanese cooperation (Aoki, 2002; Hosono *et al.*, 2016). This 'success' case in the Cerrado is now used to promote the Triangular Cooperation for Agricultural Development of the Tropical Savannah in Mozambique (ProSAVANA – JBM) project, which was signed jointly by the governments of Japan, Brazil, and Mozambique in 2009. The aim of the project is, however, somewhat obscure. The project has been criticised as being more for political, diplomatic, commercial, and publicity purposes, rather than for food security or for the needs of local inhabitants. Importantly, ProSAVANA has been criticised as a case of land-grabbing. The project has been condemned for damaging subsistence family farmers as foreign investment has embarked upon the development of industrial agriculture (Funada-Classen, 2013; Okada, 2015).

Conclusion

This chapter demonstrated how Japan has developed an intertwining structure of finance and agri-food sectors from the beginning of modernization which has tended to obscure the role of Japanese finance capital in both national economic development and, further, in the elaboration of the corporate food regime in the global agri-food economy. This hybrid structure of industrial-mercantile-financial capital resulted from the urge to rapidly develop a strong nation-state as a latecomer to the capitalist world. Accordingly, the Japanese government closely collaborated with existing Japanese finance conglomerates – *zaibatsu* – and strongly supported the government traders, those companies that would later become *sogo-shosha*. This indivisible institutional combination of industrial-mercantile-finance capital, strongly intertwined and supported by an active role of the state, contributed to the industrial and capitalist development of Japan, as well as implementing the state's imperial and militarist policies during the war and economic development in the post-World War II period. As a part of this project, these hybrid economic institutions contributed significantly to developing Japan's modern food system. It is argued that *zaibatsu* and *sogo-shosha* have been active finance actors in Japan's food system since the beginning of capitalist development – even if much of this role is obscured in the latter case due to their description as 'trading companies'.

All along this historical trajectory, *sogo-shosha* have been more than 'trading companies': they have also been highly active and strategic financial actors. First, they worked as government traders to accumulate foreign currency as a leading component of *zaibatsu*. In the post-World War II era, in addition to trading grains in the international market to procure food for Japan, they also actively invested in a wide range of food- and feed-related companies to structure the integration of various food value chains. Under global conditions set by the contemporary corporate food regime, they became more active in investment and finance, organizing new markets in the global food system. In addition to working as a part of *zaibatsu* or working closely with banks and financial institutions in the corporate groups, *sogo-shosha* themselves handled finance – in foreign currency exchange, trade-related financing and credit, investment, or organizing much larger syndicate loans, as well as trading in commodity futures markets.

Although this hybrid character makes it difficult to distinguish how much of their financial activities and investment are for their business operations and how much is directly for speculative and financial gains, based on the historical development trajectory discussed in this chapter, I argue they are significant finance actors in the current financialization of the global food system, not only in Japan but also, more recently, at a global level. These institutions have been providers of capital investment (along with many other activities) that have stabilized food regime relations and assisted transitions to new regimes both in Japan and around the Asian region and elsewhere in the world. While the global trend towards financialization is not theoretically contested, I hope this set of contextual insights – drawn from the specific economic history and geography of Japan – can contribute to future research on the specificities of financialization in particular national settings. The empirical analysis of how much Japanese financial capital and financial actors, including *sogo-shosha*, have been contributing to financialization of the global food system is only just beginning. It should continue, so that we can provide insight into better understanding how other Asian countries have also been incorporated into the global food system as latecomers to the capitalist world.

Notes

1 For example, a keyword search of 'food and financialization' (in Japanese) on Japan's dominant academic database, CiNii, resulted in no hits, while more than 10,000 articles were found in google scholar for the same search phrase in English as of April 2017.

2 Its original version is said to be 'from Ramen noodles to missiles'.

3 The Japan Foreign Trade Council, an industry association of *shosha*, says the *shosha*:

> are exploring ways to take advantage of high powered, diversified and value-added functions, built up over long years of experience in (1) information collection, analysis and market development capabilities (2) project management and risk management know-how and (3) IT (information technology), LT (logistics technology), FT (financial technology) and MT (marketing technology), etc.
>
> See www.jftc.or.jp/shosha/english/function/index.html

4 An earlier version of this section was presented at the Agrifood XXII (Agri-Food Research Network), Queenstown, NZ in December 2015, and is more fully discussed in Hiraga and Hisano (forthcoming) 'The First Food Regime in Asian Context? Japan's Capitalist Development and the Making of Soybean as a Global Commodity in the 1890s-1930s'.

5 The Japanese oil industry distinguishes between the concept of the 'mountain factory' and the 'sea factory', which reflects the different characteristics of historical development between rapeseed crushing and soybean crushing in Japan. Until around the 1960s, small-scale, locally based, 'mountain factories' were built inland, often in rural mountainous areas near rapeseed production sites. They mainly crushed domestic rapeseed for domestic consumption. In contrast, soybean crushing began with the use of imported soybeans. Large-scale 'sea factories' were built near ports, mostly with facilities for solvent-extraction and advanced chemical refinement processes, crushing imported oilseeds and supplying cheap soybean oil and soy meal for industries (Nonaka, 2013).

6 *Kombinat* is a Russian term for business groups, which is used in Japan to indicate 'combination' or 'complex'. These companies planned and built large-scale factories in industrial districts usually near the port, especially petrochemical complexes or iron-steel complexes, in the Pacific Belt of Japan.

7 The Plaza Accord was an agreement of the G5 (France, West Germany, Japan, the US and the UK), signed in 1985 at the Plaza Hotel, US, to intervene in currency markets to depreciate the US dollar against the Japanese yen and German Deutsche Mark.

References

AGRI in Asia. (2016) 'Brazil's agriculture and Japan's *sogo-shosha*' (Japanese), available at: http://agrinasia.com/archives/1920, accessed 11 July 2016

Alonso-Fradejas, A., Liu, J., Salerno, T. and Xu, Y. (2015) 'Inquiring into the political economy of oil palm as a global flex crop', *Journal of Peasant Studies*, vol 43, no 1, pp. 141–165

Ando, M. (2002) *Totally-unexpected Business Idea: Secret of the Miracle of Nissin Shokuhin (Japanese)*, Kodansha, Tokyo

Aoki, H. (2002) *The History of Offense and Defense in Brazil's Soybean: The Result of Two Decades' International Cooperation (Japanese)*, International Cooperation Publisher, Japan

Burch, D. and Lawrence, G. (2009) 'Towards a third food regime: Behind the transformation', *Agriculture and Human Values*, vol 26, no 4, pp. 267–279

Clapp, J. (2014) 'Financialization, distance and global food politics', *Journal of Peasant Studies*, vol 41, no 5, pp. 797–814

Cwiertka, K. (2006) *Modern Japanese Cuisine: Food, Power and National Identity*, London, Reaktion Books

Epstein, G. (2005) 'Introduction: Financialization and the world economy', in G. Epstein (ed.) *Financialization and the World Economy*, Cheltenham, Edward Elgar, pp. 3–16

Friedmann, H. (1982) 'The political economy of food: The rise and fall of the postwar international food order', *American Journal of Sociology*, vol 88, S248–286

Friedmann, H. (1993) 'The political economy of food: a global crisis', *New Left Review*, no 197, pp. 29–57

Friedmann, H. (2016) 'Commentary: Food regime analysis and agrarian questions: widening the conversation', *Journal of Peasant Studies*, vol 43, no 3, pp. 671–692

Friedmann, H. and McMichael, P. (1989) 'Agriculture and the state system: The rise and decline of national agricultures, 1870 to the present', *Sociologia Ruralis*, vol 29, no 2, pp. 93–117

Funada-Classen, S. (2013) *Analysis of the Discourse and Background of the ProSAVANA Programme in Mozambique – Focusing on Japan's Role*, Tokyo University of Foreign Studies, Tokyo

GPIF. (2015) *GPIF Annual Report Fiscal Year 2015 (Japanese)*, available at: www.gpif.go.jp/en/fund/pdf/annual_report_fiscal_year_2015.pdf, accessed 5 May 2017

Hall, D. (2015) 'The role of Japan's general trading companies (*Sogo-shosha*) in the global land grab', *Conference Paper No 3, Land Grabbing, Conflict and Agrarian-environmental Transformations: Perspectives from East and Southeast Asia*, 5–6 June 2015, Chiang Mai University, Thailand, pp. 1–26

Hiraga, M. (2015) 'Japanese contribution to transforming soybean into a global commodity in occupied Manchurian: A case of the First food regime in an Asian context?' *Presentation at the XXII conference of the Australasian Agri-food Research Network*, Queenstown, NZ

Honen Oil. (1963) *Honen Oil 40 years of Corporate History (Japanese)*, Honen Oil, Tokyo

Hosono, A., da Rocha, C. and Hongo, Y. (eds) (2016) *Development for Sustainable Agriculture: The Brazilian Cerrado*, Palgrave Macmillan, Hampshire UK

Integration Research Group. (1971) *Shosha Capital Expanding to Agriculture (Japanese)*, National Chamber of Agriculture, Tokyo

Itochu. (2015) *Annual Report 2015*, available at: www.itochu.co.jp/en/files/ar2015e.pdf, accessed 27 May 2017

Itochu. (2016) *Annual Report 2016*, available at: www.itochu.co.jp/en/files/ar2016E.pdf, accessed 27 May 2017

Itochu. (2017) *Annual Report 2017*, available at: www.itochu.co.jp/en/files/ar2016E.pdf, accessed 13 September 2017

Iwamura, N. (2010) *'What Must their Parents be like?' The History which Changed Japanese Families in Showa (Japanese)*, Chuokoronshinsha, Tokyo

Japan Foreign Trade Council (JFTC). (2009) *Japan's Food Strategy and Shosha (Japanese)*, Japan Foreign Trade Council, Tokyo

Japan Foreign Trade Council (JFTC). (2014) *Japan's Growth Strategy and Shosha (Japanese)*, Japan Foreign Trade Council, Tokyo

JETRO. (2015) *Japan External Trade Organization: Agrotrade Handbook 2015 (Japanese)*, JETRO, Tokyo

Kikkawa, T. (2006) 'Beyond the "lost decade": Problems confronting Japanese companies, and Solutions', *Japanese Research in Business History*, vol 23, pp. 107–126

Kozuka. (1999) *Change in Food and the Development of Food Manufacturers (Japanese)*, Agriculture and Forestry Statistics Association, Tokyo

Krippner, G. (2011) *Capitalizing on Crisis*, Harvard University Press, Boston

McMichael, P. (2000) 'A global interpretation of the rise of the East Asian food import complex', *World Development*, vol 28, no 3, pp. 409–424

McMichael, P. (2005) 'Global development and the corporate food regime', in F. Buttel and P. McMichael (eds) *New Directions in the Sociology of Global Development*, Elsevier, Amsterdam, pp. 265–299

McMichael, P. (2012) 'The land grab and corporate food regime restructuring', *Journal of Peasant Studies*, vol 29, no 3–4, pp. 681–701

McMichael, P. (2013a) 'Value-chain agriculture and debt relations: Contradictory outcomes', *Third World Quarterly*, vol 34, no 4, pp. 671–690

McMichael, P. (2013b) *Food Regimes and Agrarian Questions*, Fernwood Publishing, Halifax

Maekawa Report. (1986) *Committee on Economic Structural Adjustment for International Cooperation (Japanese)*, available at: www.komazawa-u.ac.jp/~kobamasa/lecture/japaneco/maekawarep.htm, accessed 22 May 2017

MAFF. (1956) *Oil Statistics for 1956*, Food Department, Ministry of Agriculture, Forestry and Fisheries of Japan, Tokyo

Marubeni. (2012) *Annual Report 2012*, available at: www.marubeni.com/ir/reports/annual_report/data/are12_full.pdf, accessed 22 May 2017

Marubeni. (2013) *Annual Report 2012*, available at: www.marubeni.co.jp/ir/reports/annual_report/data/ar13jp_all.pdf, accessed 13 September 2017

Marubeni. (2016) *Annual Report 2016*, available at: www.marubeni.com/ir/reports/annual_report/data/ar16en_all.pdf, accessed 22 May 2017

Mitsubishi Corporation. (1986) *Mitsubishi Corporation History (Japanese)*, Mitsubishi Corporation, Tokyo

Mitsubishi Corporation. (2013) *New: Modern Sogo-shosha Theory (Japanese)*, Waseda University Press, Tokyo

Mitsubishi Corporation. (2016) *Integrated Report 2016*, available at: www.mitsubishicorp.com/jp/en/ir/library/ar/pdf/areport/2016/all.pdf, accessed 27 May 2017

Mitsui & Co. (2016) *Annual Report 2016*, available at: www.mitsui.com/jp/en/ir/library/report/__icsFiles/afieldfile/2016/10/25/en_ar2016_all.pdf, accessed 27 May 2017

Monthly Oils. (2015) *October Issue of Monthly Oils (Japanese)*, Saiwai-Shobo, Tokyo

Murphy, S., Burch, D. and Clapp, J. (2012) *Cereal Secrets: The World's Largest Grain Traders and Global* Agriculture, Oxfam, London

Nakajima, T. (1967) 'Ch.6 Edible Oil Industry', in Nakajima, T. (ed.) *Modern Japanese Industry Development No. 18 Food (Japanese)*, The Association of Modern Japanese Industry Development, Japan

Nonaka, A. (2013) *Japan's Rapeseed: Current Situation to Perspective (Japanese)*, Showado, Tokyo

Okada, K. (2015) 'The role of Japan in overseas agricultural investment: Case of ProSAVANA project in Mozambique', Conference Paper No 82, *Land Grabbing, Conflict and Agrarian-environmental Transformation: Perspectives from East and South East Asia*, 5–6 June 2015, Chiang Mai University, Thailand, pp. 1–68

Oliveira, G. and Schneider, M. (2015) 'The politics of flexing soybeans: China, Brazil and global agroindustrial restructuring', *Journal of Peasant Studies*, vol 43, no 1, pp. 167–194

RAN. (2017) 'Every investor has a responsibility', available at: http://forestsandfinance.org/wp-content/uploads/2017/04/RAN_Every-Investor-Has-a-Responsibility_2017.pdf, accessed 5 May 2017

Sano, S. (2016) *Strategies of Japanese Trading Companies under Neoliberalism: The Case of the Grain Sector in Brazil*, Faculty of Economics Working Paper Series No 21, Toyo University, Tokyo

Steven, R. (1990) *Japan's New Imperialism*, Macmillan, London

Sugiyama, K. (1957) *My Personal History (Japanese)*, Nikkei, Tokyo

Suzuki, T. (2003) *'US Wheat Strategy' and the Diet of the Japanese People (Japanese)*, Fujiwara Shoten, Tokyo

Toyo Keizai. (2016) *Kaisha-Shikiho Industrial Map 2017 (Japanese)*, Toyo Keizai Shinpo-sha, Tokyo

9 Poultry grabs and agri-food financialization

The case of JBS of Brazil

Jonathan Almanza, Douglas H. Constance and Francisco Martínez

Introduction

This chapter combines an agri-food regimes conceptual framework (Friedmann and McMichael, 1989) with a commodity systems methodology (Friedland, 1984) to investigate the role of financialization in the emergence of JBS of Brazil as the largest meat company in the world. The combination of commodity systems analysis and agri-food regimes theory provides a terminology to interpret the internal dynamics of particular commodity sectors within the wider institutional forces that pattern the globalization of agri-food systems (Campbell and Dixon, 2009). We invoke the term 'poultry grabs' as an extension of the concept of land grabs (see Borras *et al.*, 2011; Edleman *et al.*, 2013) to illustrate the tendency for both agri-food TNCs and financial capital to invest in poultry as a way to profit from increased meat consumption by the growing middle class in developed and developing countries. The industrialization and rationalization of the poultry commodity system has reduced its market price in comparison to beef and pork, resulting in increased market share. As the middle class expands in developing countries, the 'nutrition transition' (Dixon, 2009) increases protein consumption demand, more often in the form of poultry. Global poultry demand is rising, and companies are investing proactively to meet that demand. This chapter focuses on the case of JBS of Brazil to investigate this phenomenon and thereby inform discussions on the role of financialization in the globalization of the agri-food system. Although JBS began as a beef company and has dominant positions across the protein sector, this chapter focuses on the poultry sector.

Poultry, in particular broiler chickens, was the first agri-food commodity to be industrialized (Breimyer, 1965). The US model of the broiler industry is advanced as the future model of agriculture: vertical and horizontal integration based on flexible accumulation in production and processing (Boyd and Watts, 1997; Heffernan, 2000). The combination of a sharecropping, contract, production system and non-union, minority-dominated, processing plant workers are central aspects of the 'Southern

Model' of agri-food production (Constance, 2008). Consolidation has resulted in a market structure whereby a few firms dominate production regionally and globally (Howard, 2016). The agro-industrial districts that emerged in the US South during the 1950s are the prototypes of flexible organizational forms that are a preferred model of agri-food accumulation as part of economic globalization (Boyd and Watts, 1997). This model is being adopted for other commodities and in other countries as agri-food TNCs diffuse the system globally (Heffernan, 2000; Burch, 2005; Bonanno and Constance, 2008; Martínez *et al.*, 2013).

The poultry sector will lead meat sales worldwide over the next decade as it continues to gain market share of total meat production and exports compared to beef and pork. Rising consumer demand due to the price differential – and perceived health benefits – drive the gains. Affordability and rising domestic consumption are important factors in poultry growth in developing countries worldwide. Poultry production is projected to expand from 115 million metric tons in 2016 to 131 million metric tons by 2025. Similarly, world poultry meat imports and exports are projected to increase from 12.7 million metric tons in 2016 to more than 14.1 million metric tons by 2025 (Watt Poultry, 2016; FAO, 2017).

To meet the growing demand, substantial production increases are anticipated in Brazil, the US, India, Thailand, the EU, and most other producing countries except China. China is suffering from disease outbreak problems and will greatly increase imports in the short run, accounting for almost 40 per cent of the projected expansion in the world poultry trade. Brazil will be the major beneficiary of rising international demand, with significant exports to China, as well as to Iraq, South Africa, Mexico and the Republic of Korea, followed by exports from the US and Thailand (Watt Poultry, 2016; FAO, 2017).

The chapter begins with an overview of the regimes approach to agri-food studies and ends with a contextual discussion of land grabs. The case is organized and presented from a commodity systems approach (Friedland, 1984, 2001). Commodity systems analysis employs a political-economic perspective centring upon the state-market-labour nexus to analyse the structure of agri-food production. It focuses on the role of the state in rural development and the associated structure of the particular agri-food sector. The case begins with a short review of the development of the poultry industry in the US and its diffusion globally. This section illustrates the flexible accumulation characteristics of the modern poultry industry, and its attractiveness for 'poultry grabs'. Then, a more detailed presentation of JBS's emergence and rise to dominance is provided to illustrate the role of financialization in the globalization of the agri-food system.

Agri-food regimes

Regime approaches organize history into stable global trade arrangements and their symbiotic governance mechanisms. Contradictory relations within the regime generate crises and decay that produce a period of transition to the next regime. Agri-food regimes analysis (Friedmann and McMichael, 1989) offers a structured method of analysis of the global political economy by examining the historical geo-politics that underpin the stable periods of capital accumulation and the transitional periods between regimes (Campbell and Dixon, 2009).

The First Agri-food Regime – the Settler-Colonial Regime (1870s–1930s) – was dominated by Great Britain as the core countries extracted natural resources from the periphery countries to support industrial development. This regime combined tropical imports from the colonies with grains and livestock from the settler states to provide low-cost foods for Europe's industrialization. The global depression of the 1930s combined with decolonization and World War II eroded this regime (Friedmann and McMichael, 1989; McMichael, 2005, 2009). The Second Agri-food Regime – the Agricultural Surplus Regime (1950s–1970s) – emerged after World War II as the US emphasized commodity programs over rural development and exempted agriculture from the global trade talks (via the General Agreement on Tariffs and Trade [GATT]). The US used its agricultural surplus as a food weapon in strategic postcolonial Cold War states. US food aid subsidized low wages in industrializing countries and created global supply chains that diffused the US diet and reduced food self-sufficiency in developing countries. Factors including the demise of the Bretton Woods Agreement, the delinking of the US dollar from the gold standard, the European Union's export subsidy system, and the accelerated development of emerging economies such as Brazil, China, and India, hastened the decline of the US regime (Friedmann and McMichael, 1989; McMichael, 2005, 2009).

While proponents disagree on the specific characteristics of the Third Agri-food Regime,[1] three similar proposals are useful for our analysis: the Corporate Food Regime; the Financialized Food Regime; and the Neoliberal Food Regime. The Corporate Food Regime (McMichael, 2005, 2009) focuses on the globalization project's agenda to eliminate historical forms of peasant agriculture in favour of a global agri-food system based on neoliberal restructuring and flexible accumulation. Globalization is an ideologically-driven political project designed to weaken social-democratic successes (Harvey, 2005). Neoliberalism enhances capital accumulation by removing the protections subordinate groups secured under social-democratic states during the second regime. This Regime employs supranational institutions such as the World Trade Organization (WTO), International Monetary Fund (IMF), World Bank and associated free trade agreements as mechanisms to impose agri-food restructuring on the global South. Through global banks and venture capital firms, transnational finance capital funds agri-food TNCs to create global commodities, resulting in increased market consolidation in agri-food sectors nationally and globally. Additionally, financialization-focused capital accumulation strategies are privileged over production-based models. Despite the rhetoric of free trade, the global North continues to protect its agricultural industries and dump surplus commodities into the global marketplace. Structural adjustment policies encourage a proliferation of agro-exporting from the global South so as to service debt and facilitate capital mobility. Nation-states are forced to shift their focus from national development strategies to global credit worthiness. Resistance movements such as La Via Campesina challenge the Corporate Food Regime and call for food sovereignty and the rights of indigenous peoples. Despite this, the global food security and energy crises of 2007 triggered a land grab resulting in enhanced accumulation by dispossession (McMichael, 2009).

The Financialized Food Regime (Burch and Lawrence, 2007, 2009) focuses on the radical transformation of financial relations and the food retailing sector. This transformation reorganizes global commodity chains in favour of food retailers, especially supermarkets, to the detriment of agri-food manufacturing firms. Through strategic alliances with financial institutions and creative financial instruments, global agri-food retailers organize their own flexible supply chains to service newly developed store brands and therefore avoid their historical dependence on established agri-food brands. Financialization also refers to the trend away from a focus on the long-term production of food to short-term profits generated by capital mobility through the use of investment instruments such as hedge funds, junk bonds and derivatives. Some examples are Goldman Sachs investing in poultry production in China and other forms of direct investment in agricultural land for production (land grabs). These globally focused fractions of capital have shifted the policy focus of nation-states from social safety-net programs to profit maximization where shareholder value is the key measure of performance (Lazonick and O'Sullivan, 2000; Lazonick, 2010). Profit maximization strategies based on flexible systems of production, including contracts rather than formal ownership, are diffused from the global North to the global South as part of the Financialized Regime.

The Neoliberal Food Regime (Pechlaner and Otero, 2008, 2010; Otero, 2011) focuses on the increasing inequalities between the North and South due to trade liberalization and neoliberal globalism. Focusing on the impact of NAFTA on Mexico, they amend the Corporate Food Regime to focus on the ideas of neoliberal globalism and neoregulation. Neoliberal globalism is the ideological underpinning of corporate behaviour that seeks flexible accumulation opportunities. Challenging views that minimize the role of the nation-state in the face of globalization, neoregulation re-centres it as the organizational structure that imposes the market as the self-regulating governance mechanism under the globalization project. The nation-state is seen to assume the role of neoregulator as it implements the globalization project in each country with varying results depending on the relative autonomy of the state in relation to corporate and social movement actors. Through neoregulation, the state mediates the contested formulation of public policies that constrain or enable the actions of TNCs and thereby restructures the political economy of the region.

While there are substantive differences among these three perspectives, they agree that during this transition phase to the Third Food Regime flexible forms of capital accumulation are re-orienting and re-regulating the stabilities of the previous regime. Neoliberalism provides the ideological underpinnings of these socio-economic changes as TNCs drive the process of neoliberal restructuring that has transformed agri-food supply chains in favour of capital over subordinate groups. This flexible accumulation system is diffused around the globe as new territories in the global South are integrated into the supply chains that service the global North. Agri-food TNCs, funded by transnational finance capital with the help of global governance organizations, diffuse the globalization project within nation-states which – through neoregulation – facilitate and mediate the process.

For the purposes of this chapter, we borrow the focus on financialization and market consolidation from McMichael, the focus on financialization and flexibility via creative investment instruments from Burch and Lawrence, and the focus on the role of the state in mediating neoliberalism from Otero and Pechlaner, to investigate and interpret the case of JBS and poultry grabs.

From land grabs to poultry grabs

The land grab literature emerged in the wake of the food, energy, and financial crises of 2007–8. The initial 'making sense' period (2007–12) of this research focused on large-scale land acquisitions, mostly in Africa by countries of the global North (Edelman *et al.*, 2013), which was characterized as the latest example of accumulation by dispossession (McMichael, 2009). Agricultural land was a good investment, both speculative and productive, in the face of food shortages, resource challenges, and financial recession. Land grabs are not a new phenomenon (McMichael, 2013; Magdoff, 2013) but, rather, are emblematic of mounting crises in the global agri-food system regarding ecological degradation, food and energy security, and population growth. The rush to gain access to agri-food assets – often underproductive lands – to allow intensification of production skipped the WTO free-trade regime market framework through direct ownership or leasing. The rush was fuelled by food security scares, assisted by neoliberalism, and driven by financialization and speculation (Magdoff, 2013; Margulis *et al.*, 2013).

The land grab literature expanded to move beyond the narrow focus on land and dispossession to include the various mechanisms through which asset grabbing occurs (Borras *et al.*, 2011; Edleman *et al.*, 2013). Topics such as green grabbing, water grabbing, food grabbing, industrial/urbanization grabbing, biofuels/biomass grabbing, labour grabbing, mining grabbing and infrastructure grabbing (roads and dams) entered the discourse. Local producers were often not displaced, but rather were linked to global value chains. Land, labour and capital were needed and labour was often subsumed via wage work, contract farming, and/or other partnership agreements similar to historical patterns of plantation work or sharecropping (Little and Watts, 1994). The grabbing phenomenon became much more polycentric and varied as G20 countries became involved, often in the form of collaborations between agribusiness and sympathetic governments. It represents a new territorialization and spatial restructuring of the global agri-food system (McMichael, 2013).

The modern poultry industry

The poultry industry emerged in the 1930s in the northeastern DelMarVa (Delaware, Maryland, and Virginia) region as a system of independent growers, processors, feed dealers, and distribution systems. Synthesized vitamins allowed chickens to live indoors full-time. Government and industry research improved genetics, feed conversions, and poultry health, while rural electrification modernized poultry housing. The size and concentration of 'growout' (chicken raising) facilities increased steadily (Reimund *et al.*, 1981). During World War II States in the US South (e.g. Georgia, Alabama, and Arkansas) rapidly increased production to supply growing demand. Eventually, in the South, vertical integration brought all the aspects of the supply chain (hatcheries, processing, feed, transportation) under the ownership of one firm, with chicken growers coordinated through production contracts. Integrators built a Southern Model that took advantage of small and/or marginal farmers accustomed to sharecropping and non-union and minority processing workers (Striffler, 2005; Constance, 2008; Oxfam, 2015). The modern broiler system 'possessed a distinctive southern accent from its inception' (Boyd and Watts, 1997, p. 184) and steadily replaced the independent DelMarVa system.

Vertical integration through production contracts allowed the firms to expand their operations without incurring the fixed costs of land and growout buildings. Contract growers mortgaged their land to obtain loans to build single-use growout buildings in which they grew the integrators' birds as part of a sharecropping arrangement (Heffernan, 1984; Constance, 2008). The contract system allows the integrator to take advantage of the chief assets of the family farm – cheap, 'docile', and flexible labour – without the burdens of equity or the costs of wage labour. The reliance on non-union, minority processing labour reduced wage costs (Human

Rights Watch, 2005). These poultry agro-industrial districts are early forms of flexible accumulation – the preferred model of agri-food globalization (Boyd and Watts, 1997).

Through continued vertical and horizontal integration, the industry developed regional monopsonies whereby the poultry growers often had only one firm with which to contract. This trend often results in asymmetrical power relations – called 'monopsony opportunism' – between integrators and growers, whereby the growers are kept in 'debt bondage' (Breimyer, 1965; Davis, 1980; Heffernan, 2000; Constance *et al.*, 2013). The poultry agri-TNCs have diffused the model globally and into other commodities (Constance and Heffernan, 1991; Martínez *et al.*, 2013).

Poultry sector globalization

In the 1950s and 1960s global firms such as Continental Grain, Ralston Purina, and Cargill started broiler operations in developing countries, often as a means to add value to their feedstuffs. These grain companies were followed by Tyson, Pilgrim's Pride, and other US and foreign integrators in the 1980s. By the early 1990s agri-TNC s had created a global poultry agri-food complex that linked the most favourable areas of production to profitable consumer markets (Constance and Heffernan, 1991). For example, by 2003 Tyson Foods Inc. (2005) had joint-venture poultry operations in Argentina, Brazil, China, Denmark, Indonesia, Japan, Korea, Malaysia, Panama, the Philippines, Spain, the United Kingdom, and Venezuela. After entering the Mexican market through a joint venture in 1987, by 2003 Tyson de Mexico was the largest producer of value-added chicken and was expanding into other areas of Latin America (Constance *et al.*, 2010). Following its Mexican model, Tyson expanded aggressively into China, India and Brazil. The three Brazilian broiler companies were bought to supply domestic demand and exports to Asia and Europe. The Thai firm Charoen Pokphand (CP) pursued similar patterns of vertical and horizontal integration in Southeast Asia and China (Burch, 2005; Constance *et al.*, 2014).

Pilgrim's Pride, Inc. has also been very active in Mexico. In 2007, the company became the largest broiler producer in the US and the world with its acquisition of Goldkist (Pilgrim's Pride, Inc., 2007). Pilgrim's Pride entered the Mexican market in 1995 with the purchase of five vertically integrated broiler companies from Purina and was soon the second largest broiler company in Mexico. From 2006 to 2008 Pilgrim secured US$1.6 billion in new debt, mostly to buy Goldkist and to pass Tyson as the largest broiler company in the US and the world. The debt was in unsecured notes (US$650 million) and credit (US$450 million). Pilgrim's Pride found itself caught in a combination of the global recession, a saturated domestic market, and increased feed costs due to the ethanol bubble. It was a single-commodity producer that exercised poor hedging strategies and went bankrupt. In 2009 it was bought by JBS of Brazil, which became the world's largest animal protein corporation (JBS, 2016). In 2010, Tyson de Mexico, JBS/Pilgrim's Pride, and Bachoco (a Mexican corporation) controlled 67 per cent of the Mexican poultry market (Martínez *et al.*, 2013). The vertical integration system developed in the US South around agro-industrial districts was diffused to Mexico. In 2014 JBS/Pilgrim's Pride, Tyson Foods, Inc, and CP dominated the global poultry industry with annual sales of US$40.0 billion, US$33.3 billion, and US$33.0 billion, respectively (Constance *et al.*, 2014).

A China 'poultry grab' sidebar

In China in 2008 some of the most valuable investments were in the agricultural products sector. In August 2008 Goldman Sachs bought control of ten poultry operations for US$300 million. It also owned 13 per cent of China Yurun Food Group, as well as 60 per cent of Shuanghui Investment and Development Group. In 2013 Shuanghui bought Smithfield Pork, the largest pork company in the world, with financing from Goldman Sachs, Bank of China, and Temasek Holdings Goldman Sachs is also principle owner of Burger King in China, along with the Bain and Texas Pacific Group. Meat consumption in China has quadrupled since 1980; poultry demand was expected to rise 60 per cent by 2030 (Constance *et al.*, 2014). In a similar venture, in 2014 the US private equity firm KKR invested US$400 million in Fujian Sunner Development Co. Ltd., the largest chicken products firm in China (Steger, 2014).

An Australia 'poultry grab' sidebar

In 2013 two firms – Inghams and Steggles/Balada – controlled 70 per cent of Australian poultry production. These were privately held, fully integrated, modern facilities with about 800 contract growers in 2005. Poultry farmers are some of the wealthiest farmers in Australia. Through a system called 'The Shield of Crown', contracts were negotiated through collective bargaining between farmers and government. This system – based on open records – generates good incomes, relative certainty around business planning, and continuity of supply, but it does not fit the Cairns group of countries' philosophy of open trade. In March 2013 the global private equity firm Texas Pacific Group (TPG) based in Fort Worth, Texas, bought Inghams Poultry for US$880 million. Inghams is the largest poultry company in Australia with a 33 per cent market share. TPG secured US$770 million in loan commitments from seven banks to finance the deal: Westpac Banking Corp; Macquarie Group Ltd.; Nomura Holdings; Bank of America; Merrill Lynch, Australia and New Zealand (ANZ) Banking Group, National Australia Bank Ltd.; and, HSBC Holdings Plc. In December 2013 TPG started stripping cash via a massive US$650 million sell-off of Ingham properties (Constance *et al.*, 2014).

The rise of JBS

The origins of JBS and J&F Investmentos: the beef years

In the early1950s JBS founder José Batista Sobrinho started buying cattle to resell to slaughterhouses in Anápolis in the state of Goias, Brazil. In 1953 he and his brothers began slaughter operations in Goias. In that same year they established J&F Investimentos to control the financial and strategic decisions of their business. As the business grew, Joseph's sons – Joseph Jr., Wesley, and Joesley – chose to work in the family business instead of going to college. They adopted this business philosophy: 'We focus only on what we can control, and forget the rest' (Bell and Ross, 2008, p. 1). Acquistions in Planaltina (1968) and Luziania (1970) increased cattle slaughter capacity to 500 head per day (JBS, 2016).

From the 1980s to the early 2000s JBS acquired more slaughterhouses and net sales increased from US$400 million in 1997 to US$1.2 billion in 2004. JBS' first international expansion was in 2005 when it acquired 100 per cent of Swift-Armour, S.A., the largest beef company in Argentina. The purchase was made through a judicial auction by the Argentine government. Sales rose in 2006 to US$1.9 billion and JBS continued to expand with refrigeration and slaughter acquisitions in Argentina (Bell and Ross, 2008; Torres, 2011). By 2006, JBS had 21 beef processing plants in Brazil and five in Argentina. Major international acquisitions in 2007 increased net sales fivefold from US$1.9 billion to US$12.7 billion. After these, and other major, investments its beef processing capacity reached 5,800 head per day in 2007. In 2007 JBS became the first animal protein company to go public on the Sao Paulo stock exchange (JBS, 2016).

BNDES: Brazilian Development Bank

Two factors explain this significant growth: (1) mergers and acquisitions and (2) the political and economic leverage of the Brazilian government through BNDES.[2] The BNDES is the Brazilian Development Bank, which provides financial support to Brazilian companies for domestic and global expansion (BNDES, 2017). Although Brazil embraced neoliberalism throughout the 1990s, the government continued to use incentive policies in key sectors. The main policy tool was subsidized credit to large firms that were already competitive in commodity sectors. The BNDES actively promoted the concentration and expansion in the Brazilian meat industry. BNDESpar, part of BNDES, was created by the Brazilian government to acquire stock shares. In early 2017 it owned 21.42 per cent of JBS shares, with another 4.92 per cent of shares owned by Federal Savings Bank. The Batista family is the controlling shareholder with 44.35 per cent; the remaining shares are owned by minority shareholders (JBS, 2017).

These government loans and stock investments were essential for JBS' aggressive strategy of national and global acquisitions, such as that of Swift and Co. USA in 2007 (a company with significant beef, pork and lamb operations in the US, Argentina, and Australia). It bought Smithfield Beef of the US (2008), which included Five Rivers Cattle Feeding and the Tasman Group of Australia (2008). In 2009 it bought a controlling interest in Pilgrim's Pride Inc., the second largest poultry company in the US. When it purchased Bertin (2009), the second largest beef company in Brazil, JBS became the largest cattle feeder and the largest animal protein company in the world with annual sales of US$29 billion (JBS, 2016). This was a notable event as JBS did not even appear among Brazil's 400 largest companies until 2002. With BNDES' support for expansion and acquisitions, JBS moved from sixty-first in 2006 to the fifth place in 2010, and first place among non-financial private business groups in Brazil (Almeida and Ross, 2012).

J&F Investmentos played a crucial role in the financial success of JBS. It created the JBS bank in 2008, which merged with Banco Matone in 2011, and was renamed Banco Original. Banco Original supported the agro-industrial sector by providing loans, working capital and strategic and financial consulting. Its services included retail banking, corporate credit, structured finance, investment instruments and project financing. In 2013 it bought the television station Canal Rural, the largest specialized communication platform in the agroindustrial sector in Brazil. In 2015 it bought a controlling stake in Alpargatas, the largest footwear company in Latin America, with 20,000 employees, and 150,000 points of sale in Brazil and 10,000 abroad in 106 countries. J&F Investimentos financed its Alpargatas purchase with a loan from state-owned bank Caixa Economica Federal (Reuters, 2016; J&F Investimentos, 2017).

JBS, SA: adding poultry to the portfolio

Through aggressive growth – supported by BNDES and coordinated by J&F Investmientos – JBS, SA became one of the Fortune 500 largest companies in the world. It was ranked 496 in 2010, 286 in 2012, and 185 in 2016 (Fortune, 2017). Its next acquisitions diversified the portfolio, particularly in the poultry sector. In 2009 JBS created JBS Leather, the largest leather producer and exporter in the world with the capacity to process 99,000 head of cattle per day. In December 2009 JBS bought 64 per cent control of Pilgrim's Pride Corp (PPC) for US$800 million, thus entering the poultry market and becoming the second largest producer of chicken in the world, with operations in the US, Mexico and Puerto Rico, and a capacity of 7.6 million chickens per day. PPC had filed for bankruptcy due to a production surplus, weakening demand, and increase in production costs resulting in its inability to repay the debt burden created by the acquisition of Gold Kist (JBS, 2009; El Informador, 2017).

In 2010 it made two acquisitions in Australia, one in Belgium and one in the US. With the purchase of Tatiara Meats for US$27.5 million, it instantly became the largest lamb processor in Australia, with revenues of US$183 million and slaughtering capacity of 24,500 head per day. The purchase of Rockdale Beef for US$37.2 million added to its Swift portfolio in Australia. The Toledo Group in Belgium (EU$11 million) specializes in the research, development and commercialization of baked products of bovine cattle for Western European consumers. Finally, McElhaney's Feedlot in Arizona (US$24 million) added it to its Five Rivers subsidiary (JBS, 2010; Vaz, 2010; JBS Five Rivers, 2017). JBS continued its activities in the industrialized food sector and became the second largest Brazilian platform for the production and distribution of value-added food products (JBS, 2011; Reuters America Latina, 2013). In 2012 JBS increased its stake in Pilgrim's Pride to 75.3 per cent of shares. In 2014 it bought (for US$575 million) the poultry business of Tyson Foods in Brazil and Mexico. Tyson's Mexican operations were purchased for US$400 million through the subsidiary Pilgrim's Pride, whose majority shareholder is JBS USA Holdings. Tyson's business in Brazil was acquired for US$175 million through the subsidiary JBS Foods, formerly known as Seara Brazil. JBS announced that these actions should increase annual revenues by over US$900 million (El Financiero, 2014; Infocampo, 2014; JBS, 2014).

In 2015 it acquired Moy Park for US$1.5 billion, including a US$300 million debt. Moy Park is one of the top 15 food companies in the United Kingdom, the largest private sector company in Northern Ireland and one of the leading poultry producers in Europe. Through Moy Park JBS expanded its presence in the European processed food market. In 2015 it also acquired Primo Smallgoods Group (US$1.24 billion), a leading company in the Oceania processed meats markets. It also bought the Big Frango Group poultry company in Brazil (for US$165 million) (El Economista, 2014; Jelmayer and Magalhaes, 2014). Finally, in October 2015 it purchased the assets of Cargill Pork in the US for US$1.45 billion. The deal included two meat processing plants, five feed mills, and four pig farms. In 2016 it expanded its beef operation in Belen, Paraguay and through Pilgrim's Pride, bought the GMP Company (US$35 million), a leading provider of premium brand chicken products in the Midwest with two processing plants and a value-added processing facility (Rojas, 2016; Watrous, 2016)

The reorganization plan

In 2016 JBS had operations in over 20 countries and exported to over 150 countries; it was the largest meat company in the world. Sales increased from US$0.4 billion in 1999, to US$1.9 billion in 2006 to US$48 billion in 2015 (JBS, 2016). In May 2016 it proposed a major reorganization and a move of its headquarters to Ireland to minimize taxes and maximize financial flexibility. JBS announced that a former Morgan Stanley executive would be the Chief Financial Officer for the new JBS Foods International and would lead the company's reorganization process. The goal of the restructuring was to create a platform that allowed JBS to reduce the cost of fund-raising, to optimise taxes, to serve a broad base of investors, as well as to create value for its shareholders. In addition, JBS Food International would be listed on the NYSE, which would make it more attractive for investors. JBS would still be controlled by existing Brazilian shareholders with its main shareholder J&F Investimentos, which would maintain its headquarters and strategic decisions in Sao Paulo, Brazil. But, in October 2016 the reorganization plan was cancelled due to opposition from the BNDESpar investment bank, when it exercised its veto power over the plan. Following the announcement JBS shares tumbled on the Sao Paulo stock exchange (JBS, 2016).

The fall of JBS

The corruption and 'Weak Meat' scandals

The restructuring process was linked to the political conflicts that have arisen in Brazil since 2014. JBS is implicated in one of the biggest political and business corruption scandals in the history of Brazil. According to the economic press (Freitas *et al.*, 2016), Brazil's government has for many years operated a corruption network in which a significant number of political figures have pressured executives from large companies to obtain bribes in exchange for preferential treatment. The bribes were then used to finance political campaigns. JBS and other companies were the main contributors to the 2014 election campaign. The Batistas were among the largest political donors to all parties in the 2014 elections. Henrique Meirelles, Brazilian President Michael Temer's finance minister, served as chairman of J&F Investimentos until he took the finance role in May 2016 (Leahy, 2016).

Although JBS strongly denied any wrongdoing, Joesley Batista, JBS Chairman and brother of the chief executive, was charged with alleged offences connected with Banco Original, the bank controlled by the Batista family. During the hearings, Joesley admitted that JBS had committed acts of corruption for several years through the payment of bribes to approximately 2,000 politicians, including the two former presidents of Brazil, and the current President, Michael Temer. This was done in exchange for subsidized loans and other types of clandestine operations. For his part, President Temer said that the dramatic growth experienced by JBS in recent years was due to the support provided by the government of Luiz Inacio Lula da Silva. JBS obtained its first subsidized loan from the Lula government in 2005, a relationship that allowed it to increase its sales exponentially between 2007 and 2016 (Freitas *et al.*, 2016).

In March 2017 JBS was the subject of a government investigation called 'Weak Meat', which exposed a producer network dedicated to bribing health authorities to allow the sale of contaminated or adulterated meat. Several importing countries immediately announced the ban of Brazilian meat. This investigation led two of the Batista's sons to offer their testimony in exchange for immunity from legal prosecution. Both admitted spending hundreds of millions of dollars bribing a significant number of politicians and added that, had they not done so, 'It would not have worked. It would not have been so fast' (Politica, 2017). JBS Chairman Joesley Batista and his brother, Wesley Batista (Vice Chairman of the Board), resigned from the Board of Directors. The firm was ordered to pay a US$3.2 billion fine for corruption, and later announced plans to sell billions of assets, including Five Rivers Cattle Feeding in North America and Moy Park in Europe. The day after the announcement, JBS' stock dropped 31 per cent and its credit rating was downgraded (DeFotis, 2017).

In June 2017, JBS announced the sale of all of the shares of its subsidiaries in Argentina, Paraguay and Uruguay to three companies controlled by Brazil's Minerva, for US$300 million. However, a court suspended the transaction because it considered that the decision to sell the assets could compromise investigations into President Temer and a considerable number of politicians who were caught up in the corruption scandal. The sale of assets is subject to the approval of the board of directors of the company and the prior consent of BNDESpar, which holds 21.32 per cent of the JBS shares. The company is also facing investigations of inside information in stock exchange and exchange operations, as well as for selling a significant amount of its shares in the days before the scandal (DeFotis, 2017). In September the immunity plea bargain was rescinded and Joesley and JBS executive Ricardo Saud were arrested for concealing evidence (Freitas and Margues, 2017).

Discussion and conclusions

The discussion and conclusions section is guided by three persistent themes from the agri-food regimes literature. First, the corporate food regime entails the growing power and consolidation of the agri-TNCs as they construct global value chains based on flexible accumulation and financialization, including land grabs in the wake of the 2007–8 food/energy/financial crises (McMichael, 2005, 2009, 2013, 2014). Second, there is an increased tendency for major production TNCs to develop financial subsidiaries to engage in productive and speculative investments and for venture capital to invest in the agri-food sector for speculative investment (Burch and Lawrence, 2007, 2009). And third, the state still plays a crucial role in 'neo-regulating' and managing the processes and contradictions of the globalization project, specifically, the complexities and variations of the state/TNC/social movement nexus within each country (Pechlaner and Otero, 2008, 2010; Otero, 2011).

JBS is the world's largest meat company. The two main factors that explain its dominant position are its decisions in the financial sector and the support of the Brazilian government. The strategy of maximizing the value of its shares through the creation of J&F Investimentos in coordination with BNDES and its merger and acquisitions were the basis of its expansion and diversification. The crucial function of J&F Investimentos' financial capital occurred in three phases. The first is the role of BNDES and BNDESpar in the purchases of Swift Armor in 2005 and Pilgrim's Pride in 2009, which became the model for later major acquisitions. The second is the creation of the JBS Bank and its merger with the bank Matone in 2011 and the creation of Banco Original. This allowed shareholders a 153 per cent increase in the value of assets between 2011 and 2016. The third is the portfolio diversification through the acquisitions of the pulp companies El Dorado in 2010 and Alpargatas in 2015.

These activities reveal how financialization through J&F Investimentos, combined with government support through BNDES and BNDEspar, facilitated the emergence of JBS as a major agri-food TNC. Following Tyson's diversification model, JBS developed national and regional dominance in beef packing, then expanded into Australia and the US. The 2009 purchase of Pilgrim 'grabbed' the US and Mexican markets, followed by the 2013 purchase of Seara Brazil and the 2014 purchase of Tyson's assets in Brazil and Mexico. Significantly, Pilgrim's Pride was a victim of the 2008 financial crisis and the process of financialization. It incurred US$1.1 billion in new debt to buy Goldkist, including US$650 million in unsecured notes (junk bonds) and US$450 million in credit. Pilgrim's debt almost tripled from US$500 million in 2006 to US$1.35 billion in 2017 (Simply Wall St, 2017). Pilgrim became caught as a single commodity producer in a global recession facing rising feed costs due to the ethanol boom. It was unable to pay its high interest debt and, subsequently, filed for bankruptcy. Finally, the Moy Park acquisition in 2015 'grabbed' access to the European market. JBS created a diversified poultry production platform for domestic consumption and global exports to major markets.

The JBS case exemplifies the ways state power has aided one of Brazil's largest corporations. It reaffirms how financial liberalization under neoliberalism concentrates wealth and increases the economic and political power of capitalist elites. The JBS corruption allegations reflect problems relating to linkages between large corporations and top government officials (Murphy, 2006). Trade and financial liberalization were fostered through financial contributions to the campaigns of the candidates in popular elections, allowing corporations to have more political sway. These practices of corruption were common, as revealed by JBS bribery activities with three Brazilian Presidents. The fact that the current Minister of Finance was previously President of J&F Investimentos, and previously employed at Goldman Sachs, is illustrative of the same pattern. These socio-political conditions were developed in the last 30 years as part of the same process of financial and commercial liberalization occurring in other national economies.

Joesley Batista's statement that without the (illegal) support given to the electoral campaigns JBS' rise to dominance 'would not have been so fast' illuminates the influence that these illegal actions have in the development of large agro-complexes. Illegal actions, combined with the complex networks of relations of the big corporations with the nation-state, are central elements in the consolidation of agri-food TNCs. The linkage of executives of large corporations with spheres of government not only confirms the Neoregulation Food Regime position, but also reveals the enhanced role of corporations in the struggle for political spaces within their host countries.

The allegations of corruption have occurred in the midst of an intense power struggle within the Brazilian government between right-wing and centre-left forces, which transcends the specific JBS context and extends into a geopolitical conflict beyond the national scope. In line with McMichael's Global Food Regime approach, the bottom line of the conflict is the struggle to advance the neoliberal project and limit the social democratic forces represented by the Lula regime. The alliances between political groups and large companies created in the Lula regime through BNDES were critical for the global expansion of national agri-food companies such as JBS. The incursion of meat industry CEOs into top-level government positions is a clear example of corporate dominance in the definition of the legal and institutional framework of the global meat industry. The case of JBS in Brazil provides an example of how the state supports firms via financial measure such as equity or debt, as well as through collaboration between top government and TNC representatives in investments and merger and acquisitions, globally (Costa, 2012; Hua and Carrera, 2016; Gutiérrez, 2017).

This case points to the need for more research on the state/TNC/international institutions nexus in global agri-food systems. Are we seeing the emergence of a 'neocorporatist' state – or better the neo-domination of the TNCs over the state? Global institutional rules are expanded into the international agencies on behalf of corporate interests. Finance and global trade arrangements, along with the renegotiation of free trade agreements, also need closer attention in agri-food commodity case studies. Similar comparative research on the political economy of possible poultry grabs in large emerging economics such as Russia, China, India and Brazil would be instructive (Costa, 2012; Hua and Carrera, 2016)

While the corruption scandal casts some doubts on JBS' continued dominance in the meat sector, the events reported here do reveal that financialization has facilitated its rapid rise to dominance. Although JBS rode the post-crisis financialization wave to dominance, its historical relationship to state support through BNDES limited its options for seeking better financial and tax regulations. This point highlights the continued power of the state to limit efforts of enhanced territoriality by those agri-TNCs which have structural state linkages.

The financial, food, and energy crises of 2008 bore witness to a process of global financial liberalization where capital is now more dominant than in any previous period in the history of capitalism (see Serfati, 2008, 2011, 2013; Lazonic, 2010; Stiglitz, 2012; Krugman, 2015; Bonanno, 2016: Bonanno *et al.*, 2016). While JBS embraced the financialization wave and concentrated on the animal protein sector, the creation of J&F Investimientos and its aggressive portfolio diversification into other sectors supports the Financialization Food Regime thesis. Under the Lula administration – and embedded in a culture of bribery and corruption – Brazil, a powerful G20 actor, directly subsidized the rise of JBS as its dominant actor in the global meat industry. The long list of acquisitions can be interpreted as overt 'asset grabs' to position JBS and Brazil favourably in the post-crisis era.

The poultry commodity system's grounding in flexible accumulation structures – specifically contract production and informal labour involved in processing – and its preferred status as the meat of choice for growing middle classes, make it especially attractive for agri-TNCs and venture capital investments. Although poultry is a lucrative asset grab, it is also problematic regarding ethical and human rights issues. It remains a crucial commodity system for social science analysis, especially because other agri-food commodities are following the poultry model. The case of JBS and its poultry grabs reveal the particular mechanisms of agri-food financialization, and in this case the key role of the state/TNC nexus in the process. JBS' asset grabs accelerated in the post-crises era, assisted by neoliberalism and driven by financialization and speculation (see Magdoff, 2013; Margulis *et al.*, 2013).

The sidebars on Goldman Sachs and KKR in China and Texas Pacific Group in Australia provide further evidence of poultry grabs in line with the Financialization Food Regime thesis. Goldman Sachs is a dominant, legacy venture capital firm which moved into China ahead of the nutrition transition. Its major investment in Shuanghui Investment and Development Group, which bought Smithfield Pork and instantly became the largest pork company in the world, represents a major pork grab to service both domestic and global markets. KKR is also a legacy venture capital firm. The case of TPG and Inghams Poultry in Australia is an even better example. With US$770 million in financing from a who's who list of financial banks, it bought Inghams for $880 million and then stripped out $650 million in cash. The point here is that venture capital is funding investments – asset grabs – in the protein sector beyond business-as-usual mergers and acquisitions by agri-food TNCs.

More research is needed on the specific characteristics of agri-food asset grabs in general, and poultry grabs, in particular. This case of JBS reveals a particular state/TNC nexus that allowed JBS and J&F Invesimentos to build a global meat empire. Other dominant poultry agri-TNC s such as Tyson Foods, Inc. and Charoen Pokphand continue to expand their market share through acquisitions of major national firms. Finally, we should expect more poultry grabs such as the TPG case in Australia and KKR in China as venture capital firms search for agri-food assets ahead of the nutrition transition wave.

Notes

1 Friedmann (2005) presents the Corporate-Environmental Food Regime based on a convergence of corporate greening and green agri-food social movements whereby global agri-food retailers service subdivided class diets of transnational rich and poor consumers. Dixon (2009) advances a Nutritionalization Food Regime whereby the bifurcation of class diets has distributed healthy foods to affluent consumers and increasing malnutrition to less affluent peoples. Finally, Campbell (2009) expands on the works of Friedmann (2005) regarding ecological aspects and of McMichael (2005) regarding a 'Food from Nowhere' position, to advance the 'Food from Somewhere' alternative to the corporatization of the agri-food system.

2 The Brazilian Development Bank (BNDES) is the main financing agent for development in Brazil (BNDES, 2017). Created in 1952, the BNDES played a key role in stimulating the expansion of industry and infrastructure in the country. The Bank offers several financial support mechanisms to Brazilian companies with emphasis on innovation, local development and socio-environmental development. The BNDES aligns its operations to improve Brazil's position in globalized world. The BNDES finances the expansion of national companies internationally and seeks to diversify the sources of its resources on the international market. The BNDES finances the export of Brazilian goods and services in projects carried out overseas and institutional fundraising through multilateral organizations, sharing experiences and promotion opportunities. The BNDES operates on a global scale where it consolidates its role as financing agent to the economic and social development of Brazil.

References

Almeida, M. and Ross, B. (2012) 'Globalization, democratization, and the challenges of industrial policy in Brazil', in C. Lim and R. Locke (eds) *Industry Policy*, World Bank, Washington DC

Bell, D. and Ross, C. (2008) 'JBS Swift & Co: Case study', available at: https://hbr.org/product/JBS-Swift–Co-/an/509021-PDF-ENG, accessed 16 March 2017

BNDES (Brazilian Development Bank). (2017) 'The Brazilian Development Bank', available at: www.bndes.gov.br/SiteBNDES/bndes/bndes_en/Institucional/The_BNDES/, accessed 22 June 2017

Bonanno, A. (2016) 'La financiarización de la agricultura y alimentos en el contexto de la reestructuración neoliberal: Rasgos principales y contradicciones fundamentales', *Estudios Reales*, vol 5, no 10, pp. 1–17

Bonanno, A. and Constance, D. (2008) *Stories of Globalization: Transnational Corporations, Resistance, and the State*, Pennsylvania State University Press, Pennsylvania

Bonanno, A., Martínez, F. and Aboites, G. (2016) 'El Neoliberalismo: Un momento de la globalización', *Editorial Fontamara y Editorial de la Universidad Autónoma de Coahuila*, México

Borras, S., Hall, R., Scoones, I., White, B. and Wolford, W. (2011) 'Towards a better understanding of global land grabbing: An editorial introduction', *Journal of Peasant Studies*, vol 38, no 2, pp. 209–216

Boyd, W. and Watts, M. (1997) 'Agro-industrial just-in-time: The chicken industry and postwar capitalism', in D. Goodman and M. Watts (eds) *Globalizing Food: Agrarian Questions and Global Restructuring*, Routledge, London, pp. 192–225

Breimyer, H. (1965) *Individual Freedom and the Economic Organization of Agriculture*, University of Illinois Press, Urbana

Burch, D. (2005) 'Production, consumption and trade in poultry: Corporate linkages and North-South supply chains', in N. Fold and B. Pritchard (eds) *Cross-continental Agri-food Chains*, Routledge, London, pp. 166–178

Burch, D. and Lawrence, G. (eds) (2007) *Supermarkets and Agri-food Supply Chains: Transformations in the Production and Consumption of Food*, Edward Elgar, Cheltenham, UK

Burch, D. and Lawrence, G. (2009) 'Towards a third food regime: Behind the transformation', *Agriculture and Human Values*, vol 26, no 4, pp. 267–273

Campbell, H. (2009) 'Breaking new ground in food regime theory: Corporate environmentalism, ecological feedbacks and the "food from nowhere" regime', *Agriculture and Human Values*, vol 26, no 4, pp. 309–319

Campbell, H. and Dixon, J. (2009) 'Introduction to the special symposium: Reflecting on twenty years of the food regimes approach in agri-food studies', *Agriculture and Human Values*, vol 26, no 4, pp. 261–265

Constance, D. (2008) 'The southern model of broiler production and its global implications', *Culture and Agriculture*, vol 30, no 1, pp. 17–31

Constance, D. and Heffernan, W. (1991) 'The global poultry agri-food complex', *International Journal of Sociology of Agriculture and Food*, vol 1, pp. 126–141

Constance, D., Martínez, F. and Aboites, G. (2010) 'The globalization of the poultry industry: Tyson Foods and Pilgrim's Pride in Mexico', in A. Bonanno, H. Bakker, R. Jussaume, Y. Kawamura, and M. Shucksmith (eds) *From Community to Consumption: New and Classical Statements in Rural Sociological Research*, Emerald, Bingley, pp. 59–76

Constance, D., Martínez, F., Aboites, G. and Bonanno, A. (2013) 'The Problems with Poultry Production and Processing', in H. James (ed.) *The Ethics and Economics of Agri-food Competition*, Springer, The Netherlands, pp. 155–175

Constance, D., Konefal, J., Dixon, J., Martínez, F. and Aboites, G. (2014) 'A financialization regime analysis of poultry grabs in Mexico, China, and Australia', *Paper Presented at the Annual Meeting of the Rural Sociological Society*, New Orleans, August

Costa, F. (2012) 'Capitalismo de estado neocorporativista. Texto para Discussão', Instituto de Economia da Unicamp, Campinas

Davis, J. (1980) 'Capitalist agricultural development and the exploitation of the propertied labourer', in F. Buttel and H. Newby (eds) *The Rural Sociology of the Advanced Societies*, Allenheld, Montclair, pp. 133–154

DeFotis, D. (2017) 'Brazil bribery beef: JBS Sinks 31 per cent, Pilgrim's Pride Safe?' available at: Barron's. May 22. www.barrons.com/articles/brazil-bribery-beef-jbs-sinks-31-pilgrims-pride-safe-1495493779, accessed 26 June 2017

Dixon, J. (2009) 'From the imperial to the empty calorie: How nutrition relations underpin food regime transitions', *Agriculture and Human Values*, vol 26, no 4, pp. 321–333

Edelman, M., Oya, C. and Borras, S. (2013) 'Global land grabs: Historical processes, theoretical and methodological implications and current trajectories', *Third World Quarterly*, vol 34, no 9, pp. 1517–1531

El Economista. (2014) 'JBS compra Primo-smallgoods', available at: http://eleconomista.com.mx/industria-global/2014/11/21/jbs-compra-primo-smallgoods-1250-mdd, accessed 24 June 2017

El Financiero. (2014) 'JBS comprará operaciones de Tyson Foods en México y Brasil', available at: www.elfinanciero.com.mx/empresas/jbs-comprara-operaciones-de-tyson-foods-en-mexico-y-brasil.html, accessed 24 June 2017

El Informador. (2017) 'Se declara en quiebra productor de pollo Pilgrim's Pride', available at: www.informador.com.mx/economia/2008/58958/6/se-declara-en-quiebra-productor-de-pollo- pilgrims-pride.htm, accessed 24 June 2017

FAO (United Nations Food and Agricultural Organization). (2017) *Food Outlook Report*, available at: www.fao.org/3/a-i7343e.pdf, accessed 22 June 2017

Fortune. (2017) 'Fortune 500', available at: http://fortune.com/fortune500/, accessed 12 12 April 2017

Freitas G., Freitas, T. and Godoy, D. (2016) 'JBS feels blow as Brazil's Batista Boys mired in a new probe', available at: www.bloomberg.com/news/articles/2016-09-14/batista-boys-become-liability-with-jbs-mired-in-criminal-probe, accessed 26 June 2017

Freitas, T. and Margues, F. (2017) 'JBS's Batista turns himself into police after arrest order', available at: www.bloomberg.com/news/articles/2017-09-10/jbs-batista-closer-to-jail-as-supreme-court-orders-arrest, accessed 12 September 2017

Friedland, W. (1984) 'Commodity systems analysis', in H. Schwarzweller (ed.) *Research in Rural Sociology: Vol. 1*, JAI Press, Greenwich, pp. 221–235

Friedland, W. (2001) 'Reprise on commodity systems methodology', *International Journal of Sociology of Agriculture and Food*, vol 9, pp. 82–103

Friedmann, H. (2005) 'From colonialism to green capitalism: Social movements and the emergence of food regimes', in F. Buttel and P. McMichael (eds) *New Directions in the Sociology of Global Development: Research in Rural Sociology and Development: Volume 11*, Elsevier Press, Oxford, pp. 227–264

Friedmann, H. and McMichael, P. (1989) 'Agriculture and the state system: The rise and fall of national agriculture', *Sociologia Ruralis*, vol 29, no 2, pp. 93–117

Gutiérrez, T. (2017) 'El Proceso de Financialización y la Consolidación de las Empresas Transnacionales Avícolas: Los Casos de Bachoco y JBS (2005–2014)', *Tesis de Maestría en Economía Regional. Centro de Investigaciones Socióeconomicas*, Universidad Autónoma de Coahuila, México

Harvey, D. (2005) *A Brief History of Neoliberalism*, Oxford University Press, Oxford

Heffernan, W. (1984) 'Constraints in the poultry industry', in H. Schwarzweller (ed.) *Research in Rural Sociology and Development: Vol 1*, JAI Press, Greenwich, pp. 237–260

Heffernan, W. (2000) 'Concentration of ownership in agriculture', in F. Magdoff, J. Foster and F. Buttel (eds) *Hungry for Profit: The Agribusiness Threat to Farmers, Food, and the Environment,* Monthly Review Press, New York, pp. 61–76

Howard, P. (2016) *Concentration and Power in the Food System: Who Controls What We Eat?* Bloomsbury Academic, New York

Hua, S. and Carrera, J. (2016) 'Leading Brazilian multinational enterprises: Trends in an era of significant uncertainties and challenges', available at: http://ccsi.columbia.edu/files/2013/10/EMGP-Brazil-Report-2015-Jan-27-final.pdf, accessed 15 May 2017

Human Rights Watch. (2005) 'Blood, sweat and fear: Workers' rights in US meat and poultry plants', available at: www.hrw.org/en/reports/2005/01/24/blood-sweat-and-fear, accessed 13 November 2008

Infocampo. (2014) 'Noticias Ganaderas. JBS y Pilgrim´s Pride compran negocio de aves en México y Brasil', available at: www.infocampo.com.ar/jbs-y-pilgrim-s-pride-compran-negocio-de-aves-en-mexico-y-brasil/, accessed 24 June 2017

J&F Investimentos. (2017) 'J&F Investimentos', available at: http://jfinvest.com.br/who-we-are/our-history/?lang=en, accessed 24 June 2017

JBS. (2009) 'JBS SA announces the acquisition of Pilgrim's Pride Corp. and the Association with Bertin SA', www.slideshare.net/jbsri/acquisition-of-pilgrims-pride-and-association-with-bertin-sa, accessed 23 April 2010

JBS. (2010) *Annual Report 2010,* available at: http://quote.morningstar.com/stock-filing/Annual-report/2010/12/31/t.aspx?t=XBSP:JBSS3&ft=&d=a637d810e2d107b125a98d3e07ac0c7f, accessed 23 June 2017

JBS. (2011) *Annual Report 2011,* available at: http://quote.morningstar.com/stock-filing/Annual-Report/2011/12/31/t.aspx?t=XBSP:JBSS3&ft=&d=9d96129c2922f5ce2b7d1e9c556be569, accessed 24 June 2017

JBS. (2014) *Annual Report 2014,* available at: http://quote.morningstar.com/stock-filing/Annual-Report/2014/12/31/t.aspx?t=XBSP:JBSS3&ft=&d=e07878d6b1f2c1401e6d71b6fc1ba670, accessed 26 June 2017

JBS. (2016) *JBS History,* available at: http://jbssa.com/about/history/, accessed 20 March 2017

JBS. (2017) 'Composição Acionária e Societária', available at: http://jbss.infoinvest.com.br/static/ptb/composicao-acionaria-e-societaria.asp, accessed 29 March 2017

JBS Five Rivers. (2017) *Company History. JBS Five Rivers Cattle Feeding LLC,* available at: https://fiveriverscattle.com/pages/companyhistory.aspx, accessed 24 June 2017

Jelmayer, R. and Magalhaes, L. (2014) 'JBS meat producer build in acquisitions', available at: www.wsj.com/articles/meat-producer-jbs-continues-expanding-1416593212, accessed 27 June 2017

Krugman P. (2015) 'What the 1 per cent do not want you to know', available at: www.youtube.com/watch?v=QzQYA9Qjsi0ntervie, accessed 23 June 2017

Lazonick, W. (2010) 'Marketization, globalization, financialization: The fragility of the US Economy in an era of global change', available at: www.ie.ufrj.br/intranet/ie/userintranet/hpp/arquivos/lazonick.pdf, accessed 24 June 2017

Lazonick, W. and O'Sullivan, M. (2000) 'Maximising shareholder value: A new ideology for corporate governance', *Economy and Society,* vol 29, pp. 13–35

Leahy, J. (2016) 'Brazilian corruption probe spreads to meatpacker JBS', available at: www.ft.com/content/af269be6-3fa3-11e6-93e1-426e2e7bb8ad, accessed 27 March 2017

Little, P. and Watts, M. (eds) (1994) *Living Under Contract*, University of Wisconsin Press, Madison, WI

McMichael, P. (2005) 'Global development and the corporate food regime', in F. Buttel and P. McMichael (eds) *New Directions in the Sociology of Global Development, Research in Rural Sociology and Development: Volume 11*, Elsevier Press, Oxford, pp. 265–300

McMichael, P. (2009) 'A food regime genealogy', *Journal of Peasant Studies*, vol 36, no 1, pp. 139–169

McMichael, P. (2013) 'Land grabbing as security mercantilism in international relations', *Globalizations*, vol 10, no 1, pp. 47–64

McMichael, P. (2014) 'Rethinking the land grab ontology', *Rural Sociology*, vol 79, no 1, pp. 34–55

Madgoff, F. (2013) 'Twenty-First Century land grabs: Accumulation by agricultural disposession', available at: https://monthlyreview.org/2013/11/01/mr-065-06-2013-10_0/, accessed 13 August 2017

Margulis, M., McKeon, N. and Borras, S. (2013) 'Land grabbing and global governance: Critical perspectives', *Globalizations*, vol 10, no 1, pp. 1–23

Martínez, F., Aboites, G. and Constance, D. (2013) 'Neoliberal restructuring, neoregulation, and the Mexican poultry industry', *Agriculture and Human Values*, vol 30, no 4, pp. 495–510

Murphy, S. (2006) 'Concentrated market power and agriculture trade', available at: www.iatp.org/files/451_2_89014.pdf, accessed 22 June 2017

Otero, G. (2011) 'Neoliberal globalization, NAFTA, and migration: Mexico's loss of food and labour sovereignty', *Journal of Poverty*, vol 15, no 4, pp. 384–402

Oxfam. (2015) *Lives on the Line: The Human Cost of Cheap Chicken*, available at: www.Oxfamamerica.org/livesontheline, accessed 14 March 2016

Pechlaner, G. and Otero, G. (2008) 'The third regime: Neoliberal globalism and agricultural biotechnology in North America', *Sociologia Ruralis*, vol 48, no 4, pp. 351–371

Pechlaner, G. and Otero, G. (2010) 'The neoliberal food regime: Neoregulation of the new division of labour in North America, *Rural Sociology*, vol 75, no 2, pp. 179–209

Pilgrim's Pride, Inc. (2007) 'Pilgrim's Pride completes acquisition of Goldkist', available at: www.pilgrimspride.com/investors/GoldKist/jan09announcement.aspx, accessed 13 July 2007

Politica. (2017) 'Delação da JBS: Veja todos os vídeos e os principais pontos', available at: https://g1.globo.com/politica/noticia/delacao-da-jbs-veja-os-videos-dos-depoimentos.ghtml, accessed 24 June 2017

Reimund, D., Martin, J. and Moore, C. (1981) *Structural Change in Agriculture: The Experience for Broilers, Fed Cattle, and Processing Vegetables*, Technical Bulletin No. 1648, ERS/USDA, Washington

Reuters America Latina. (2013) 'Datos, adquisiciones de la empresa brasileña de carnes JBS', available at: http://lta.reuters.com/article/domesticNews/idLTASIE95902I20130610, accessed 6 October 2017

Reuters. (2016) 'Brazil's J&F financed Alpargatas deal with government bank loan-paper', available at: www.reuters.com/article/alpargatas-sa-jf-investimentos-idUSL2N1740RU, accessed 28 March 2017

Rojas, A. (2016) 'En Belen comenzó a operar moderno frigorífico de JBS. 2 de octubre del 2016', available at: www.abc.com.py/edicion-impresa/economia/

en-belen-comenzo-a-operar-moderno-frigorifico-de-jbs-1524143.html, accessed 28 June 2017

Serfati, C. (2008) 'Dimensiones Financieras de la empresa transnacional: Cadena global de valor e innovación tecnológica, *Revista ola Financiera*, vol 2, no 4, pp. 111–149

Serfati, C. (2011) 'Transnational corporations as financial groups', available at: www.researchgate.net/publication/290150267_Transnational_corporations_as_financial_groups, accessed 29 June 2017

Serfati, C. (2013) 'La lógica financiero – rentista de las sociedades transnacionales', available at: www.mundosigloxxi.ciecas.ipn.mx/pdf/v08/29/01.pdf, accessed 6 October 2017

Simply Wall St. (2017) 'Is Pilgrim's Pride Corporation's (PPC) financial health a matter of concern?', available at: https://simplywall.st/news/2017/07/12/is-pilgrims-pride-corporations-ppc-financial-health-a-matter-of-concern/m, accessed 6 October 2017

Steger, I. (2014) 'KKR makes investment in Chinese chicken producer,' *Wall Street Journal*, available at www.wsj.com/articles/kkr-makes-investment-in-chinese-chicken-producer-1409054932, accessed 16 March 2018

Stiglitz, J. (2012) 'El precio de la desigualdad', available at: www.elcultural.com/noticias/letras/El-precio-de-la-desigualdad/3921, accessed 22 June 2017

Striffler, S. (2005) *Chicken: The Dangerous Transformation of America's Favorite Food*, Yale University Press, New Haven, CT

Torres, G. (2011) 'El rol de las pequeñas y medianas empresas de la industria avícola en el proceso de la globalización: El caso de la zona metropolitana de Saltillo (1958–2010)', available at: www.cise.uadec.mx/downloads/tesis/tesis 2008-2010/TGAJ.PDF, accessed 22 June 2017

Tyson Foods, Inc. (2005) 'About Tyson International,' available at: www.tyson-foodsinc.com/international, accessed 13 May 2005

Vaz, T. (2010) 'JBS compra Grupo Toledo por 11 millones de Euros', http://exame.abril.com.br/negocios/jbs-compra-grupo-toledo-11-milhoes-euros-588896/, accessed 22 June 2017

Watrous, M. (2016) 'Pilgrim's Pride to acquire GNP Company for 350 Million', available at: www.foodbusinessnews.net/articles/7244-pilgrim-s-pride-to-acquire-gnp-company-for-350-million, accessed 28 June 2017

Watt Poultry. (2016) 'Poultry Trends Highlights', available at: www.wattglobal media.com/news/2016-watt-poultry-trends-highlights-new-exclusive-data-poultry-egg-markets/, accessed 14 July 2017.

10 Between cooperative and private equity capital in the Norwegian food sector

How do the corporate, cultural and political environments make a difference?

Reidar Almås

Introduction

In this chapter, I analyse the role and significance of private equity firms and other capital companies in agri-food supply chains, focusing on the food manufacturing sector in Norway. Burch[1] and Lawrence (2007a, 2007b) have argued that power in supply chains has shifted from the food manufacturers to the supermarkets in most industrialized societies, with significant impacts on farmers, processors and consumers. However, I will raise a number of questions about the nature of this transformation, and suggest that it did not necessarily involve a permanent shift in control. I will argue that while the retail sector could, in Winson's (1993) terms, be characterized as 'the Masters of the Food System', there were other players emerging, who were in a position to challenge that dominant role. I have noted the growing importance of private equity companies which, in purchasing supermarkets, food suppliers, and fast food companies, were able to significantly restructure the scale, the spread of ownership, and the range of operations of companies in the food processing sector. The private equity companies – such as the Texas Pacific Group (TPG), Kohlberg, Kravis and Roberts (KKR), Blackstone, Bain Capital and many others – came to be known as the 'Masters of the Universe'.[2] Conceptually the 'Masters of the Universe' may trump the 'Masters of the Food System'. However, I doubt if that is the case in Scandinavia, with its special regimes of regulated capitalism (Barth *et al.*, 2014). The question may be to what degree financialization is happening from country to country, and from one business and regulatory environment to another.

In taking control of any company, whether it is in the food sector or not, private equity companies have only limited aims. They are not, for example, interested in taking control of food processors in order to better manage these companies in the medium to long term, or to compete more effectively with other food manufacturers. In fact, private equity companies are not particularly interested in the mechanics of production at all; they do not take control in order to expand their holdings in any expanding sector. Rather, private equity groups see any company they take over as 'a bundle of assets to be deployed or redeployed depending on the short-run rates of return that can be earned' (Fligstein and Markowitz, 1993, p. 187). Their intent is to 'realize shareholder value' in order to generate high rates of return to their investment in the shortest possible time.

Finance capital, as represented by the private equity companies, claims to be able to shake off and renew capital in other sectors by restructuring under-performing companies, and by realizing and/or liquidating assets that can be more profitably invested in other, more resilient or profitable, areas. However, there are undoubtedly significant impacts all along the agri-food supply chain arising from the acquisition by private equity companies of food manufacturers. This chapter, therefore, represents an attempt to describe and evaluate some of these impacts by analysing two cases in the Norwegian food sector. These examples are:

1 The chicken cluster around Støren in Gauldal
2 The agriculture cluster at Kviamarka, Jæren.

Before proceeding to the analysis it is necessary to clarify a number of conceptual and other issues relating to the growth and operations of the private equity companies, which have come to play such an important role in the operations of the capitalist system in recent years.

Methodological and conceptual approaches to restructuring of the Norwegian food industry

In order to come to the core of what is happening in the competition and restructuring of the Norwegian food industry, the most important insight is contextual knowledge of the socio-cultural conditions and political constellations of its national framework, together with current knowledge of what is happening on the ground in the food industry. There has been a drive towards deregulation in Norway, although the core Norwegian regulatory regime in the food sector remains intact. Norway as well as Sweden and Denmark have had one of the most regulated systems welfare capitalism in the world since the mid-1930s. Coming out of the capitalist crisis of that time, the Scandinavian mode of welfare capitalism, or the Scandinavian model (Barth *et al.*, 2003) has combined social security and capitalist dynamics in a special way of organizing society, and is referred to as the Nordic or Scandinavian model.

The Scandinavian model of capitalism, as defined by Barth *et al.* (2014) is characterized by strong unions, strong employers' organizations, major welfare priorities and expenditures, and routine centrally coordinated collective bargaining between the parties in the workplace. As a consequence, the Scandinavian countries are characterized by small pay differentials, universal welfare schemes that apply to everyone, and a commitment to full employment by the authorities (Esping-Andersen, 1990; Barth *et al.*, 2014). The Scandinavian economies are small, open, economies thus adapted to a system of free international trade, where the conditions in export industries are crucial for wage growth throughout the economy.

In contrast, what can be termed the 'American model' is characterized by a less regulated capitalism with a lean welfare state (Galbraith, 1967; Hall and Soskice, 2001). Even if orthodox Marxist like Miliband (1973) would characterize both models as belonging to the 'capitalist mode of production', I would argue that we have at least two modes of capitalism: the American, which I will call model 1.0, and the Scandinavian, which I will call model 2.0. A chart on the relation between wage compression and welfare generosity introduced by Barth *et al.* (2014) illustrates the difference between model 1.0 (the American) and model 2.0 (the Scandinavian model).

The vertical axis shows Welfare Generosity as measured by the Overall Generosity Index from the Comparative Welfare Entitlements Data set developed by Lyle Scruggs, University of Connecticut (see Scruggs and Allan, 2006). Wage Dispersion is measured by the ratio of the ninth to the

Figure 10.1 Wage dispersion and welfare generosity in selected capitalist countries.
Source: Barth *et al.*, 2015, p. 62.

first decile of gross hourly wages from the OECD earnings data base. Both series are averages over available years from 1976 to 2002.

As this figure shows, the outcomes of those two models or varieties of capitalism are quite different when it comes to wage compression and welfare generosity. In the lower right corner you find the US and United Kingdom (less wage dispersion) and Canada (more welfare) not far away from the US, while you find the three Scandinavian countries in the upper-left corner, with lowest wage dispersion and highest welfare generosity in Norway and Sweden and a little more wage dispersed and less generous welfare in Denmark, close by. Barth *et al.* (2014, pp. 63–65) explain these outcomes by pointing to the certain class compromise between labour and capital in Scandinavia and the institutionalized collective bargaining system and three part collaboration between the opposing interests in the labour market and the state.

It is interesting to investigate how this Norwegian model of welfare capitalism influences activities in the food sector. In my research I have focused upon two regional clusters in the Norwegian food industry. Why study such clusters, and not just individual companies? In a cluster study, the industrial perspective is extended to the local community and the region with the resources, population base and political and economic dynamics. Power use and changed relationships, which are central themes when it comes to financialization, are taking place to varying degrees both between companies and between public and private sectors. By extending the analysis to the regional level, one achieves a geographical and sociocultural overview and insight, while risking the loss of some of the overall national and global perspectives. A regional demarcation was adopted to see if it helps in an understanding of the strategic positioning and power struggle between enterprises and government agencies in the food sector. This chapter does not aim to provide a complete overview of financialization in the Norwegian food chain, but provides a close-up study of two clusters during an important phase of contemporary change in Norway's food industry.

This close-up study of two different food regions provides two facets of a larger picture of today's Norwegian food industry. It is based on a qualitative, contextualized, and historically grounded approach. Specifically, this chapter builds on facts and information in various media sources from the two regions over a 20-year period, deposited with the Norwegian Retriever database.[3] A number of researchers at the Centre for Rural Research have followed these two regions in various research projects (Almås, 1984, 1985, 2002; Brustad and Bjørkhaug, 2015, Frisvoll, 2003; Frisvoll and Rønningen, 2009; Klimek and Bjørkhaug, 2017), making it possible to utilize a great deal of contextual knowledge. During the last research project period from 2010–2014, field studies, company visits and regular interviews were conducted with industry actors, board members, local authorities and politicians. Interviews have been conducted with 28 informants in the two regions (Almås, 2015). In October 2013, a food industry dialogue conference was held at Jæren between the researchers and officers from the municipality and company management. During the entire project period, local and regional media have been carefully monitored.

In what degree has private equity financialization been one of the driving forces behind the restructuring of the Norwegian food industry in the early 2000s? The main idea of private equity acquisitions is to create growth through organizational restructuring over a relatively short period of time and then sell out to gain quick profits (Kaplan and Strömberg, 2009; Burch and Lawrence, 2013; Klimek and Bjørkhaug, 2017). The term private equity is defined as the substitution of private capital for the share capital which underwrites the activities of public companies listed on the stock exchange. Companies operating in most of the developed countries are either privately owned by individuals or cooperatives, or are publicly listed. A privately owned company does not have a formal structure of shareholdings, but may be owned by an individual, a family or a cooperative which finances its operations through accumulated savings, profits or bank borrowings. A publicly listed company, in contrast, finances its operations in part by the sale of shares in the company, which will be traded on the stock market. A private equity take-over usually involves the purchase of all or most of the shares in a publicly listed company, and removal of these shares from the public arena.

As noted earlier, the aim of a takeover by a private equity partnership is not to become engaged in the long-term management of a company, but to leverage the assets of a company in the short term in order to realise potential capital gains to the members of the private equity consortium. They do not seek long-term control, nor are they particularly interested in the long-term viability of any company they take over. In some instances, productive capital (and the jobs that go with this) may simply be eliminated, or stripped of its assets, even where a company may be making a profit, if this is in the short-term interests of the private equity owners (Rossman and Greenfield, 2006, pp. 4–5; Burch and Lawrence, 2013). There is no reason why a private equity company should commit itself to a profitable enterprise in the medium term, if the private equity partners are able to make larger profits from a series of short-term operations.

Another key feature of the private equity model is that the new owners are not subject to the kinds of regulatory requirements to which a publicly listed company must adhere. When a private equity partnership acquires a company listed on the stock exchange, it means that shareholders no longer own tradeable equity in the company. Shareholdings are liquidated and the company de-listed. This means that, as a private company, it is no longer required to produce annual reports, or audited accounts, or to make public its financial position. This means that the new owners of a company have complete control over the disposal of assets and earnings.

Stakeholders versus the shareholders

One major tension within capitalism is the separation of ownership from control. The emergence of publicly listed companies and the growth of the number of shareholders have made available vast amounts of capital for investment. However, it has also led to the emergence of the 'company manager', or chief executive officer, who controls the corporation but does not own it. This notion of the separation of ownership from control and its effects – the tensions it creates between the owners of capital (shareholders) and the managers of the enterprise (the executives) – has a long scientific tradition exemplified by works of Bell (1960), Burnham (1962), Galbraith (1967), and Miliband (1973). Shareholders prefer companies to give back leftover cash through increased dividends, while management often prefers to spend profits on new investments. The academic conceptualization of this dynamic is 'agency theory' (Jensen, 1986; Eisenhardt, 1989).

In an early analysis of the relationship between the governance of the publicly listed company and the emergence of private equity activity, Michael Jensen (1989, p. 5) wrote that: 'The public corporation is a social invention of vast historical importance. Its genius is rooted in its capacity to spread financial risk over the diversified portfolios of millions of individuals and institutions'. By diversifying risks that would otherwise be borne by the owner-entrepreneurs the public corporation can lower the cost of capital. However, these risk-bearing benefits can come at a cost – creating fundamental conflicts of interest between those who bear the risk (shareholders) and those who manage risk (the executives) (Jensen, 1989). Jensen argues that the publicly listed corporation had outlived its usefulness in many sectors of the economy and is being eclipsed. New organizations are emerging in its place, organizations that are not owned by households but large institutions and entrepreneurs: 'Takeovers, corporate breakups, divisional spin-offs, leverages buyouts, and going-private are the most visible manifestations of a massive organizational change' (Jensen, 1989, p. 61). Do we see an end of shareholder capitalism, as private equity buys up publicly listed companies? And what of the farmer-owned cooperative food firms, which are so important in the Norwegian dairy and meat industries? One of the main aims of a private equity owner, after having extracted 'shareholder value', is to return a company to the stock market and realize further profit. However, private equity companies may perform another role, in that they may serve to regenerate and reinvigorate capitalist enterprises. There is a significant difference in the way publicly listed and cooperative companies act in their varying capacities to respond to changing circumstances, and especially in their capacity to innovate.

Cooperatives and management-run stakeholder corporations, in which a wide range of interests are accommodated – such as employees, customers, suppliers, the environment, the wider community, and government – can have their hands tied when it comes to restructuring and innovation. The immediate financial interests of the shareholders may become only one of a number of considerations that have to be taken into account under a regime of good governance and corporate social responsibility. Moreover, the capacity to innovate and introduce flexibility in structuring operations in the use of capital, labour and resources, may be diminished in a context in which farmers, trade unions and governments have to be consulted. In this context, public-equity companies and acquisition funds find a fertile ground to enter and enrich, if they are welcomed.

Innovation, shareholders' and other stakeholders' interests in the short and long run

The issue of innovation is critically important in the modern flexible system of production in which the capacity to respond quickly to changing needs can determine whether or not a company survives. The Schumpeterian (1943) notion of the creative individual who guides the corporation at a time of rapid change, suggests that the privately owned and managed corporation is more likely to be able to manage such change. However, as the cry for shareholder value emerged in the 1990s, a series of 'buy-outs' and takeovers, characterized by asset-stripping, downsizing, and outsourcing, were initiated, described as 'realizing shareholder value'. This 'movement' can be characterized as a process of financialization, with private equity activity key to the reintegration of ownership and control in the period of late capitalism.

Private equity owners, in undertaking tasks like asset-stripping and downsizing, are not driven only by a need to reform capitalism; they are, above all, driven by the vision of a major return on borrowed capital. In the process they may also succeed in regenerating and re-energize companies which may have become inflexible and moribund. To this extent, they also serve to re-energize the capitalist system, increasing profitability and return on investment, while at the same time generating conflicts with suppliers, labour, communities and other competitors, in general.

Another development that challenged the capacity to 'realize shareholder value' in the post-World War II period was the squeeze on profits which arose out of the long period of economic growth and full employment. According to Glyn and Sutcliffe (1972), the Keynesian consensus which governed economic management in the US and Western Europe in the post-World War II period – manifested in high levels of welfare and a strong commitment to public expenditure – led to the situation in which workers were able to gain significant wage increases. This resulted in a reduction in the share of national output going to capital in relative terms as long as the 'Keynesian accommodation' generated full employment. Workers were able to obtain a greater share of company profits as a consequence of their bargaining power. Shareholders, in contrast, found themselves in a relatively weak position vis-à-vis the corporate executives who preferred to retain some of the company's profits for investment, rather than to increase dividends.

Such a threat to capital could only be resolved by a radical transformation that would reverse the gains made by labour. This was achieved by allowing unemployment to grow and reducing welfare expenditures and public spending. This shift to a neoliberal agenda, the task undertaken by 'Thatcherism' in the UK, and 'Reaganomics' in the US, undermined the so-called Keynesian consensus of the 1950s and 1960s. The labour and employment policies introduced by private equity companies, and the historical time frame within which private equity activity was established, are consistent with the neoliberal agenda and with the attempts to restore shareholder value by an attack on working standards and conditions, including the 'downsizing' of the workforce and outsourcing of key activities.

These two great post-war developments – the 'eclipse' of the public corporation and the victory of neoliberalism over Keynesianism in order to reverse the decline in profits – are clearly a part of the same phenomenon: the need to re-assert the primacy of capital over labour, and shareholders over managers, in an attempt to restore and revitalize the capitalist system. However, they do not necessarily operate to the same degree in all countries and situations. On the one hand, in the UK, where the labour force was relatively well-organized, the main focus of the Thatcherist neoliberal agenda was to reduce the power of the trade union movement and to weaken organized labour. In the United States, on the other hand, where labour was not well organized, the emphasis has been on 'a pattern of accumulation in which profit-making occurs increasingly through financial channels rather than through trade and commodity production' (Krippner, 2005, p. 174). In the Scandinavian countries, neoliberalism never fully took hold, even though Keynesian policies came under pressure and the welfare state was transformed, but not rolled back (Barth *et al.*, 2003). The transformative role of private equity has been well illustrated by developments in the agri-food industry, and it is to this sector that our attention is now directed.

Private equity, cooperatives and the agri-food sector

It has been suggested that the private equity sector is particularly attracted to food manufacturers and the retail sector, largely because both of these sectors exhibit strong cash flows which can be accessed in order to repay debt or provide the working capital needed. In addition, many companies in the food sector have significant property holdings, usually in prime 'high street' locations. These property holdings have been seen by private equity companies as 'under-performing assets', whose real value is not usually reflected in the profits or in generating new capital for investment. Private equity operators are always looking for ways to generate ongoing returns on their investment, and there are three main avenues that have been put in place to achieve this goal (Burch and Lawrence, 2013, p. 248).

First, the leader in any private equity consortium is highly exposed, having taken on a high level of debt in order to take over a company. Consequently the leader has a strong incentive to reduce that debt and the levels of interest it carries. This is usually done by selling some of the existing assets of the company, using the returns to reduce debt (Burch and Lawrence, 2013, p. 253) Second, there is a need to make savings and introduce 'efficiencies' in the remaining company operations in order to generate increased profits to make the company more attractive when it comes to refloating it on the stock exchange. Finally, there is the capacity to generate new streams of income going to the private equity partners. This might involve, for example, a retail property being sold to a property company and then leased back (Burch and Lawrence, 2013, p. 254). These three approaches provide a conceptual framework for the analysis of private equity companies and their effects on agri-food companies in the food sector. The three main strategies adopted by the private equity owners as they seek to 'realize shareholder value' can be summarized thus:

1 Divesting existing assets to repay debt;
2 Making savings and introducing 'efficiencies' in company operations;
3 Leveraging existing assets and generating new income streams.

(Burch and Lawrence, 2013, p. 252)

As will be seen in the cases that follow, not all of the above strategies are adopted in every instance. What strategy is adopted, and with what effect, will in large part depend upon the particular opportunities available in each takeover. Not all food companies own or control large amounts of property, so the capacity to leverage that asset in order to generate new income streams will vary. Similarly, the scope for generating efficiencies within any particular company will depend upon a range of factors: are the company's operations labour-intensive or capital-intensive; do workers require high skill levels; or, can a job be done with unskilled, foreign labour? Such considerations are important in determining how a private equity company will operate in any particular situation. I now turn to a consideration of the effect of these strategies in the two case examples from Norway.

Two cases from Norway

So far, we have seen that 'financialization' is not a single phenomenon: There may be several types of capital restructuring and private equity takeovers, as well as structural changes led by cooperative companies or even driven by government agencies and changing policies. Finance companies may act differently from case to case, depending on the firms they approach and the business environment in which they act, as is illustrated below.

Case I: chicken cluster around Støren in Gauldal

When Agnar Østhus came back from the US to his former hometown in order to launch his plans for a chicken industry in 1990, it was a series of lucky circumstances that made his start-up enterprise grow extraordinarily quickly. Midtre Gauldal authorities at the municipality centre Støren, 50 km south of Trondheim, had a decline in population and desperately needed new industries. Throughout the 1900s, the municipality was dominated by primary industries with traditional production and processing of food, wood and wood products as key elements. From the 1970s, both forestry and wood processing had undergone a tremendous rationalization, with a strong decline in employment, both in the primary stage and the processing industry. Midtre Gauldal and surrounding municipalities are comprised of multiple forest and farming villages. Støren is a gateway to two valleys and traffic hub for both the railroad and the E6 highway. After the agricultural heydays in the 1970s and 1980s, workplaces in agriculture and food processing were greatly reduced (Almås, 1984, 2004).

The decline in employment in agriculture and the agricultural processing industry had coincided with the decline in employment in the timber industry. Closures in the farmer-owned food industry in Gauldalen created opportunities for the ambitious entrepreneur, Agnar Østhus. Arriving in 1990, he found supportive politicians and change-willing farmers. In a short time Østhus established Norsk Kylling (Norwegian Chicken) in close cooperation with the Midtre Gauldal municipal authorities – the latter providing support in terms of regulations and planning. This green light from the municipal authorities also motivated farmers in the valley to move into chicken production.

In 1991, Agnar Østhus established Norsk Kylling in an abandoned clothing factory at Støren. Østhus set up franchise agreements to farm chickens with a network of farmers in neighbouring villages, in particular in the agricultural village Soknedal, where he was born. This happened in a period when chain formation in the Norwegian retail trade was gathering pace. The Norsk Kylling slaughterhouse at Støren expanded rapidly, and in 1994 Norsk Kylling became the main supplier to Rema 1000, the fastest growing supermarket chain at that time.[4] Consumption of chicken in Norway had doubled each decade from 1970, and the 1990s were not an exception, as Norwegian chicken production increased from 19,700 tonnes in 1989 to 42,300 tonnes in 2001.[5]

The supply of local labour in Støren was not large enough for the quick expansion of Norsk Kylling, but the influx of refugees and labour immigrants from Eastern Europe created by EU expansion in the 1990s helped to solve this problem. Norsk Kylling actively recruited labour, mostly from Lithuania, Poland and Slovakia. Most people were hired through recruitment companies, and many of them worked for short seasons. This was in the business owner's interest, because of the rapid expansion and seasonal variations in production. But also many workers liked the rotation shifts, which lasted six weeks in Norway and six weeks in their home country. Unionization was actively opposed by corporate leadership and language barriers made it difficult for the Norwegian Labour Union (LO) to engage in dialogue with the workers.[6]

Shortly after their establishment, Norsk Kylling came into the government's spotlight. Several cases of conflict unfolded in the media. These cases that were exposed in the media were negative stories about the company because it broke laws and regulations, and were spread over a period of 20 years, from June 1995 and up to 2015. The cases may be grouped into four general themes:[7]

1 Conflicts with the Labour Inspectorate and the LO about wages and working conditions
2 Conflict with agricultural authorities about chickens licensing
3 Conflicts with the FSA on animal health, animal welfare and food safety
4 Conflicts with the environmental authorities on emissions to air and rivers.

Østhus usually rejected all complaints, claiming that allegations of failures were false or other actors' responsibilities. For example, that the municipality had the responsibility for emissions and pollutions of the waterways. The owner company was taken to court several times by the Ministry of Agriculture and Food, the Labour Inspection Authority, the Food and Safety Authority, and the County of Sør-Trøndelag, due to violations of the law. Even though the public reputation of Norsk Kylling was frayed, this fact did not appear to hurt the consumers' judgement of the company's products. Both the councillors and mayors of Midtre Gauldal municipality defended the enterprise and, as late as in October 2003, the main customer retail chain – Rema 1000 – characterized Norsk Kylling as a model company.[8]

In 2008, however, when the Food and Safety Authority forecast daily fines of NOK10,000 because of unlawful killing methods, purchasing manager Martin Klafstad in Rema 1000 threatened to halt all trade with Norsk Kylling if their practices were not changed immediately.[9] Rema 1000 sought better control of Norsk Kylling, not least out of consideration for the reputation of Rema 1000 and their product brands (see also Brustad and Bjørkhaug, 2015).

From founder capitalism through private equity capital to Rema retail chain industries

After 10–15 years of growth, Norsk Kylling had become a cornerstone company in Midtre Gauldal, with a workforce of 90 persons. Turkey was included in the assortment and, at the turn of 2005/6, Norsk Kylling sales were 6,000 tonnes of chicken and 1,000 tonnes of turkey to a first-hand value of NOK290 million.[10] From its earliest days Østhus ran Norsk Kylling as majority shareholder in a limited company. Because of the quick growth, more capital was needed, and in March 2005 Norsk Kylling was sold to the Finnish private equity company CapMan. Østhus remained in the top management of the company. CapMan's primary backing was capital from Finnish and international pension funds.[11] CapMan was constructed as a classic private risk fund in order to buy up companies considered to be ripe for restructuring and which, after 5–7 years, could be sold for high profit. CapMan then established Cardinal Foods as an umbrella owner company for all their white meat and egg investments in Norway. In addition to the majority owner CapMan, the main shareholders were Synnøve Finden, a private dairy company and the principal owner and founder, Agnar Østhus. Cardinal Foods established their own brand company: 'Den Stolte Hane' ('The Proud Rooster') which became the main supplier to the ICA retail chain RIMI, one of the four retail chains in Norway at that time.

From 2007, the collaboration between cooperatives and government to regulate prices in the Norwegian chicken industry was suspended. This meant that the big producers at the time – Nortura, Jærkylling and Norsk Kylling – had to regulate caging of day-old chicks, which they expected to sell during the next months. In 2010, CapMan started a process to sell themselves out of Norsk Kylling. Cardinal Foods sold its shares in Norsk Kylling to Rema Industries in 2011. In 2012 Agnar Østhus bought out of Norsk Kylling, with Rema Industries taking over 100 per cent of the share capital. During this separation from the Cardinal Foods brand, Rema 1000 developed its own brand, Solvinge, at the same time that the chicken factory at Støren made Kløver chicken for the Bunnpris chain.

Agnar Østhus had, together with another local entrepreneur, Roar Aarhaug, established the Hugaas Group, which consists of a major cluster of companies in farming, food processing, and agricultural construction. After the acquisition of Norsk Kylling, Rema 1000 also purchased 50 per cent of the Hugaas group, and at the present time, the retail chain Rema 1000 stands out as the controlling integrator of the chicken and farm building cluster in Trøndelag County. Today Norsk Kylling and the Hugaas Group, owned by Rema 1000, controls the whole value chain of chicken production except genetics and feed production. In several rounds, both the principal owner Agnar Østhus and the private equity capital firm CapMan have removed themselves completely from the Norsk

Kylling value chain. While the original owner violated environmental and labour regulations, these practices continued with the private equity owners. Shortly after the retail chain takeover, they have had to 'clean up' the company, due to the emergence of new violations of hygiene, payment to turkey farmers and labour legislation.[12]

Case II: the agri-food cluster in Kviamarka at Jæren

The agri-food cluster at Kviamarka, Jæren is known for its agricultural history and was instrumental in modern agronomic breakthroughs in Norway (Almås, 2002). Jæren is noted for its stone-covered and meager soils, occupied by many small farms. Iconic fiction author Arne Garborg (1892, p. 1) described the peasants there as:

> A strong, heavy people who dig through life with pondering and toil, chop in the ground and examine the Scriptures, grind grain from the gravel and hope out of their dreams, believe in the pennies and comfort themselves to God.
>
> (author's translation)

But this barren land and Lutheran pietism were an advantage when new tractor technology, fertilizers and compound feed gathered pace in Norwegian agriculture during the 1950s (Almås, 2002, pp. 131–137). In the northern European context, Jæren, as far as natural history is concerned, is a part of East Anglia in Britain and Jutland in Denmark, which is easy to detect when flying into this part of Norway by plane from England or Denmark. Culturally, the peasant society in southwestern Norway belongs culturally more to Anglo-American farming where belief in God and in hard work were part of a protestant ethic (Weber, 1995).

In the 1950s and 1960s the farmers at Jæren were among the first to employ new technologies, improving the formerly barren land into a food export region. Therefore, it was not a coincidence that the largest and most dynamic agricultural-related industry in Norway was established in this area. This was in Kviamarka in the country's largest farming municipality, Hå.

How the food industry in Jæren was restructured and relocated

In 2001, the cooperative farmer-owned company Prior (later renamed Nortura), began planning one of the world's most modern poultry slaughterhouse at Kviamarka on a site readied for construction. The Norwegian poultry industry, with only a small number of private actors, was under reorganization at this time. Late in 2002 an agreement was signed between Prior and the private operator Jærkylling about building a joint operating company and slaughter facility. The slaughterhouse was located next door to the two companies at Kviamarka. Slaughtered chickens would be transported on conveyor belts through tunnels from the commonly-owned slaughterhouse and over to the processing and packing sections of the two companies. Everything looked to be a fruitful cooperation between a cooperative and a share company, with a very benevolent municipality as facilitator.

The main factor contributing to the large cluster of food-producing firms at Kviamarka was the decision, in 2002, by the cooperative dairy company TINE to replace four dairy plants at Jæren with one.[13] The decision to build a TINE dairy in Jæren sparked disagreement between the municipalities at Jæren. The final TINE board decision – to add the large dairy plant to Kviamarka – came in 2007.[14] The Hå municipality had then been active on many levels for several years, developing plans for food industry expansion in Kviamarka. TINE based themselves in Kviamarka as a consequence of this municipal leadership and vision.

When this decision was taken, a large nursery – formerly to be based in Jæren – was relocated to Kviamarka, adjacent to the dairy. In that way, significant environmental benefits could be achieved through the use of heating and carbon dioxide from the dairy plant. The establishment of the environmental nursery in 2011, was a direct spin-off of the decision to establish the large dairy plant. The dairy had surplus energy that could be harnessed in the energy-intensive greenhouse production of vegetables, and the nursery could use heated spillwater from the dairy. A two-way pipe system was built, eventually to serve all the food industries at Kviamarka through recycling of energy and carbon dioxide. Both environmentally and economically, this was beneficial for several companies. A heating system for the entire industrial area was built, and the industrial area was connected to a pipe system with access to North Sea gas from Kårstø.

Collaboration and contest between private and co-operative capital

The collaboration in chicken slaughtering functioned well for a few years, while the chicken market increased and synergies gave substantial economies of scale.[15] In 2007 Nortura, of which Prior now was a part, was enjoying growth of 12 per cent each year in the chicken market. The private competitor, Jærkylling – which in 2006 had been acquired by private equity company CapMan that owned the brand Cardinal Foods – experienced an equivalent growth, with the consumption of chicken meat in Norway expected to pass that of beef in 2011.[16] However, private equity-owned Cardinal Food invested heavily in their processing plant at Kviamarka, planning to acquire more of the national chicken market. This expansion of Cardinal Foods resulted in a conflict over access to raw materials from the slaughterhouse. Jærkylling had too few primary producers supplying them and had to access chicken carcasses from the Nortura producers. In the consumer market, the Nortura brand name Prior was in fierce competition with Cardinal Foods brand Den stolte hane. These two brands battled to sign agreements with consumer chains. The cooperative agreement was now questioned and in the spring of 2010 the two parties entered a legal battle in the courtroom.[17]

Jærkylling succeeded in the District Court and the two companies reached a settlement in July 2011, stating that for a transitional period of four to seven years Nortura would supply Jærkylling with chickens from their own producers, with Jærkylling seeking to secure its own producers.[18] Jærkylling planned to sign up approximately 20 new producers each year, which should not be at the expense of Nortura and its own 120 chicken producers.[19] Jærkylling expected to take over two-thirds of the annual growth in the Norwegian chicken market, given that at this time 30 new chicken barns a year were erected nationwide. The agreement of 2011 reduced the conflict between Nortura and Jærkylling, allowing for cooperation in chicken slaughtering.[20] In the last couple of years, however, conflict has been reignited as the Norwegian chicken market has levelled off because of public health concerns, and animal welfare criticisms, of the industry.[21]

Changing ownership in the chicken sector with private equity capital as the driving force

Production of chicken and turkey in Norway has grown significantly during the last 25 years, driven by technological, financial and organizational changes. Demand for chicken was small in Norway in the 1960s, and it was not before the late 1970s that chicken consumption in Norway accelerated (Almås, 2002, pp. 213–216). In 1977 the first processing facility for chicken meat was built by the cooperative Prior, but it was not until the 1990s that chicken meat regularly appeared on the Norwegian dinner table. In the market place, where the cooperative Prior was accepted as the dominant player, two new industrial actors began gaining ground: Norsk Kylling and Jærkylling. The restructuring process on the private side started in 2004 when a holding company owned by the Finnish private equity fund CapMan, established 'The Proud Rooster', a brand still sold in shops today.

More than NOK2.2 billion has been invested in Kviamarka – an area of 24 hectares where employment of food workers has passed 700. Further growth in the number of businesses in Kviamarka is virtually exhausted for spatial reasons. Yet, synergistic effects remain. For example, the farmer cooperative Nortura has signed an agreement to acquire Prima's slaughterhouse and building site for NOK600 million, with Nortura planning to build the nation's largest slaughterhouse.[22] Prima Group and Nortura established, at the same time, a joint cooperation company for meat cutting (although they will still compete in finished goods consumer markets).[23] The collaboration between farmers' cooperatives, private actors and governments at Jæren has been unique and productive. Yet, the interactions have not been conflict-free, and there is still strong competition among market participants about supply from primary producers and access to retail chains. The relationship between the farmer-owned company and the private, local rival Jærkylling were mostly cooperative, but there was both conflict and cooperation. After building the common slaughterhouse they attempted to squeeze out the cooperative Nortura by first expanding its operations, and then 'stealing' supply from its rival's network of farmers. These aggressive tactics were maintained after the private equity takeover of Jærkylling.

The emergence of the Norsk Kylling cluster in Gauldal is a recent example of how opening up a closed, national, segment of food production can occur under globalization. It is a variant of the Nordic or Scandinavian model of capitalism. However, after the retail chain Rema 1000 took control of Norsk Kylling in 2011, the company management started to move back to the Nordic/Scandinavian model. True enough, the acquisition of Norsk Kylling was also part of an overall strategy of Rema 1000 to integrate backward in value chain. However, through the rapid expansion period while the company Norsk Kylling had broken a series of laws and developed a bad public reputation, Rema 1000 had to adapt its management practices to the Norwegian model if they wanted to avoid more troubles with government authorities.

Making sense of these changes

From the beginning, Østhus and his companion Aarhaug, had neglected or circumvented Norwegian license rules, ignored notifications of pollution to rivers and air, and had broken Labour Codes through illegal use of overtime, unlawful working conditions and use of surveillance cameras. They had also exploited and neglected Norwegian food laws.[24] Since its formation in 1991, until Rema took over in 2011, the owner had either ignored, or had confronted, the demands of a series of Norwegian governments.

Why did Rema 1000 finally purchase 100 per cent of the shares in Norsk Kylling? Did they do it for profit? This is not likely. As an expression of social responsibility? Again, not likely. To ensure stable supply of raw materials of an acceptable quality? Certainly, but not just because of that. It appears that Rema 1000 primarily purchased Norsk Kylling – and inserted a reputable industrial leader with a solid background from the from Norwegian food industry – in order to build a better reputation and show customers and governments that Norsk Kylling would readjust to Norwegian environmental regulations and labour rights. Rema 1000 could no longer allow Norsk Kylling to be at war with the Norwegian model.

Two previous CEOs lost their positions, despite the fact that their case could end in court, which it did. In this way, Norsk Kylling was brought back into the Norwegian fold.

What does the turnaround to the Norwegian model mean for those working in Norsk Kylling? There are certainly some positives. The enterprise could now be organized by the Norwegian Employers' Organization (NHO). It also meant that the most of the 325 employees were employed directly by Norsk Kylling and could be organized in the labour union.[25] What does this turnaround mean for those 200 producers supplying the company with chicken? Norsk Kylling must listen to farmer producers and treat them with respect. What does the turnaround mean for the Norwegian food industry? The history of Norsk Kylling's first 25 years may be a learning piece about what is acceptable, and what is not acceptable, when it comes to the Norwegian model of food production.

The return to the Norwegian model of food production does not mean that the pressure on competitors and on agricultural cooperatives is diminished, quite the contrary. The turnaround means neither less power nor less profit to Rema 1000 as owner and integrator in the long run. The return to the Norwegian model means more concentration of power and more vertical integration of the value chain for chicken and turkey. Vertical integration under a retail chain is so far the most enduring innovation by Norsk Kylling. Even though Norsk Kylling posted losses of NOK255 million after the clearing process in 2013 and 2014,[26] Rema 1000 gained control of the quality assurance in the production chain. They can push suppliers to improve quality and they do so, in order to make their value chain more safe and efficient. However, in the event of an overproduction crisis, which occurs regularly in the poultry industry, the integrator has the power and control to limit the delivery of day-old chicks to primary producers.

Norsk Kylling provides a good example of the tendency toward 'Americanization' in the mode of food production in Norwegian agriculture. However, this case review of the recent local economic history of Norsk Kylling shows that this mode of production must be adapted to Norwegian food and environmental regulations, labour laws and agreements and the government's requirements in Norway. What we have seen here doesn't verify the claim that globalization leads to the same, Americanized, mode of production everywhere in the world (Cheshire *et al.*, 2007; Constance, 2008). Governmental regulations, communal conditions and the level of organizational strength among farmers and workers are of crucial significance in determining the development path of Norwegian food industries. This case study shows that private equity failed to make inroads in the Norwegian retail sector, such as Burch and Lawrence (2013) indicate with evidence from other parts of the world. On the contrary, the case with Norsk Kylling shows that a retail company is able to fight back against private equity capital and integrate backward in the food supply chain.

The retail chain had to install its own management, different from that under the original ownership and the joint private equity/founder ownership. This case also shows the Norwegian regulatory regime under model two welfare capitalism, (Barth *et al.*, 2014) is able to partly control the development of the food industry during globalization, and even force retail chain owners and private equity capital investors to contribute to descent work relations and to adhere to the mandate of the Food Safety Authorities.

Corporate social responsibility, transparency and workers' rights

It is difficult to avoid the conclusion that in many cases, the leaders of private equity companies have little concern for corporate social responsibility, transparency, and the rights and the livelihoods of their employees. The short-term focus on high returns to capital will often involve a reduction in employment and/or a restructuring of the conditions of the existing workforce. It should come as no surprise that the senior managers and investors in the private equity sector do not have any particular commitment to the notion of transparency and social equality. In the Nordic countries there has been a countervailing force in the form of the strong trade unions, representing workers and other staff members. However, as we have seen in the case studies, it is possible to undermine the activities of organized labour.

A business regime of good governance and corporate social responsibility means that government bodies and trade unions have to be consulted and suppliers, customers and communities considered within a consensual decision-making framework. As we have seen in the Norsk Kylling case, the Rema family-owned retail company bought out both the founder and private equity capital. It took full ownership and control of the company when conflicts relating to food safety and labour rights escalated – a time when there was a real chance that the chicken slaughterhouse would be closed. As a result, the buy-out cost Rema dearly and resulted in great net losses both in 2014 and 2015. However, today all employees are directly employed by Norsk Kylling, the company is a member of the Norwegian employers' organization (NHO) and most of the workers are unionized. Both environmental and safety authorities are satisfied with the operations. The Norwegian model 'won' for now, at the privately owned Norsk Kylling company, in spite of excessive competition and attempts at acquisitions from private equity capital in a stagnating chicken market. Klimek and Bjørkhaug (2017, p187) argue that 'entry barriers protecting the Norwegian market are of particular interest since they reduce and restrict import competition'. They also argue that favourable industry trends in the Norwegian food market with duopolistic growth opportunities, in which one of the dominating cooperatives are challenged, emerges as one important explanation for the private equity investments: 'Private equity strategists focus on top-ten categories in the market, i.e. those with bigger sales volumes ... and maintenance of well-known local brands is a characteristic that makes an agri-food company an attractive target of private equity capital' (Klimek and Bjørkhaug, 2017, p. 187). Do we see signs of an end to shareholder capitalism in Norway, as private equity companies buy up publicly listed companies, as predicted by Jensen (1989) and Eisenhardt (1989)? A review of just two cases does not lend enough evidence for such a conclusion. What may be the case, however, is that Norwegian food companies are not easy targets for private equity buy-ups, as we also have seen in other case studies. A serious attempt by KKR to buy the very large Norwegian publicly listed share company ORKLA in the early 2000s failed because shareholders sided with management to defend the company against unfriendly takeover attacks (Almås, 2013). The interest in buying was also dampened among the potential buyers when knowledge about the significant restrictions on ownership of hydro power resources, and the Norwegian tariffs on raw materials for the food industry, became apparent (Almås, 2013).

From a regulatory point of view, it is necessary to control the 'Masters of the Universe' as the private equity companies are called by some observers. As Clapp (2014) argues, given globalization and the development of a neoliberal agri-food model, an international system of governance is necessary to protect farmers, workers and consumers against the negative impacts of neoliberalism. It is also necessary that there exist legitimate national systems of governance that are able to negotiate national interests and rights, mediating between the development of global trade and local value creation (Bjørkhaug *et al.*, 2015).

There is little sign of the penetration of an American model of capitalism (Constance *et al.*, 2012) penetrating Norwegian food industries. To the contrary we may see a further development of the Scandinavian mode of welfare capitalism, let us call it model 2.1. In addition to being a welfare capitalism with a certain compromise between state and capital, the Norwegian model of welfare capitalism has a distinct compromise between the urban and rural interests. Ever since the crisis settlement between the Labor party and the Farmers' party in 1935, this distinctive compromise between town and countryside characterized the capitalist mode of production in Norway.

It may be argued that Norway still is influenced by Keynesian ideas which continue to impact upon policy by giving strength to the state and labour. However, Keynesianism came to Norway, not just from Western capitalism, but as a heavy influence from the planning system in Soviet Union as well (Colbjørnsen and Sømme, 1933). Introduced by the philosopher Hans Ebbing, the Norwegian model of the double compromise is not only based on a class compromise between labour and capital, but in addition to that also on a compromise between town and countryside interests (Ebbing, 2016). My empirical study lends empirical evidence to the view that the Norwegian model 2.1 deviates from the Swedish and Danish version of Scandinavian welfare capitalism. Findings from a recently published dissertation by Bjørn Klimek also support the thinking of a decisive difference between the Norwegian and Danish model of Scandinavian welfare capitalism (Klimek, 2017, pp. 52–57).

Conclusion

How could it be that Norsk Kylling and Jærkylling/ Den stolte hane (The Proud Rooster) – two food industry companies with the same products – evolved so differently? Norsk Kylling ended up being owned by food retail chain Rema 1000, while 'The Proud Rooster' is owned by a private equity company. But both collaborate with a Norwegian farmer-owned cooperative. The special Norwegian political economy of the agro-food sector is important. Because of the regulatory regime, which was only partly deregulated, private equity capital has lost some interest in the Norwegian food industry. While the chicken price regulation was abolished in 2007, and later the same happened with beef, the high tariff levels protecting the Norwegian production remained. The special regulation regime for agricultural products in Norway supports the thesis that Norwegian welfare capitalism is a special variant of the Scandinavian model. The double compromise between work and capital and between city and country is a characteristic of Norwegian capitalism that strongly affects the way private equity capital is able to penetrate the Norwegian food industry.

I will not propose any final conclusion about the impact of financialization on the Norwegian agri-food sector. There is no doubt that private equity companies can have a major impact on the organization, ownership and operations of all activities along the food chain, from farm to plate. But it is important to note that private equity companies have only a limited interest in this – or any other – particular sector. Private equity partnerships can, however, transform the operations of a company over the long term, and long after the private equity partners have moved on.

The cases above indicate that – apart from making the partners in any private equity company large amounts of money – the most significant role of private equity within twenty-first-century capitalism is to renew and restructure a system which, in the view of private equity, has lost sight of its core business of creating value. Private equity provides an opportunity for capital to become renewed and recharged, with all that that implies for workers, small businesses, farmers and consumers.

However, when it comes to issues such as environmental sustainability, social responsibility, workers' rights, community acceptance, and so on, private equity interests collide with Scandinavian values, social practices and political institutions. Scandinavian labour laws, trade regulation and natural resource management regulation seem to delimit the operation of private equity capital to a certain extent. Equity acquisitions and buy-outs in the food sector are not always successful, partly because of the peculiar Norwegian trade regulation and natural resource management regimes. The presence of cooperatives in both production and retail sectors also seems to have some influence, both as real actors playing the same game, but also because there is less capital and fewer private companies that can ultimately be purchased. In Norway, one-third of the retail food trade is controlled by the semi-Nordic consumer cooperative Coop. In food processing, some 90 per cent of the dairy industry is controlled by farmers' cooperative Tine, while about half of the meat industry is run by farmers' cooperative Nortura. While Sweden and, to an even greater extent, Denmark has deviated somewhat from the 'clean' Scandinavian model in recent years, it appears that the distinctive Scandinavian welfare capitalism in Norway has developed further on the basis of the two compromises between labour and capital and between city and country. The fact that Norway in 1994 remained outside the European Union has helped to consolidate and further develop the Norwegian model, which I in this chapter have called welfare capitalism model 2.1.

Notes

1 I thank Emeritus Professor David Burch for introducing me to this research field during my sabbatical stay in Brisbane in 2007–8 and for his insights into the operations of private equity firms in global agri-food industries.
2 'The new masters of the universe', *Wall Street Journal*, 27 July 2005.
3 See: www.retriever.no/?utm_campaign=Brand&utm_source=google&utm_medium=ppc&utm_term=retriever&utm_content=1048672x858264979648623667.
4 Newspaper Dagbladet 05.11.1996.
5 Budsjettnemnda for jordbruket.
6 Interview with leader of the food industry union Jan-Egil Pedersen (Norsk Nærings- og Nytelsesmiddelarbeiderforbund [NNN]) 11.11.2011.
7 See overview in the newspaper Adresseavisen 09.11. 2007 and 28.10.2013.
8 Newspaper Adresseavisen 10.10. 2003 and 25.10.2003.
9 Newspaper Adresseavisen 19.04.2008.
10 Newspaper Fosna-Folket 11.11.2005.
11 Newspaper Dagens Næringsliv 23.03.2005.
12 Newspaper Adresseavisen 28.12. 2012
13 Newspaper Stavanger Aftenblad 18.08.2002.
14 Newspaper Stavanger Aftenblad 20.11.2007.
15 Terje Haavardsholm, Jærkylling, interview 15.05.2013.
16 Newspaper Stavanger Aftenblad 15.12.2007.
17 Newspaper Stavanger Aftenblad 20.04.2010.
18 Kjøttbransjen (The Meat Industry Journal) 20.04.2012.
19 Newspaper Jærbladet 24.08.2011.
20 Terje Haavardsholm, Jærkylling, interview 15.05.2013.
21 Terje Haavardsholm, Jærkylling, interview 15.05.2013.
22 Newspaper Nationen 05.04.2014.
23 Newspaper Stavanger Aftenblad 22.10. 2014. www.aftenbladet.no/nyheter/lokalt/jaeren/Konkurranse-ja-til-Nortura-og-Prima-3541950.html

24 This has been well documented through court proceedings held by the earlier workmen (top executive) against Norwegian Chicken. See for example Addresseavisen in the period 29.10 2013 to 20.11.2013.

25 CEO Frode S. Martinsen, interview 26.11.2012.

26 Newspaper Adresseavisen 05.03.2015.

References

Almås, R. (1984) Budalsrapporten: Eksempel på lokal utvikling i eit avgrensa bygdesamfunn, Norsk senter for bygdeforskning, Trondheim

Almås, R. (1985) Evaluering av lokalt utviklingsarbeid: Erfaringer frå Budal, Mausundvær og Salsbruket, Norsk senter for bygdeforskning, Trondheim

Almås, R. (2002) *Norges landbrukshistorie*, Samlaget, Oslo

Almås, R. (ed.) (2004) *Norwegian Agricultural History*, Tapir Academic Press, Trondheim

Almås, R. (2013) 'Private equity in the Nordic food sector: How do the corporate, cultural and political environments make a difference?' Paper presented at the Australasian Agri-Food Research Network Conference, Melbourne, 3 December

Almås, R. (2015) 'Regional samhandling, nyskaping og utviklingsbanar i tre matklynger: Gauldalen, Frøya og Kviamarka', in H. Bjørkhaug, R. Almås and J. Vik (eds) *Norsk Matmakt i Endrin*, Fagbokforlaget, Bergen, pp. 391–411

Barth, E., Moene, K. and Wallerstein, M. (2003) *Likhet under press. Utfordringer for den skandinaviske velferdsmodellen.* Oslo: Gyldendal Akademisk

Barth, E., Moene, K. and Willumsen, F. (2014) 'The Scandinavian model – an interpretation', *Journal of Public Economics*, vol 117, September 2014, pp. 60–72

Bell, D. (1960) *The End of Ideology*, The Free Press, New York

Bjørkhaug, H., Almås, R. and Vik, J. (2015) Norsk Matmakt i Endring: Nyskaping og Endrede Maktforhold i den Norske Matvarekjeden, Fagbokforlaget, Bergen

Brustad, S. and Bjørkhaug, H. (2015) 'Omdømme, makt og mulighetsrom for bærekraftig styring – om Norsk kylling i media', in H. Bjørkhaug, R. Almås and J. Vik (eds) *Norsk Matmakt i Endring*, Fagbokforlaget, Bergen, pp. 175–198

Burch, D. and Lawrence, G. (eds) (2007a) *Supermarkets and Agri-food Supply Chains: Transformations in the Production and Consumption of Foods*, Edward Elgar, Cheltenham

Burch, D. and Lawrence, G. (2007b) 'Supermarket own brands, supply chains, and the transformation of the agri-food system', *International Journal of Sociology of Agriculture and Food*, vol 13, no 1, pp. 1–18

Burch, D. and Lawrence, G. (2013) 'Financialisation in agri-food supply chains: Private equity and the transformation of the retail sector, *Agriculture and Human Values*, vol 30, no 2, pp. 247–258

Burnham, J. (1962) *The Managerial Revolution*, Indiana University Press, Bloomington

Cheshire, L., Higgins, V. and Lawrence, G. (eds) (2007) *Rural Governance: International Perspectives*, Routledge, London

Clapp, J. (2014) 'Financialisation, distance and global food politics', *Journal of Peasant Studies*, vol 41, no 5, pp. 797–814

Colbjørnsen, O. and Sømme, A. (1933) *En norsk treårsplan*, Det Norske Arbeiderpartis Forlag, Oslo

Constance, D. (2008) 'The Southern model of broiler production and its global implications', *Culture and Agriculture*, vol 30, nos 1–2, pp. 17–31

Constance, D., Konefal, J., Heffernan, W. and Rainey, A. (2012) 'Monopsony opportunism and debt slavery in the poultry industry: The origins of flexible accumulation in agri-food', Paper presented at the *XIII World Congress of Rural Sociology*, Lisbon, Portugal, 31 July

Ebbing, H. (2016) 'Liberaliseringen av Norge – Den norske modellen» og dens forutsetninger,' *Vardøger*, vol 36

Eisenhardt, K. (1989) 'Agency theory: An assessment and review', *Academy of Management Review*, vol 1, no 4, pp. 57–74

Esping-Andersen, G. (1990) *The Three Worlds of Welfare Capitalism*, Princeton University Press, Princeton NJ

Fligstein, N. and Markowitz, L. (1993) 'Financial reorganisation of American corporations in the 1980s', in W. Wilson (ed.) *Sociology and the Public Agenda*, Sage, Newbury Park, pp. 185–206

Frisvoll, S. (2003) *Beslutningsprosesser i havbruket – lokale eieres valg i møte med globale og lokale forhold*, Rapport 07/03, Norsk senter for bygdeforskning, Trondheim

Frisvoll, S. and Rønningen, K. (2009) *Kampen om gråsonene: Oppdrett i Froan landskapsvernområde*, Rapport 08/09, Norsk senter for bygdeforskning, Trondheim

Galbraith, J. (1967) *The New Industrial State*, Hamish Hamilton, London

Garborg, A. (1892) *Fred*, Aschehoug, Oslo

Glyn, A. and Sutcliffe, R. (1972) *British Capitalism, Workers and the Profits Squeeze*, Penguin Books, Harmondsworth

Hall, P. and Soskice, D. (2001) 'An introduction to varieties of capitalism', in P. Hall and D. Soskice (eds) *Varieties of Capitalism: The Institutional Foundations of Comparative Advantage*, Oxford University Press, Oxford, pp. 1–6

Jensen, M. (1986) 'Agency costs of free cash flow, corporate finance, and takeovers', *American Economic Review*, vol 76, no 2, pp. 323–329

Jensen, M. (1989) 'The eclipse of the public corporation', *Harvard Business Review*, vol 67, no 5, pp. 61–74

Kaplan, S. and Strömberg, P. (2009) 'Leveraged buyouts and private equity', *Journal of Economic Perspectives*, vol 23, no 1, pp. 121–146

Klimek, B. (2017) 'Food industry structure in Norway and Denmark since the 1990s. Path dependence and institutional trajectories in Nordic food markets' PhD dissertation, University of Oslo, Oslo

Klimek, B. and Bjørkhaug, H. (2017) 'Norwegian agri-food attracting Private equity capital: Varieties of capitalism – varieties of financialisation? ' *Sociologia Ruralis*, vol 57, no 2, pp. 171–190

Krippner, G. (2005) 'The financialisation of the American economy', *Socio-Economic Review*, vol 3, pp. 173–208

Miliband, R. (1973) *The State in Capitalist Society*, Quartet Books, London

Rossman, P. and Greenfield, G. (2006) *Financialization: New Routes to Profit, New Challenges for Trade Unions*, Global Labour Institute and the International Union of Food, Agricultural, Hotel, Restaurant, Catering, Tobacco and Allied Workers' Associations, Geneva

Schumpeter, J. (1943) *Capitalism, Socialism and Democracy*, Routledge, London

Scruggs, L. and Allan, J. (2006) 'Welfare-state decommodification in 18 OECD countries: A replication and revision', *Journal of European Social Policy*, vol 16, no 1, pp. 55–72

Weber, M. (1995) *Den protestantiske etikk og kapitalismens ånd*, Pax, Oslo

Winson, A. (1993) *The Intimate Commodity*, Garamond Press, Toronto.

Part III

Financialization of farming and forestry

11 Old roots, new shoots

Thickening the local histories of agri-food financialization

Melanie Sommerville

Introduction: two debates about agri-food financialization

This chapter begins at the crossroads of two debates in recent scholarship on the financialization of agri-food systems. The first debate emerged in a dialogue between Brett Christophers (2015a), a frequent contributor to the broader academic literature around financialization, and three agri-food scholars: Geoffrey Lawrence (2015), Madeleine Fairbairn (2015), and Stefan Ouma (2015). In his piece, entitled 'The Limits to Financialization', Christophers raises two questions that his respondents subsequently consider. Christophers' first question regards the degree to which financialization as a concept generates new insights into the cultural and political economic dynamics of contemporary capitalism – in the author's words, 'where does financialization … sit on the spectrum between powerful and innovative theory at one extreme and superficial and redundant label at the other?' (Christophers, 2015a, p. 187). Christophers' second question, in turn, is whether the (sometimes divergent) empirical phenomena that scholars refer to as financialization are in fact 'unique' to our current conjuncture, or may rather be preceded by 'historic parallels and (dis)continuities' (Christophers, 2015a, pp. 191–192). Christophers expresses some doubt on each of these counts. He suggests, among other points, that scholarly presumptions as to the theoretical utility of financialization may be undermined by the temporally restricted lens through which many researchers have assessed these phenomena. Our agri-food studies colleagues diverge in their responses to Christophers' qualms. Lawrence (2015, pp. 201–203, and see Chapter 2 this volume) finds financialization 'indispensable' to theorizing insofar as the concept has helped to identify new 'actors', 'products', and 'instruments' in agri-food systems as well as new appropriations of 'power and wealth' therein. Fairbairn (2015, p. 211) takes a similar stance, acknowledging that while our current situation has some 'parallels' with past instances of financial expansion, it also has distinctive elements that can helpfully be clarified and shorthanded using the concept. Ouma (2015, p. 227), by contrast, is more cautious, suggesting that many recent works have employed an 'indiscriminate' use of financialization that risks overstating finance's interest in agri-food systems. He points to the need for more careful examinations of finance's interweaving with these systems, including by focusing on the 'practical activities of financial economization' and on the barriers that finance may encounter therein (Ouma, 2015, p. 225).

Ouma's response to Christophers picks up threads from a larger critique that he has leveraged in pieces published in the years surrounding the aforementioned dialogue (Ouma, 2014, 2016 and see Chapter 6 this volume). These pieces are interlocutors in the second debate that this chapter aims to reflect upon and extend. In his critique, Ouma suggests that early accounts of agri-food financialization have tended towards determinism, a problem he links to the structuralist political economy framing of many early accounts. Drawing on the work of Gillian Hart (2004), Ouma (2014, pp. 163–164) states that this approach risks perpetuating an 'impact' model of financialization that frames the phenomenon as a set of processes imposed 'from above'. One result is that finance is portrayed as a sector that is 'external' to agriculture when in fact the two have been entangled for some time (Ouma, 2016, p. 83). Ouma (2014, p. 164) argues in favour of more nuanced accounts that attend to the 'more technical, everyday dimensions' of financialization, including the heterogeneous agency (and resistance) of farmers among other agricultural actors.[1] He suggests that agri-food scholars might usefully explore a 'middle ground', a recommendation also made by Fairbairn in her reflection on the historical novelty of financialization. For Ouma (2015, p. 228), such ground would incorporate both the abstract and quotidian elements and practices of financialization. For Fairbairn (2015, p. 213), the midpoint exists between focusing 'myopically on the events of the last few decades and examining recent financial developments as carbon copies of their historical counterparts'. In his response to these commentators, Christophers largely agrees. While he remains sceptical about Lawrence's insistence that financialization is 'crucial to understanding particular, historical–geographical developments', he suggests that the 'worth of [the concept] lies in its application' (Christophers, 2015b, p. 229). Christophers (2015b, pp. 229–230) suggests that, following Fairbairn, such application ought to be 'comparative and connective rather than isolated and disconnected' and, following Ouma, should avoid 'black box[ing] the micro-practices involved'.

In this chapter, I aim to further these debates by reviewing and extending recent research into the financialization of farmland and agriculture in the Canadian prairie province of Saskatchewan. My contention is that the emerging body of work that focuses on this province has begun to sketch out an historically nuanced and grounded approach to financialization, such as Christophers, Fairbairn and Ouma suggest ought to be prioritized in future studies. My own work alongside that of other prairie researchers has carefully traced both the political reforms and economic restructuring processes that undergird a longer running interpenetration of farming and finance in Saskatchewan, while highlighting the ways that these shifts have opened the door to a new round of investor activity in the province. The studies in question have provided both a sense of the scope of new investment activities and detailed accounts of the new financial vehicles that have emerged over the past 15 years as well as their underlying assumptions, profit models, and operational mechanics. I argue that while this work has begun to embed financialization within ongoing trajectories of agrarian transition, it could still go further towards providing a thickened local history of financialization that recognizes farmers as among its key agents. I make a modest start at this task by considering the shifting cultures and subjectivities of prairie farmers and the ways that recent developments may re-substantiate – as well as challenge – current ideas about the 'best practices' of farming and what makes for an 'outstanding farmer'. Part synthesis of existing research and part identification of areas where further clarification is needed, this chapter aims to contribute to lasting theorizations of agrarian change in Saskatchewan at a time when some surmise that investment activity in the province is beginning to taper off (see Magnan and Sunley, 2017). I close the chapter by considering how the prairie approach might be extended beyond its current focus on land access and production in the grain and oilseeds sector, including by examining different agricultural sectors, by exploring farmers' broader set of interactions with finance during agricultural operations, by integrating still-longer time scales, and through scholarly work in different geographic areas.

Financialization, farmland and primary agriculture in Saskatchewan

Agriculture has long been a mainstay of the Saskatchewan economy. The province contains 47 per cent of Canada's crop land and is dominated by grain and oilseeds production (comprising 62 per cent of all farms and 80 per cent of gross farm receipts) (Statistics Canada, 2017). Saskatchewan's agri-food systems became something of a darling among investors in the mid-to-late 2000s, when rising commodity prices seemed to foretell a new era for agriculture (Sparling and Uzea, 2012) that could boost the province's political economic importance from fly over region to Canada's 'big middle' (Pitts, 2011). Canola, Saskatchewan's leading crop, was pronounced the seed of 'the other oil boom' (Pitts, 2011) and agriculture the seat of a new group of potential 'soil tycoons' (Lappano, 2009). Investment capital flocked to the province amid these reports, drawn by the province's large quantity and high quality of farmland, its affordability relative to neighbouring provinces and states, its strong processing and transportation infrastructure, and its favourable location relative to regional markets (Sommerville and Magnan, 2015).

The diverse investment configurations that have emerged in Saskatchewan since the mid-2000s have involved both farmland ownership and participation in agricultural production. At one end of the range lie variations on what Fairbairn (2014) calls the 'own-lease out' model, wherein investors acquire a diversified portfolio of farmland and lease it to tenant farmers on anticipation of rental payments and capital gains through land appreciation. Among the most well-known practitioners of this model are a set of private equity farmland investment funds managed by three companies: Assiniboia Farmland Limited Partnership (established in 2005), AgCapita Farmland Investment (2005), and Bonnefield Financial Inc. (2010) (Sommerville and Magnan, 2015). While Assiniboia sold its full land portfolio to the Canadian Pension Plan Investment Board (CPPIB) in 2014 (see Assiniboia, 2013),[2] Bonnefield continues to operate several investment partnerships. Similarly, by 2015 AgCapita had sold off two of its portfolios to another investment fund called TopSoil Farm Land Management, but continues to manage other partnerships (Magnan and Sunley, 2017). Operating alongside these funds in farmland markets are certain wealthy individuals. Based on their detailed analysis of land titles in the province, Annette Desmarais and colleagues identify an Alberta resident named Robert Andjelic as Saskatchewan's single largest farmland owner in the province (Desmarais *et al.*, 2015).[3] Practitioners of the 'own-lease out' model are not typically involved in farming their amassed land, although some do exercise extensive oversight with their tenants (Sommerville and Magnan, 2015). Whether the underlying capital originates with institutional or individual sources, 'own-lease out' configurations are typically considered lower-risk investment ventures since farmland values in Saskatchewan have historically been more stable than have commodity prices and are less influenced by production factors such as weather and pests.

At the other end of the investment range are those configurations that are actively involved in farming the land that they acquire. These investment models are similar to those that Fairbairn (2014) calls 'own-operate' except that – in keeping with an important feature of farming in Saskatchewan that I will return to later in this chapter – the farms in question often rely on a blend of owned and leased land. The most widely recognized examples of this arrangement are One Earth Farms (OEF, founded in 2009) and Broadacre Agriculture Inc. (2010), which each established very large farms spread across multiple locations in the province. OEF, which farmed almost exclusively on land leased from First Nations, was established by Sprott Resource Corp., a natural resource-focused private equity fund traded on the Toronto Stock Exchange (Magnan, 2012; Sommerville and Magnan, 2015; Sommerville, 2018). The company subsequently attracted capital from several regional agribusinesses as well as the CAPE Fund, a private equity fund focused on First Nations development. Broadacre was founded by Pike Management Group (PMG), a consultancy business that expanded into farm management and farmland investment (Magnan, 2015). The company grew in part by absorbing another large farming operation called Wigmore Farms in 2012 (Rowland, 2008). Wigmore had itself started life as a family farm before expanding to incorporate the land of two other family operations (see Magnan, 2012), one of which had a sale-lease back agreement with Assiniboia (Rowland, 2008). By 2014, Broadacre owned about 13 per cent of its land base and leased the other 87 per cent (see Marshall, 2014), with the leased land including properties owned by PMG (Magnan, 2015).[4] Investors in these models – which I will call 'access-operate' configurations[5] – aim to benefit from large-scale commodity production, appreciation on the land they own, and/or capital gains on the eventual sale of their business as a going concern. 'Access-operate' models are typically considered higher risk than the previously discussed 'own-lease out' ventures because investors have greater exposure to commodity prices and production risks. While this assessment would seem to be borne out by One Earth Farm's exit from crop operations by 2014 – and Broadacre's filing for creditor protection that same year – analysts have, instead, tied each company's failure to a broader range of factors, including management problems (Pratt, 2014; Guenther, 2015; Magnan, 2015; Sommerville, 2018).

This second category of 'access-operate' investment arrangements has a blurry boundary with a third group of investors operating in Saskatchewan. Desmarais *et al.* (2015, p. 35) call these investors 'farmer/investor hybrids', describing them as 'large, rapidly growing entities that have only recently bought land' in the areas studied by the authors. While these investors 'appear to be taking a role in managing the farming operations on their lands', the authors distinguish them from 'long-established, usually smaller and slower expanding family farms', suggesting that in addition to farming revenues, 'farmer/investor hybrids' are likely also motivated by the potential for 'long-term returns on their sizable land investments' (Desmarais *et al.*, 2015, p. 35). This third investment model, which I will return to later in the chapter, highlights a particular strength of prairie financialization research: its exploration of the relationships between farmers (among other agri-food interests) and recent investment models. As James Williams (2014, p. 407) has pointed out, many early agri-food financialization analyses have tended to construe farmers as 'not only passive victims of the financialized food system … but also as a singular and undifferentiated group' which 'overlooks important differences among farmers [as well as] their relative assertions of agency'. In contrast, the research program developing in Saskatchewan has emphasized both the role of farmers as participants within – or partners to – financial vehicles, as well as certain connections and parallels between these vehicles and larger changes in the provincial farming sector.

But before turning to that matter, it is important to provide some inkling of the scale of the recent agri-investment activity in the province.

Here, we can look to Desmarais *et al.* (2016, p. 157) whose analysis reveals a 16-fold increase in investor-owned land between 2002 and 2014 to encompass approximately 338,700 hectares. This land is distributed unevenly across the province. In three rural municipalities (RMs) examined by the same authors, investors owned between 8 per cent and 13 per cent of the farmland (Desmarais *et al.*, 2015, p. 32). In turn, André Magnan and Sean Sunley (2017, p. 98) calculate that, between 2003 and 2014, 'investor purchases accounted for 15 per cent or more of arms-length farmland transactions [in terms of area of land acquired] in five out of sixteen RMs [that were] examined'. While these accounts offer a necessarily partial view of investment activity in Saskatchewan – insofar as they do not include the area leased by certain investors (see above) – they nevertheless help to contextualize the discussion that follows.

Embedding financialization in longer trajectories of agricultural restructuring

Any account of contemporary agrarian change in Saskatchewan needs to consider the origin of rural landholdings and prairie farming in processes of colonization and settlement that dispossessed First Nations peoples. Desmarais *et al.* (2016) trace existing patterns of family farming to the 1872 *Dominion Lands Act*, which privatized farmland and granted it to European settlers willing to clear it for agriculture. The signing of the *Numbered Treaties* (1874–1906) and passing of the 1876 *Indian Act* created the reserve system comprising native lands. Although First Nations had hoped that treaties would enable them to pursue farming, their agricultural ambitions were hampered by discriminatory federal policies extending until after World War II (see Carter, 1990 for a detailed overview). First Nations farmers have continued to face barriers in the decades since. Particularly relevant for financialization scholars are the difficulties that First Nations face obtaining agricultural credit given the collective and inalienable form of land title existing on reserves. Together with deskilling, rural racism, and challenges in accessing state supports for farming, these difficulties have differentially exposed these farmers to restructuring pressures in recent decades, even as First Nations' have increased their rural landholdings through land claims settlements (Sommerville, 2018). The result is that where on-reserve farms continue to exist, many struggle with viability, and in many cases First Nations have exited farming and now lease their farmland to non-native tenants. This history of dispossession and discrimination is a condition of possibility of all recent financial ventures involving farmland in the province, whether on-reserve or off. But it also created particular opportunities for OEF, which presented itself as offering a form of redress for First Nations in order to secure reserve land and, to a lesser degree, labour from these communities as well as capital from would-be investors (Sommerville, 2018).

While Saskatchewan's farm sector remains dominated by family-run operations, it has undergone considerable consolidation over the past seven decades. Farm numbers have declined steadily and farm sizes increased as some families have exited the sector and others have absorbed available land (Magnan, 2012; Sommerville and Magnan, 2015). Although higher commodity prices over the last several years have improved revenues, this followed three decades of low and sometimes negative farm incomes, when prices stagnated and Canada's increasingly concentrated farm input and processing sectors continued to extract profits (NFU, 2005). Historically, Saskatchewan has been prone to droughts, but in more recent years flooding and excess soil moisture have troubled flatlander farmers, preventing large swaths of the province from being seeded. A neoliberal shift in policies since the 1980s has accentuated many farmers' difficulties. This shift has included the implementation of free trade (NAFTA, WTO), the elimination of grain transportation subsidies (the Crow Rate), several overhauls of farm income stabilization programs, the privatization of farmer-owned grain handling cooperatives (Saskatchewan Wheat Pool), and the removal of single desk wheat and barley marketing through the Canadian Wheat Board (Desmarais *et al.*, 2015, p. 21).

Despite these changes, farmland values in Saskatchewan remained relatively stable, posting moderate but steady increases amid favourable taxation policies and low interest rates over much of the past 30 years.[6] Farmland values have also been strongly conditioned by the *Saskatchewan Farm Security Act*, which regulates the ownership and leasing of farmland in the province (see Desmarais *et al.*, 2016 for a comprehensive review). Briefly, restrictions put in place in 1974 in response to a run up in farmland prices restricted ownership and leasing of large land parcels to provincial residents. These regulations were partially loosened in 2002, opening the farmland market to Canadian citizens and permanent residents. As Desmarais *et al.* (2016, p. 149) point out, this shift from a stance of relative protection to one of being 'open for business' prioritized financial investment in farmland over the broader range of social goals that farming has historically serviced. The fact that farmland values remained lower than surrounding areas, even after the partial liberalization, enticed investors who anticipated a 'catch up' in regional markets (Magnan and Sunley, 2017, p. 95; see also Sommerville and Magnan, 2015). This prediction seems to have been borne out, with Saskatchewan farmland appreciating 261 per cent between year-end 2005 and 2015 (FCC, 2017), a rate that Magnan and Sunley (2017) link to both stronger crop receipts and, at least in some regions, the presence of investors. First-mover investors thus saw their speculative play handsomely rewarded, with firms such as Assiniboia and AgCapita enjoying returns that more than doubled their initial investments (Sommerville and Magnan, 2015). Despite revisions in 2015 that prohibited pension plans from acquiring additional farmland, the *Act* still allows investment by wealthy individuals. Farmland prices continue to rise although growth is slowing (Arnason, 2017). The impacts of these rising values on farmers are complex, as I will discuss below. For First Nations, the exemption of reserve lands from the Act – based on the fact that such lands are federally, rather than provincially, regulated – facilitated the large-scale leasing undertaken by OEF (Sommerville, 2018).

In order to understand the implications of rising land prices and investment activity for Saskatchewan farmers, one must return to the key restructuring pressures identified above, namely: the ongoing consolidation of farming; a prolonged period of low (though recently improving) farm incomes; and, intermittent weather events – all of which unfolded amid a neoliberal erosion of farm supports. Together, these factors have contributed to three large-scale shifts. First, alongside their efforts to intensify their operations, farmers have expanded their farms. In some cases, farmers have proceeded by buying additional land, while in others they have leased it, typically from neighbours that have exited the sector, their descendants, or from neighbouring First Nations. Indeed, in common with the model adapted by Broadacre and One Earth Farms (which were particularly lease heavy), census data indicate that by 2006, 54 per cent of Saskatchewan farm operations leased some portion of their land base. The census also shows a gradual shift towards leasing for the province as a whole – for example, between 1981 and 2006 the area of reporting farms that was owned by farmers decreased by 10.8 per cent while that leased by farmers increased by 23.1 per cent.[7] This could imply a decrease in farmers' overall tenure security and farm resilience (since owned land can be a valuable source of equity in years of slim margins). Furthermore, farmers' growing reliance on leased land suggests that it is increasingly landlords, rather than prairie farmers, who benefit from land appreciation.

The second major aspect of restructuring is that farmers have relied more and more heavily on credit, meaning that farm debt climbed from C$18 billion in 1981 to C$96 billion by 2016.[8] While Saskatchewan farmers retain access to subsidized borrowing via Farm Credit Canada (FCC), a Crown corporation that currently controls a third of the lending market, FCC's operations have become more market-oriented since the 1980s (see Martin and Clapp, 2015) and its preferential access to the perceived top tier of farmers is under attack from other mainstream lenders (Bergevin and Poschmann, 2013; Dorosh, 2014). Moreover, FCC has also provided loans to both 'own-lease out' investment vehicles such as Assiniboia (Assiniboia, 2010) and 'access-operate' vehicles including Broadacre (Marshall, 2014). While farm balance sheets currently appear rosy due to strong farmland prices, many farmers could be seriously impacted by the Bank of Canada's decision to increase interest rates in both July and September 2017 and again in January 2018 (with further increases expected to follow). Moreover, some commentators are arguing for the incorporation of land assets into new measures of farm wealth, which could result in government supports to farmers being cut back accordingly (Martin, 2010; Sparling and Uzea, 2012).

Finally, the third challenge emerging amid ongoing restructuring relates to the issue of farm succession. Canada's farmers are aging, and many of their sons and daughters are leaving to pursue jobs in cities. Even where children choose to stay and take over the farm, succession involves many pragmatic challenges. Parents may need to liberate capital to fund their retirement given the absence of private pensions and the fact that many farmers opt not to pay into Canada's public pension plan (the same plan that recently acquired a huge tract of Saskatchewan farmland). When on-farm capital is consistently sunk back into land costs to maintain or expand the farm, farmers often find themselves in a situation of being 'dirt rich, cash poor'. The children remaining on the farm, in turn, are likely to be under pressure to further expand operations to maintain viability. Meanwhile, new farmers have trouble getting started without a land base transferred from the previous generation or other assets to rely upon as a source of equity.

Investment vehicles have positioned themselves as responding to these problems in their attempts to curry favour with government officials, the general public and, of course, farmers. Investors suggest that they offer an alternative source of capital for an investment-hungry sector. 'Own-lease out' firms argue that they are a welcome land buyer for farmers who are highly indebted and need to deleverage to reduce risk – as well as for late career producers or farm descendants who are ready to dispose of their landholdings (Sommerville and Magnan, 2015). These investors present themselves as responsible landlords who can help new farmers start farming and more established ones to optimize their efficiency and achieve economies of scale (Sommerville and Magnan, 2015). Through sale-lease back arrangements, these investors also claim that they can facilitate intergenerational transfer and address the cash crunch that frequently accompanies farm succession. 'Access-operate' investors, in turn, suggest that they will realize the same efficiency and scale-related gains that their 'own-lease out' counterparts aim to help other farmers grasp. They propose that their presence will benefit the public by operationalizing a new model of farming that will similarly address declining turnover in the sector. But even as restructuring trends have helped to bolster investors' 'license to invest' (Knoepfel and Imbert, 2011), they have also conditioned resistance among farmers. For example, the Agriculture Producers Association of Saskatchewan (the main farm association in the province) long refrained from speaking out against farmland investors given that their membership includes both older farmers who benefit from investor interest and higher land prices, and younger producers who are having trouble entering agriculture or expanding their land base. Indeed, many a producer's situation is still more ambivalent: where farmers operate on a blended land base, they benefit from the appreciation of land that they own but are negatively impacted by climbing rents on that which they lease.[9] Farmers may also find themselves landlocked or priced out of expansion due to the intensity of regional market activity. The situation presents similar challenges for First Nations, whose reserve land is appreciating (albeit with somewhat different dynamics due to its institutional peculiarities), but who are facing difficulties acquiring further lands with the monies granted in claims settlements given elevated land values (Sommerville, 2018).

The overall picture that has emerged in Saskatchewan is one where the 'new' shoots of emerging financial vehicles are rather thoroughly entangled with the 'old' roots of agrarian transition. Yet even as the contemporary financialization of the prairie farming sector is underpinned by pre-existing and ongoing restructuring trends, it may also further coagulate, accentuate, and accelerate these trends (Sommerville and Magnan, 2015; Desmarais *et al.*, 2016). Interestingly, it seems that the recent spate of investment activity in Saskatchewan has helped to focus scholarly attention on the province, leading to the documentation of underlying and parallel developments in the broader farm sector. This is a perhaps unexpected benefit of deploying financialization as a concept and descriptor, even as prairie researchers have insisted that this label does not tell the whole story. Nonetheless, the situation remains one where there are 'more questions than answers' and where the 'complex realities' revealed by recent work require further analysis (Desmarais *et al.*, 2015, p. 41). One of the most compelling questions regards '[t]he differential impact of family farms expanding and consolidating their land base versus new, "outsider" investors buying [and/or farming] comparable amounts of land' (Desmarais *et al.*, 2015, p. 39). Another relates to the labour relations underlying the different models of investing and family farming in the province (Desmarais *et al.*, 2015). Still another is that of the transfer of shares or other financial interest in existing farms, given the corporate structure of many investments and a growing number of family farms. A final concern is how productive and financial revenue streams interact within these different models, especially where investors or farmers rely on a blend of both owned and leased land. Indeed, examining how this blend differs across farms and geographic regions in the province could help to clarify the implications of both recent investor activity and the larger sectoral shifts underway in Saskatchewan.

Accounting for shifts in farm culture and farmer subjectivities

By connecting the emergence of new agri-food investment vehicles to longer-running agricultural restructuring processes, the research described above goes some distance towards embedding financialization in Canada's prairie farming sector. Ongoing farm consolidation pressures, an increasing reliance on credit, and a worsening farm succession crisis are all key aspects of the 'local context' in which global shifts such as the '[t]he mobility of capital [and] the volatility of financial markets … are interpreted and acted on' (Desmarais *et al.*, 2016, p. 162). The prairie approach thus offers a useful counter to those studies that, by suggesting that financialization marks a rupture or break from past processes of agrarian-change-as-it-has-actually-occurred-in-specific-locations, may have tended to portray it as an 'impact' (see Ouma, 2014, p. 164). Nonetheless, there remains a tendency in some accounts to stop short of fully grounding financialization in shifts in prairie farming approaches. More progress could be made by broadening the prairie research focus to consider the cultural and subjective aspects of financialization alongside the political and economic ones. James Williams (2014, p. 421) has argued that in recent decades financial logics have reshaped 'sensibilities, expectations and interests' in prairie 'ag space', alongside other areas of cultural and economic life. In this view, financialization proceeds in part through the hearts and minds of rural actors, including, of course, farmers. Such an approach would help illuminate some of the quotidian aspects of financialization that Ouma (2014, 2016) is concerned to account for, while also helping to differentiate the responses of farmers (Williams, 2014) to both the recent investment surge and the larger restructuring trends noted above. Indeed, Christophers (2015a) suggests that among the different approaches to financialization that scholars have developed to date, it is those emphasizing finance's expanding sphere of influence over cultures, identities, and everyday life that provide the strongest analytical bite.

Williams (2014, p. 413, drawing on Pryke and Allen, 2000) argues that recent decades have witnessed the extension of a certain 'financial imaginary' into 'the local worlds of farmers', who are increasingly viewed as 'market actors' that are expected to both monitor and execute appropriate market transactions as a means of managing risk. He traces the emergence of the 'cultural figure of the farmer as a prudent investor', a characterization that he suggests 'is powerful as it appeals to farmers' sense of independence and self-sufficiency' (Williams, 2014, p. 413). This image is continually relayed to farmers, whether through education (with young farmers attending university at record rates), training programs, or farm advisory services (Williams, 2014). We can add to Williams' analysis state agricultural support programs following the neoliberal reforms described previously. For example, Hedley (2015, p. 24) describes a shift in farm safety net programs after 1990 'based on the premise that farmers should take greater responsibility for stabilizing their own operation', which he suggests marked a 'significant departure' from earlier policies that relied exclusively on government funds for income stabilization. The resulting programs – the Net Income Stabilization Account (NISA, 1991–2002) and its successor AgriInvest (2007–present) – have allowed farmers to bank a proportion of their net sales revenues alongside federal and provincial government contributions, and to withdraw such funds when they encounter slumps in net farm, individual, or farm household incomes (Hedley, 2015). While farmers had, of course, long drawn on their own bank savings in tight years and had also relied on banks for the provision of credit and crop advances, the NISA and AgriInvest programs are the first to weave personal investing into the delivery of farm supports. Elsewhere, I have argued that a similar invocation to 'live by finance' (Martin, 2002, p. 3) is playing out among First Nations, who are being encouraged to pursue new forms of financial 'inclusion' and leveraging as a means of addressing their racialized marginalization in prairie society (Sommerville, 2018; see also, Haiven, 2017).

A key question is how this acculturation process is conditioning individual decisions on farms and in First Nations communities regarding engagement with Saskatchewan's new financial actors. Although Williams' (2014) main focus is on prairie farmers' use of futures markets (which he suggests contributed to the collapse of the Canadian Wheat Board), he also notes the arrangements made between Bonnefield, one of the 'own-lease out' farmland investment funds and its tenant farmers. Indeed, Bonnefield explicitly markets itself to 'progressive, growth oriented family farm operators', suggesting that the company can help these farmers to 'optimize' their operations and serve to mitigate the risk associated with a rise in interest rates (see Bonnefield, 2014, pp. 2–4). While research is still emerging, there is evidence that this and the other claims of social benefits that investors use (see pp. 3–5) find resonance with some farmers. For example, Magnan (2012, drawing on Pritchard *et al.*, 2007) has suggested that a form of 'entrepreneurialism' has been at work in Wigmore Farms, the vertically integrated, family-owned, 'mega-farm' whose land Broadacre subsequently incorporated into its 'access-operate' model. As he notes, Wigmore itself expanded by absorbing the land base of two other family farms. One of these farms was managed by a couple, Laura and Norm Shoemaker, who met studying agriculture at the University of Saskatchewan and went on to farm Norm's family's land (OYFC, 2007). The Shoemakers won both provincial and national 'Outstanding Young Farmers' competitions in 2007, on the basis of their excellent 'business planning', which had led them to incorporate their farm and establish various value-added capacities as a succession plan for their two young children (OYFC, 2007).[10] As mentioned above, the couple sold the majority of their land to Assiniboia in 2007 before becoming minority shareholders in Wigmore in 2008, which subsequently farmed the land on which the shoemakers had previously operated (Briere, 2008; Rowland, 2008). While the model employed by the Shoemakers and their successors is clearly only one of many in Saskatchewan, it helpfully highlights the ways that shifting ideas and cultural values around the best practices of farming – and of being an outstanding farmer – can serve as a site of 'hybridity' (Desmarais *et al.*, 2015, 2016) fostering links between family farms and newer investment models. To be sure, this is despite the differential 'moral legitimacy' (Fairbairn, 2015) that may be afforded to farmers relative to outside investors when they embrace such practices.

Future research could usefully explore this topic further, including by examining the ways that farmers' shifting relationships with finance may be reworking the 'productivist' orientation to agriculture that has historically prevailed in Saskatchewan as in other locations globally (see Wilson, 2001; Sippel *et al.*, 2017). As a complement to existing and ongoing research on restructuring, this more culturally attentive work could also help to build theories of agrarian change that will extend – both spatially and temporally – beyond the apparent recent decline in agri-food financialization in Saskatchewan (see Magnan and Sunley, 2017). For example, such research could gather farmers' differing perspectives on two current debates that are unfolding in the prairie farm press and wherever farmers gather to talk. The first debate concerns the aforementioned trend towards farm expansion – commonly pursued as a means of improving revenues, but also for the 'social symbolism' (Burton, 2004) that size conveys. While farmers have been quick to dismiss the super-scale ambitions of certain 'access-operate' investors as a matter of hubris – which, interestingly, Engelen *et al.* (2012) also diagnose among financial agents – they simultaneously view those 'Big Time Operators' (BTOs) who have (or are perceived as having) deeper prairie roots with a mix of fear and admiration. Meanwhile, some suggest that alternative strategies such as increasing yields or cost-trimming provide a more advisable path forward (Rance, 2014). The second debate regards whether farmers are better off pursuing expansion by renting or by purchasing additional land. On the one hand, farmers are being chided that, given the strong performance of farmland financially, they need to think more as though they were 'investors *in* the land' rather than 'producers *on* the land' (Oltmans, 2007, cited in Painter, 2009; emphasis in original) – a recommendation that directly resonates with Williams' (2014) assessments. On the other hand, the strategy of farming a larger, leased land-base rather than a smaller, owned one has clearly gained traction among farmers such as the Shoemakers and the Wigmores, who argue that capital 'works harder' on the operations side of farming and that there need no longer be a link between farmland ownership and farming in Saskatchewan (McClinton, 2006; Hursh, 2013). A comparative analysis of the assumptions and values underlying this agrarian twist on the so-called 'opco-propco' split – a tactic that Christophers (2010) has analysed in other sectors – could yield valuable insights.

Discussion and conclusion: extending the prairie approach

The research program on agri-food financialization in Saskatchewan that underpins this chapter remains emergent and the situation 'on the ground' remains in flux. Nonetheless, this research lends important insights into the debates that served as the chapter's entry point. Responding to arguments raised by Christophers (2015a), the first of these debates concerns the 'novelty' of the current conjuncture in agri-food systems (Fairbairn, 2015, p. 210) and, relatedly, the utility of financialization as a conceptual descriptor for processes operating therein. Here, prairie scholars have identified a range of new financial vehicles investing in farmland and productive agriculture in Saskatchewan, while also assessing a sizeable increase in the scope of these investors' activities through analyses of farmland titles and sales transactions. Perhaps more pointedly, researchers have also considered the new alignments of power resulting from these investments (Lawrence, 2015, p. 202) – but crucially, they have done so without excusing farmers and other agricultural actors from the frame. Indeed, a major strength of recent prairie research is that it has been relatively more farm-centric when compared with accounts of agri-food system financialization elsewhere in world. Recognizing the long-standing entanglements of agriculture and finance in the region, prairie scholars have insisted that understanding rural transformation in Saskatchewan requires attention to the dynamism in *each* of these realms. After all, amid recent appeals to more practice-oriented accounts of financialization (see Ouma, 2014, 2016) – the second debate in which this chapter intervenes – researchers ought not to neglect shifts in the practice of farming.

With this in mind, this chapter also identifies several points of continuity between recent financial activities in Saskatchewan and pre-existing processes of agrarian change. These include the ongoing dispossession of, and discrimination against, First Nations farmers as well as the neoliberalization of agricultural policy in Saskatchewan since 1980s, which has changed who is allowed to own farmland in the province and rejigged the landscape of opportunity for farmers. I also describe three larger restructuring trends that have helped to lay the groundwork for recent investments, namely farmers' reliance on an increasingly lease-heavy model of farm expansion and consolidation, their deepening dependence on credit from a growing range of providers, and an ongoing farm succession crisis that is challenging new and established farmers alike. New investment models position themselves as helping farmers to respond to these trends and some farmers engage them as such, thereby constraining resistance against many models over the province as a whole. Yet, understanding farmers' perspectives on the new models and/or their on-farm financial decision-making more generally demands attention to the cultural and subjective aspects of financialization alongside the political and economic ones. Farmers themselves are regularly exercising financial logics and identities, which tangle with historic regimes of productivism in education, training programs, farm advisory services and state support programs, provoking constant debates around the most efficient, profitable and respected models of farming. In accordance with recent work coming out of Australia, then, the Saskatchewan experience suggests that financialization is best understood as a process of 'local negotiation' (Sippel *et al.*, 2017, p. 253), rather than one of outright determinism as some scholars suggest characterized early accounts of the phenomenon (see Ouma, 2014, 2016). Accounting for farmers' roles in this negotiation is central to providing a more grounded, embedded, and differentiated approach to the concept.

The prairie approach has strong potential for generating lasting theorizations of agrarian transition that stretch beyond the rapid rise and now apparent decline of investment activity in Saskatchewan. Financialization ought not be seen as 'a steamroller that advances into communities regardless of their historical circumstances' (Sommerville and Magnan, 2015, p. 139). Rather, researchers are effectively writing a thickened local history of agri-food financialization in the province, which presumes neither a clear ontological separation of finance from agriculture, nor the passivity of prairie farmers and other agricultural actors. This approach avoids portraying Saskatchewan as mere backdrop to financialization, instead implicating the province's farmers in both contouring the ground for, and augmenting, both new and old financial flows. Researchers will doubtless disagree about the extent to which farmers actions reflect – to use Karl Marx's (1867) famous phrasing – the 'dull compulsion of economic relations' versus individual enterprise and entrepreneurial innovation, a blend that likely varies between operations and in any case, can be subject to rapid change. Doubtless certain financiers have indeed been perceived by certain farmers as a kind of 'big bad wolf' in recent years (Ouma, 2014, p. 163) – or will be now that interest rates are climbing – but it would appear that we have actually been suckling the cub for some time. In turn, this chapter can also help to advance accounts of finance-driven land and value-chain grabbing, suggesting that rather than recent acquisitions marking 'the *beginning* of a process of gaining or grabbing access' (Peluso and Lund, 2011, p. 669, emphasis added), such acquisitions might better be conceptualized as something in the *middle* of a long chain of events for which it is difficult to mark a clear beginning or an end.

The body of work that has blossomed in Saskatchewan under the label of financialization continues to grow. In a sense, the concept has served a strategic value, helping to draw attention to agriculture in the region and to forge or strengthen relationships between emerging and established academics, farm organizations, and the farm press. This bodes well for future work in the province and beyond, which in addition to the specific questions raised across this chapter, might usefully be extended to address four broad areas. The first extension regards attending to other commodities beyond grains and oilseeds. The second area where more work could be done regards the operation of financialization in other aspects of the production process. Prairie farmers depend on a growing array of financial arrangements, including mortgage lending from traditional lenders and farmland lease agreements with the existing landlord class. But they also rely on crop and production insurance, input and equipment financing, forward contracts and contract insurance, and agricultural futures and derivatives. Can traditional models of family farming prosper under these conditions, especially if a prolonged period of low commodity prices, a decrease in farm income supports, further increases in interest rates, and/or the bursting of a possible bubble in farmland values come to pass?

A further opportunity relates to the possibility of extending the temporal scale that has yet been applied to situate financialization within still longer run histories. Returning to Christophers' (2015a) question regarding the historical novelty of financialization and as summarized above, prairie researchers have helpfully traced out the financial continuities spanning much of the last four decades. Yet, it is clear that our efforts fall well shy of fully exhausting the different features and periods of the past. For example, what parallels and divergences exist between the current conjuncture and earlier land booms in the region, most recently in the 1970s? Fairbairn (2015) has suggested that the companion boom in US farmland prices yielded a first unsuccessful attempt at institutional investment that has been reversed in recent years. A comparative historical analysis considering Canada, where producers were somewhat shielded from rising interest rates and the subsequent decline in land values (Preville, 2003), and where the *Saskatchewan Farm Security Act* worked to hamper similar financial innovation, could prove fascinating.

Finally, there are also opportunities for extending the approach that has been taken in Saskatchewan to encompass additional geographic locales, whether through comparative research or the work of independent scholars. With respect to the former, Magnan (2015) has compared agri-food financialization in the prairie grain and oilseeds sector to that in two Australian states. In South Africa, the rapid liberalization of the fruit sector at the end of Apartheid has helped to shape the nation's land reform programs, which have become increasingly focused on cultivating a class of 'entrepreneurial' black farmers. While the de-racialization imperative has created new opportunities for attracting capital under the rubric of 'responsible' investment, my own research (Sommerville, 2016) suggests that results have been less than transformative. Rather, financialization has underpinned instances of land concentration, value-chain grabbing, and 'elite capture' of state benefits by white-dominated financial interests and multinational businesses.

Notes

1 A similar critique is made by Williams (2014, pp. 408–409), who suggests the need for greater emphasis on the 'contingency and contextuality of finance, [and] on the diverse assortment of practices, devices, theorems and technologies through which finance is regularly constituted and performed'.

2 The CPPIB is the manager of Canada's public pension plan.

3 Alberta is the province bordering Saskatchewan to the west.

4 Whether there is further overlap between the land bases of 'own-lease out' and 'access-operate' investment models remains unclear, but the question is worthy of further attention.
5 Although admittedly imperfect, the term 'access-operate' is clearer than the obvious alternatives, i.e. 'lease/own-operate' or 'own/lease-operate'. In choosing this term, I am drawing on the work of Ribot and Peluso (2003, p. 153), who define access as 'the ability to benefit from things – including material objects, persons, institutions and symbols'.
6 Hudson (2010) argues that low tax regimes tend to be capitalized into land values. While property tax rates in Saskatchewan vary by year and municipality, farmers generally benefit from a lower proportion of the assessed value of farmland being incorporated into tax assessments relative to residential or commercial/industrial land. Additionally, farmers in Saskatchewan benefit from a C$1 million lifetime capital gains tax exemption upon sale of their property (or interest in a farm business) under the federal *Income Tax Act*.
7 Unfortunately, revisions to the land tenure section of the census as well as shifts in data processing make it impossible to compare either the importance or growth of leasing in 2011 and 2016 with data from previous census years. See: CANSIM Table 002–0204, 'Census of Agriculture, tenure of land owned, leased, rented, crop-shared, used through other arrangements or used by others' for 2011 and 2016 and Statistics Canada (2007) for 1976 through 2006.
8 See: CANSIM Table 002–0008, *Farm debt, classified by lender*.
9 In Saskatchewan, rent is frequently set at a fixed percentage of assessed land value. Of course, the ambivalent effects of rising farmland values also relates to certain farmers' access to subsidized credit as well as the low interest rates that have prevailed since the early 1990s.
10 Canada's 'Outstanding Young Farmers Program' is an annual competition to 'recognize farmers that exemplify excellence in their profession and promote the tremendous contribution of agriculture' (OYFC, 2007). The project is sponsored by a range of financial and agribusiness interests as well as Agriculture and Agri-Food Canada.

References

Arnason, R. (2017) 'Farmland price boom may be ending: FCC', available at www.producer.com/2017/09/farmland-price-boom-may-be-ending-fcc/, accessed 15 January 2018

Assiniboia Farmland Limited Partnership (Assiniboia). (2010) 'Management's discussion and analysis for the year ended December 31, 2009', 16 April 2010

Assiniboia Farmland Limited Partnership (Assiniboia). (2013) 'Assiniboia Farmland Limited Partnership enters into an agreement for the sale of its farmland portfolio', available at: www.newswire.ca/news-releases/assiniboia-farmland-limited-partnership-enters-into-an-agreement-for-the-sale-of-its-farmland-portfolio-513432241.html, accessed 16 July 2017

Bergevin, P. and Poschmann, F. (2013) 'Reining in the risks: Rethinking the role of Crown Financial Corporations in Canada', available at: www.cdhowe.org/reining-risks-rethinking-role-crown-financial-corporations-canada, accessed 16 July 2017

Bonnefield Financial Inc. (Bonnefield). (2014) 'Farmland lease financing: How it can help you reach your goals', available at: www.flipsnack.com/aflannery/farmland-lease-financing-how-it-can-help-you-reach-your-goals.html, accessed 7 August 2017

Briere, K. (2008) 'Entrepreneur develops new farming model', available at: www.producer.com/2008/08/entrepreneur-develops-new-farming-model/, accessed 7 August 2017

Burton, R. (2004) 'Seeing through the "good farmer's" eyes: Towards developing an understanding of the social symbolic value of "productivist" behaviour', *Sociologia Ruralis*, vol 44, no 2, pp. 195–215

Carter, S. (1990) *Lost Harvests: Prairie Indian Reserve Farmers and Government Policy*, McGill-Queen's University Press, Montreal

Christophers, B. (2010) 'On voodoo economics: Theorizing relations of property, value, and contemporary capitalism', *Transactions of the Institute of British Geographers*, vol 35, pp. 94–108

Christophers, B. (2015a) 'The limits to financialization', *Dialogues in Human Geography*, vol 5, no 2, pp. 183–200

Christophers, B. (2015b) 'From financialization to finance: For "de-financialization" ', *Dialogues in Human Geography*, vol 5, no 2, pp. 229–232

Desmarais, A., Qualman, D., Magnan, A. and Wiebe, N. (2015) 'Land grabbing and land concentration: Mapping changing patterns of farmland ownership in three rural municipalities in Saskatchewan, Canada', *Canadian Food Studies*, vol 2, no 1, pp. 16–47

Desmarais, A., Qualman, D., Magnan, A. and Wiebe, N. (2016) 'Investor ownership or social investment? Changing farmland ownership in Saskatchewan, Canada', *Agriculture and Human Values*, vol 34, no 1, pp. 149–166

Dorosh, M. (2014) 'Overgrown farm credit giant: In Canada, the government entity serves the most credit-worthy', available at: http://busin ess.financialpost.com/opinion/overgrown-farm-credit-giant/wcm/f7c53cfe-1963–4f5f-ad1e-65a874b3125e, accessed 16 July 2017

Engelen, E., Ertürk, I., Froud, J., Johal, S., Leaver, A., Moran, M. and Williams, K. (2012) 'Misrule of experts? The financial crisis as elite debacle', *Economy and Society*, vol 41, no 3, pp. 360–382

Fairbairn, M. (2014) ' "Like gold with yield": Evolving intersections between farmland and finance', *Journal of Peasant Studies*, vol 41, no 5, pp. 777–795

Fairbairn, M. (2015) 'Reinventing the wheel? Or adding new air to old tires?' *Dialogues in Human Geography*, vol 5, no 2, pp. 210–213

Farm Credit Canada (FCC). (2017) 'Historic farmland values: 1985–2015', available at: www.fcc-fac.ca/fcc/about-fcc/corporate-profile/reports/farmland-values/farmland-values-historic.pdf, accessed 16 July 2017

Guenther, L. (2015) 'Big bankruptcies make big headlines but smaller farms are under pressure too', *Country Guide*, 12 February, available at www.country-guide.ca/2015/02/12/farm-failures/45816/, accessed 7 August 2017

Haiven, M. (2017) 'The uses of financial literacy: Financialization, the radical imagination, and the unpayable debts of settler colonialism', *Cultural Politics*, vol 13, no 3, pp. 348–369

Hart, G. (2004) 'Geography and development: Critical ethnographies', *Progress in Human Geography*, vol 28, no 1, pp. 91–100

Hedley, D. (2015) *The Evolution of Agricultural Support Policies in Canada*, available at: https://caes.usask.ca/meetings/the-evolution-of-agricultural-support-policy-in-canada.pdf, accessed 16 July 2017

Hudson, M. (2010) 'From Marx to Goldman Sachs: The fictions of fictitious capital, and the financialization of industry', Critique: Journal of Sociologist Theory, vol 38, no 3, pp. 419–444

Hursh, K. (2013) 'Should you buy or rent the land you farm?' *Farm Credit Canada AgriSuccess Journal*, March 2013

Knoepfel, I. and Imbert, D. (2011) *The Responsible Investor's Guide to Commodities*, available at: www.onvalues.ch/images/publications/ri per cent20guide per cent 20to per cent20commodities_2011.pdf, accessed 16 July 2017

Lappano, J. (2009) 'Growing capital', available at: www.corporateknights.com/channels/food-beverage/growing-capital-12522509/, accessed 7 August 2017

Lawrence, G. (2015) 'Defending financialization', *Dialogues in Human Geography*, vol 5, no 2, pp. 201–205

McClinton, L. (2006) 'Farmland: Buy or rent', *Farm Credit Canada AgriSuccess Journal*, January 2006

Magnan, A. (2012) 'New avenues of farm corporatization in the prairie grains sector: Farm family entrepreneurs and the case of One Earth Farms', *Agriculture and Human Values*, vol 29, pp. 161–175

Magnan, A. (2015) 'The financialization of agri-food in Canada and Australia: Corporate farmland and farm ownership in the grains and oilseed sector', *Journal of Rural Studies*, vol 41, pp. 1–12

Magnan, A. and Sunley, S. (2017) 'Farmland investment and financialization in Saskatchewan, 2003–2014: An empirical analysis of farmland transactions', *Journal of Rural Studies*, vol 49, pp. 92–103

Martin, R. (2002) *Financialization of Daily Life*, Temple University Press, Philadelphia

Martin, L. (2010) 'Farm incomes and farm wealth', in George Morris Centre (ed.) *Advancing a Policy Dialogue: A Series of Papers Prepared for the Canadian Agri-Food Policy Institute*, Canadian Agri-food Policy Institute, Ottawa, pp. 15–17

Martin, S. and Clapp, J. (2015) 'Finance for agriculture or agriculture for finance?', *Journal of Agrarian Change*, vol 15, pp. 549–559

Marshall, A (2014) Affidavit filed with the Court of Queen's Bench, Alberta, available at: www.pwc.com/ca/en/car/broadacre/assets/broadacre-002_110514.pdf, accessed 7 August 2017

Marx, K. (1867 [1976]) *Capital: Volume 1*, Penguin Books, London

National Farmers Union (NFU). (2005) 'The farm crisis: Its causes and solutions'. Submission to the Ministers of Agriculture Meeting, 20 June 2015, available at: www.nfu.ca/sites/www.nfu.ca/files/Ministers_of_Ag_brief_FOUR.pdf, accessed 7 August 2017

Oltmans, A. (2007) 'A new approach in farm business analysis to fit a changing farmland investment market', available at: http://ifmaonline.org/contents/a-new-approach-in-farm-business-analysis-to-fit-a-changing-farmland-investment-market-ap/, accessed 7 August 2017

Ouma, S. (2014) 'Situating global finance in the Land Rush Debate: A critical review', *Geoforum*, vol 57, pp. 162–166

Ouma, S. (2015) 'Getting in between M and M[1] or: How farmland further debunks financialization', *Dialogues in Human Geography*, vol 5, no 2, pp. 225–228

Ouma, S. (2016) 'From financialization to operations of capital: Historicizing and disentangling the finance–farmland-nexus', *Geoforum*, vol 72, pp. 82–93

Outstanding Young Farmers Canada (OYFC). (2007) 'Diversified crop operation chosen 2007 Outstanding Young Farmers for Saskatchewan', available at: www.

saskoyf.ca/documents/2007_SKOYF_winners_news_release.pdf, accessed 7 August 2017

Painter, M. (2009) 'The farmland investment market in Canada', available at: http://citeseerx.ist.psu.edu/viewdoc/download?doi=10.1.1.586.4807&rep=rep1&type=pdf, accessed 7 August 2017

Peluso, N. and Lund, C. (2011) 'New frontiers of land control: Introduction', *Journal of Peasant Studies*, vol 38, no 4, pp. 667–681

Pitts, G. (2011) 'Canola: Western Canada's other oil boom', available at: www.theglobeandmail.com/report-on-business/economy/canola-western-canadas-other-oil-boom/article595307/, accessed 7 August 2017

Pratt, S. (2014) 'Farm owner blames ownership rules: Broadacre Agriculture seeks bankruptcy protection', available at: www.producer.com/2014/11/farm-owner-blames-ownership-rules/, accessed 7 August 2017

Preville, E. (2003) *The Financial Structure of Agriculture in the Canadian and American Prairies, 1971–2001*, Library of Parliament Parliamentary Research Branch, Ottawa

Pritchard, W., Burch, D. and Lawrence, G. (2007) 'Neither "family" nor "corporate" farming: Australian tomato growers as farm family entrepreneurs', *Journal of Rural Studies*, vol 23, pp. 75–87

Pryke, M. and Allen, J. (2000) '"Monetized time-space: Derivatives – money's "new imaginary"?', *Economy and Society*, vol 29, no 2, pp. 264–284

Rance, L. (2014) 'Bigger is definitely better — if it's yield and not acreage', available at: www.winnipegfreepress.com/business/bigger-is-definitely-better–if-its-yield-and-not-acreage-284204551.html, accessed 7 August 2017

Ribot, J. and Peluso, N. (2003) 'A theory of access', *Rural Sociology*, vol 68, pp. 153–181

Rowland, K. (2008) 'Farming outside the box', available at: http://farmforum.ca/article/farming-outside-the-box-2/, accessed 7 August 2017

Sippel, S., Larder, N. and Lawrence, G. (2017) 'Grounding the financialization of farmland: Perspectives on financial actors as new land owners in rural Australia', *Agriculture and Human Values*, vol 34, no 2, pp. 251–265

Sommerville, M. (2016) 'The more things change, the more they stay the same? Finance capital, land reform, and "transformation" politics in South Africa's commercial farming sector'. Paper presented at the 'Rethinking agriculture in South Africa: Constraints and opportunities' conference. Institute for Poverty, Land and Agrarian Studies, Cape Town, South Africa. 11–13 October 2016.

Sommerville, M. (2018) '"Naturalizing finance, financializing natives": Indigeneity, race, and "responsible" agricultural investment in Canada', *Antipode*, DOI:10.1111/anti.12395

Sommerville, M. and Magnan, A. (2015) '"Pinstripes on the prairies": Examining the financialization of farming systems in the Canadian Prairie Provinces', *Journal of Peasant Studies*, vol 42, no 1, pp. 119–144

Sparling, D. and Uzea, N. (2012) *Six Years that Changed the Future for Agriculture: Impact on Farm Income in Canada 2005–2010*, Richard Ivey School of Business, London

Statistics Canada. (2007) 'Saskatchewan remains the breadbasket of Canada', available at: www.statcan.gc.ca/pub/95-640-x/2016001/article/14807-eng.htm, accessed 16 July 2017

Williams, J. (2014) 'Feeding finance: A critical account of the shifting relationships between finance, food and farming', *Economy and Society*, vol 43, no 3, pp. 401–431

Wilson, G. (2001) 'From productivism to post-productivism … and back again? Exploring the (un)changed natural and mental landscapes of European agriculture', *Transactions of the Institute of British Geographers*, vol 26, no 1, pp. 77–102.

12 Canadian supply management as a bulwark against financialization

Bruce Muirhead and Jodey Nurse-Gupta

> The fault, dear Brutus, is not in our stars,
> But in ourselves, that we are underlings.
>
> *Julius Caesar (I, ii, 140–141)*

The financialization of agriculture has been an important topic of analysis, especially since the food price spikes of 2007–8. What is this financialization beast, and why is the financialization of agriculture especially worrisome for individuals concerned with sustainable and secure food systems? Greta Krippner notes that, broadly speaking, financialization is 'the tendency for profit making in the economy to occur increasingly through financial channels rather than through productive activities', which she argues results in a far more crisis (and angst) ridden economy (Krippner, 2011, p. 4). In the agricultural context, scholars such as Geoffrey Lawrence point towards the recent 'intrusion of banks and other financial entities such as private equity firms in the purchase of farmland and the acquisition of food supplies' as particularly disruptive because of their connection to food price distortions and agricultural land speculation (Lawrence, 2014). A recent example of this behaviour is the sale by JBS USA of Five Rivers Cattle Feeding, the largest such station in the world, to affiliates of Pinnacle Asset Management LP for US$200 million. This does not mean, however, that land or assets are held merely for speculative purposes; investment syndicates increasingly appreciate their value as an income generator through rental agreements with farmers, or an annual dividend through a business investment. In short, as Madeleine Fairbairn suggests, 'it is a productive asset that moonlights as a financial asset' (Fairbairn, 2014).

There is a parallel stream to this that merits some consideration, one that has increasingly infected jurisdictions where financial pressures force so-called family farms to sell their labour to a much larger entity. Indeed, to some extent, farmers have become mere employees of large agribusiness corporations. Richard Longworth has noted that:

> Many critics rail against 'corporate farms' as though corporations owned the farms. This is the wrong idea. Often, corporations control but do not own. Most farms are still owned by farm families, but these families work for corporations. In other words, ownership may be personal but control is corporate.
>
> (Longworth, 2008, p. 65)

In short, as the title of a provocative article published more than two decades ago, in 1994, accurately asked 'why own the farm if you can own the farmer (and the crop)?'[1] (Hamilton, 1994).

How has this played out in Canada? That is difficult to determine given that this is a hugely understudied area in the Canadian context. Only a handful of academics have turned their attention to the issue to date and it has not yet garnered critical mass (Ross, 2016).[2] However, trends are suggesting that sufficient mass could be developed in the near future. For example, since the Conservative government ended the monopoly of the Canadian Wheat Board (CWB) to market Western wheat and barley in 2012, foreign, often financial, firms have moved into the vacuum. China's COFCO is now active in buying wheat, and Swiss-based Glencore has purchased Canadian company Viterra's business and is building new facilities in Western Canada. Similarly, PLC Bunge and the state-owned Saudi Agricultural and Livestock Investment Company purchased old CWB assets and have established a presence in the Canadian market. Farmers are increasingly beholden to these conglomerates. While some, like Errol Anderson, the president of ProMarket Wire in Calgary, welcome the 'newcomers looking at the Canadian market', which, in his mind, is 'making the market healthier and providing farmers with more choices' others are of a different mind (Skerritt, 2016). More choice, of course, does not guarantee farmer welfare when the companies involved are mega-industries. As one critic commented in response to a *Western Producer* article on the subject, 'We now have freedom of choice – pick who you want to get screwed by' (White, 2017). That could well be on Quebec farmers' minds as they contemplate the activities of Pangea Terres Agricoles, a company designed to 'establish business partnerships with Quebec farmers to make underused agricultural land more productive and allow the farmers to earn more money' (van Praet, 2014).

In Saskatchewan, Canada's most productive agriculturally productive province, André Magnan and Annette Desmarais calculated that the percentage of farmland owned by pension funds and other private sector investors totalled 1.44 per cent of the province's total farmland area of about 60 million acres (24.281 million ha). Despite the significant increase from approximately 52,000 (21,000 ha) to more than 830,000 acres (336,000 ha) between 2003 and 2014, in absolute terms the percentage remains small (Magnan and Desmarais, 2017, p. 8). That does not mean, however, that farm-size is static (Somerville and Magnan, 2015, p. 134).

Between 2011 and 2016, the average family farm in Saskatchewan expanded in area by 7 per cent. The largest 17 per cent of Saskatchewan farms generate 83 per cent of all industry revenue. Farmland is also increasingly rented in the province but, for the most part, these agreements continue between farmers – private equity firms do not significantly influence this trend (Statistics Canada, 2017). While the transfer of farms into more corporate holdings is tempered by provincial legislation that restricts foreign equity funds from purchasing more than ten acres (4.04 ha) of land and excludes public pension plans from investing in farmland, private equity firms can and have begun to buy into the province's agricultural land base. Still, as noted, the amount they have under their control remains small and Saskatchewan – a province with 40 per cent of Canada's farmland – remains the preserve of the family farm albeit with larger family farm units, now totalling on average about 1,750 acres (708 ha).

Supply management as anti-financialization

While certain provinces, like Saskatchewan, are keen to protect farmland ownership which serves to mitigate some of the worst effects of sector financialization, another important reason is Canada's system of supply management (SM), which has protected many farmers from being forced from their lands and livelihoods. *Supply management* is credited with providing a bedrock of financial stability in rural Canadian communities, serving family-run farms in Canada's dairy, eggs and poultry sectors. The importance of supply management as a solution for maintaining smaller-scale family farms has been advocated by organizations such as the National Farmers Union (NFU), who recognize that as 'corporate control of agriculture increases, the livelihoods of farmers and farm workers become increasingly precarious' (National Farmers Union, 2016).

The term family farm may be debated, but Elizabeth Garner and Ana Paula de la O Campos have concluded, after reviewing 36 descriptions and uses of the term, that the majority of these cases 'recognize the role of family labour and the role of the family in managing the farm operation'. However, the notion of family farming seems to go beyond farming capacity, size and orientation. It is sometimes also used 'to capture ecological, social, cultural and environmental objectives and therefore has close ties to the local culture and the rural community' (Garner and O Campos, 2014, p. 1). They propose a concept of family farming that emphasizes a recurring aspect of the term – the reliance on family labour – that informs the discussion of financialization of farmland and rural restructuring.[3] Further, they note that family farmers have close ties to the local culture and community, presumably attachments that are often absent in the case of agribusiness given its demand for massive volumes of inputs, often from afar. By this definition, supply-managed farmers remain resolutely in the family category, where speculative commodity flows, uncertain pricing and futures markets do little to disturb their relatively stable world.

The intimate connection between family farming and supply management is the result of a model that does not allow non-family corporate ownership of quota, thereby removing a corporation's ability to vertically integrate and 'cut the family farm out' (Bruce Sargent as quoted in CBC News, 2017). Canada's system of supply management is based on three pillars – producer pricing, market discipline and import management – and matches domestic demand with domestic supply while negotiating a sustainable price among stakeholders. This negotiation considers 23 variables such as the cost of feed, insurance, equipment, fertilizer and herbicides, and labour (CDC, 2017). The system is highly regulated and operated by farmers in the sectors in which it prevails.

The system of supply management that exists today in Canada has undergone significant transformations over the years. The national systems of supply management were instituted during the 1970s, although some provincial marketing boards had already been established and had begun implementing some of the pillars of supply management. During this process, the conflicts that resulted in response to the system tended to reflect ideological uneasiness with how the system influenced 'free markets' and 'enterprising capitalists' in agriculture. While some believed market controls offered rational and fair regulations to these agricultural sectors, critics charged that such interventions supported small-scale farmers at the expense of larger, supposedly more efficient producers and restricted individual farmers from acting independently.

When the Ontario Milk Marketing Board introduced a system of controlled production in the province that included production controls in the form of quotas and formulated pricing for fluid milk producers, they defended their policies for bringing stability to an industry where none had existed previously (Nurse-Gupta, forthcoming). Part of the reason that supply management has been successful in creating stability, and functioned so well in the sectors it serves is that the products are not widely traded internationally, especially in the case of dairy and eggs, With the former, only about 7 per cent of total global production is traded across borders, while for eggs, it is about 3 per cent. This caveat is critical. Generally, there is less incentive in such cases to extend the market beyond national borders. Moreover, competition tends to be stiff for that 7 per cent. About 80 per cent of it is controlled by only three countries, Australia (5 per cent), New Zealand (32 per cent) and the United States (17 per cent), and one region, the EU-27 (28 per cent). Canada is a higher cost dairy producer, being a northern country where expensive shelter and feed is necessary to house cows in the winter. By comparison, in the case of the US, California is the largest dairy state, and the industry has generally migrated to warmer climes over the past 20 years with devastating effects on the dairy farming communities of the upper Midwestern and Northeastern United States.

The paradigm of supply management flies in the face of modern agricultural organization where the emphasis since the late 1990s has been on deregulation and the development of more market orientation with respect to global agriculture. As has been pointed out, this 'approach [is] in line with the core objective of the WTO Agreement on Agriculture: establishing fundamental reforms to achieve substantial, progressive reductions in support and protection' (Hafemeister, 2018). However, the Canadian government remains committed to the system as best reflecting Canada's food sovereignty and security in that it reflects a specifically *Canadian* method of organization. Today, the country is the only one in the world that supports such a system and is likely to remain so – as Ontario's Minister of Agriculture, Food and Rural Affairs, Jeff Leal, insists 'supply management yesterday, supply management today, supply management tomorrow' (Bender, 2015).

Ontario is an influential province with a vested interest in protecting the farmers producing under supply management who contribute significantly to the province's rural economy. *Supply management* allows the stability and security of incomes that contribute to the support of local business, which builds strong rural communities. This is in contrast to those whose agricultural residents rely on fluctuating cash crop markets and the highly volatile and globally exposed beef and hog industries to earn a living. The importance of the local economic multiplier effect on rural sustainability is one reason why the conservative government of Stephen Harper did not repeal the enabling federal legislation; it recognized supply management's beneficial effect on rural communities and realized the extreme hardship that many communities would face if it were no longer in place (anonymous interview). Those relatively small, supply management farms do their business locally – they buy, sell, live and play locally. On the other hand, it is also the case that 'no mega-farmer buys locally anymore' nor do corporations purchase product from small farmers. (Longworth, 2008, p. 68).

As already noted, politicians have long recognized the benefits of supply management and the stability it creates in rural communities. Since the effective implementation of supply managed principles in the 1960s and 1970s, provincial and federal agricultural ministers have supported the system given its stabilizing effect of production control and fair farmer pricing. Minister Leal and other provincial ministers of agriculture have remained steadfast in their support of supply management during the recent North American Free Trade Agreement renegotiations. Despite American demands that the system be phased out over a ten-year period, Leal has insisted that 'Ontario is not going to cede one inch when it comes to supply management' (Johnson, 2017).

Canadian politicians appreciate supply management for a number of reasons, including that it does not rely on government assistance. It is an outlier in terms of other systems. Both the Europeans and the Americans provide public assistance when market conditions deteriorate to a certain point. The EU's Common Agriculture Policy (CAP) rewards farmers for the amount of land farmed through the Single Farm Payment, an annual distribution that inequitably rewards the largest producers disproportionately. The Queen of England, for example, is a subsidy millionaire given the vast numbers of hectares that she owns. In the US, various subsidy programs kick in when price falls. A recent report has noted that US direct and indirect expenditures benefiting US dairy producers are equivalent to about 73 per cent of revenue from the market place (*Congress*, 2018).

Supply management was designed to end the continual appeal to government in terms of providing assistance to farmers when commodity prices collapsed, as they have done on a regular basis. In the 1960s the volatility of many agricultural industries, but especially the dairy industry, prompted the federal government in Ottawa to take action. In the postwar period, various provincial and national studies were done by Canadian governments concerned that dairy demands for subsidies in view of depressed global prices and over-production was counter-productive. As the Minister of Agriculture, Douglas Harkness, noted in April 1960, 'a cost of some C$50 million or C$55 million would seriously upset the balance of the budget' (PCO Cabinet Conclusion, 1960). A few days later, cabinet concluded that any action 'to support dairy product prices would thrust a burden on the budget' that could not be considered. In short, the federal government could not afford such a payment. The solution, or so Harkness pointed out, to this on-going problem was to investigate some system of quotas and production control under farmer guidance, which became supply management by 1966 in the dairy sector. In short, the federal government would thrust the burden of support on farmers themselves and not on the state. The policy objective also addressed issues of power asymmetries between producer and processor and helped to ease price volatility in the products concerned.

Over the years, Canadian governments have appreciated supply-managed sectors' ability to self-manage and not require the government subsidies demanded in other sectors. Politicians and agricultural leaders support supply management because they understand that the system works to protect their farming communities and provide farmers fair farm prices, unlike other types of agricultural supports, including agricultural subsidies. Reporting on American agricultural subsidies has shown that these payments tend to benefit 'elite, commercial farmers' at the expense of smaller operations. For most taxpayers, the idea of supporting family farms is one thing, but 'It's quite another to subsidize the expansion of a mega-farm operation that puts family farmers out of business'. The goals of American farm bills today depart from the price-stabilization policies that used to concern policymakers, and instead, the 'subsidy explosion' in the United States tends to advantage the largest agribusinesses (Imhoff, 2012).

Supply management's ability to benefit all producers, rather than simply the largest and most powerful, is one way in which the system has stifled financialization, income inequality, and wage stagnation. It does not elevate the significance of the financial sector over the real sector or transfer income from the latter to the former. Indeed, by keeping prices stable and negotiating a profit for the farmer, it does not contribute to increased income inequality and wage stagnation that has been experienced by almost all farm sectors that participate in the international economy (Palley, 2007, p. 3). To the degree that financialization increases the importance of investment capital and adds volatility in agricultural markets (Clapp, 2014), small-scale farmers are usually hit hard, which then impels the movement into corporate ownership or control as the future recedes. As Isakson (2013) observes, this is due to price variations affecting the livelihood of farm families, and because ordinary farmers often lack the financial muscle to compete with corporate finance and the institutional investor. When this 'financial muscle' is rendered redundant as it is by SM, financialization does not become an issue.

As has already been discussed, the emphasis is on smaller farms – the average across the country works out to about 85 milking cows per farm in 2016 which very much makes the industry one where farms of between 200–900 acres (80–360 ha) predominate. In the egg sector, the average Canadian farm hosts about 22,400 layers with the largest housing slightly more than 400,000. By comparison with US producers and many others around the world, Canada's farms are much smaller and more appropriate to being family-owned and operated. Of the 20 largest egg producers in the world, 12 are based in the United States, and hold about 152 million layers (Clements, 2015) of the 351 million housed in the US in 2017. The entire layer population in Canada is approximately 24 million birds, a number that meets the needs of Canadian consumers only.

The national umbrella organization, Egg Farmers of Canada, administers a country-wide system for eggs, and one of its most important roles is the prevention of an excess of production by regulating quota allocations. Each province (and the one territory) that hosts an egg industry has a provincial board that is responsible for regulating regional production and marketing. For example, Egg Farmers of Ontario 'allocates [that province's] share of the domestic egg supply to producers and sets the price paid by graders to producers' (OMAFRA, 2014). Retailers then purchase those eggs from graders at a negotiated price. Clearly, the entire sector is highly regulated with rules developed that respond to quintessentially Canadian conditions, disregarding financial forces emanating from outside the country, and establishing greater power symmetry between farmers and processors/supermarkets.

Similarly, prices with respect to dairy Canada are regulated by the Canadian Dairy Commission (CDC) which was established in 1966 with a mandate to coordinate federal and provincial dairy policies and create 'a control mechanism for milk production which would help stabilize revenues and avoid costly surpluses' (CDC, 2017). The CDC has controlled two of the three pillars underpinning the system: support prices and market sharing quotas. Annually, the Canadian Dairy Commission sets the support price of butter and skimmed milk powder 'following consultations with industry stakeholders'. What results is system-wide stability and adequate income flows to dairy farmers to permit them to operate without the necessity of getting continuously bigger to satisfy the demands of the market as is the case in most other jurisdictions. A family dairy or egg farm really is a family farm in Canada.

Challenges to supply management

Supply management benefits certain Canadian farmers by providing them with financial solidity in their agricultural pursuits. They do so through a market largely closed to imports beyond a tariff rate quota, and production controls in the form of quota that can be passed down to succeeding generations, or sold. It is also a system that has aroused the ire of exporting countries in dairy and eggs, in particular New Zealand and the United States, with respect to milk, and the US in the case of eggs, who believe that "the free market" should decide commodity flows and prices. What those affected groups and their countries do not understand is that there is always a regulator, whether it be the supermarkets which buy farm commodities or the financial sector that metes out its own form of sustenance that allow agriculture to continue.

In the recent Trans Pacific Partnership negotiations, the two countries targeted Canada's system, demanding widespread access to dairy and egg markets. In the case of dairy, concessions granting an additional 3.25 per cent were agreed to, while for eggs, it was 2.3 per cent of production (and for chicken and turkey, 2 per cent). However, pressure was most intense in the case of dairy, where the US had initially demanded access to an additional 15 per cent of the Canadian market. Dairy production in the United States has remained very high, with chronic surplus production causing significant distress in the industry. With few additional foreign markets to absorb this surplus, Americans see their northern neighbour as ripe for dairy exports if serious inroads could be made into the supply managed system.

A similar case pertains with US eggs where, in 2017, supply greatly exceeded demand leading to ever-lower prices paid to farmers and to graders and retailers. Cal-Maine Foods, the largest shell egg producer in the US, lost US$74 million in the fiscal year ending June 2017. As the company's CEO noted:

> egg markets have been affected by increased production levels as producers repopulated their flocks after the 2015 avian influenza-related laying hen losses, and the younger, more productive hen population has produced a higher number of eggs. Overall, market demand trends have not kept pace with these production levels.
>
> (Welshans, 2017)

The reality is that chronic surplus production in the dairy and egg sectors in the United States, Europe, and elsewhere, has not been aided by the current policy of providing extensive agricultural subsidies. Only controlling production has meant an effective solution to the boom and bust cycles so often experienced in these agricultural sectors.

In this context of global overproduction, however, export sales to Canada take on new meaning, especially in the United States whose dairy and egg sectors dwarf their Canadian counterparts. American producers recognize this. Chad Gregory, president of United Egg Producers, the largest egg organization in the US, told a meeting of Canadian egg farmers in July 2016, it was a good thing they operated under supply management. Otherwise, he offered, 'we would clobber you' through the export of cheaper American eggs to Canada (Anonymous, 2016). And despite American dairy farmers' desire for a larger market in which to supply their chronic overproduction of milk, they recognize that Canada's system is not to blame, but rather their own inability to address their overproduction issues. In fact, farmers in Wisconsin who are experiencing a significant exodus in their dairy farming community, especially among smaller producers who have been squeezed out by financial pressures, are looking to Canada's system of supply management as a possible solution (Hunter, 2017) The Wisconsin Dairy Farmers Union has recently come out in favour of adopting a dairy supply management system similar to Canada's, and Union leaders have admitted that there have been 'a lot of myths' about what American farmers are told about the Canadian system, and that a system that promotes cooperation rather than competition and conflict is needed (Lee, 2018).

Both dairy and egg producers in the United States are feeling the effects of over-production, but many farm groups continue to be reluctant to take action to slow down the rate of growth and are instead intent on prying open more of Canada's markets as competition for sales in South and East Asia has heated up, in particular with the EU-27 and NZ. And that creates challenges given the influence of the dairy and egg lobbies in Washington. The 2017–18 North American Free Trade Area renegotiation reflects these tensions in that Americans have made what could only be described as outrageous demands of Canada with respect to supply management. The US Trade Representative has demanded that supply management, a cornerstone of Canadian agricultural policy, be phased out over the next decade. While this is unlikely to happen, it has created significant difficulty for Canada's negotiators. To accede to this demand would surely constitute a bad deal and, as the vast majority of Canadians have emphasized in an EKOS poll, the government should 'walk away' from such bad deal. Indeed, the federal Minister of Agriculture and Agri-Food, Lawrence MacAuley, announced at the beginning of renegotiation that 'our government is going to fight to make sure [supply management] stays in place. To deal with anything else is simply a non-starter' (Panetta, 2017). While Washington may be keen to see the end of the Canadian supply management system, few Canadian politicians would support such a move considering the broad support the system receives from both farmers and consumers in Canada.

Despite the support the system generally receives in Canada, some Canadians are opposed to it, largely for ideological reasons. Both national newspapers in the country, which espouse a right of centre perspective, the influential *Globe and Mail* and the *National Post*, have adopted editorial policies in opposition. As well, some private sector think tanks, like the Montreal Economic Institute (MEI) and the Canada West Foundation (CWF) intensely dislike the regulated nature of the paradigm, believing that farmers in the managed sectors should be able to free their inner entrepreneur and participate in a global market. Indeed, the CWF cites export opportunities as a reason to be rid of the system: 'Global markets beckon the Canadian dairy industry, in particular the rapidly growing Asian markets with millions of people now able to afford, and who want, the high-quality food Canadians produce' (Hall Findlay, 2017). However, as exporters know and despite Findlay's rosy projection, there is declining demand in that region for dairy imports as countries like China enhance their own production. With respect to the MEI, it has offered without any evidence that 'If 92% of farms [the percent in Canada not supply managed] manage to succeed without supply management, there's no reason to believe that most of the others could not do so as well' (Moreau, 2017). It maintains this position with no evidence other than a perfervid belief that a completely deregulated world is the best option. Moreover, Moreau's assertion presupposes that the other 92 per cent are doing well, which objective evidence disputes.

Domestic critics also maintain that supply managed dairy and eggs are more expensive than competing products from the US with that country being, to their minds, the natural measure. A 2015 analysis by agricultural economists from the University of Manitoba purported to demonstrate that poorer Canadians pay C$339 more per year than they would in a so-called free market environment.[4] Using a suspect methodology, the study was amplified by the CWF, the MEI and the two national newspapers. It again demonstrates the depth of ideological hostility to a regulated system, and little else.

What was the methodology? First, the article compared only three years of Canadian-American price comparisons for dairy and eggs, 2009–12. The significance of this is that the short time frame distorts the long-term official exchange rate in favour of their argument asserting supply management is a regressive tax on poorer Canadians. If Ottawa would simply get rid of the model in favour of the so-called free market, poorer Canadians would be C$339 better off in an average year. The exchange rate that they employed in their analysis was for the three years noted, during which time the Canadian and American dollars were at roughly a unique balance of US$1 = C$1.05. Over a 27-year period from 31 December 1990 to 31 December 2017, the exchange rate works out to US$1 = C$1.26. Had they used the latter, more realistic and historically accurate figure, supply management would have resulted in a net benefit to poorer Canadians rendering this seminal criticism of supply management among Canadian critics as specious.

Similarly, they compare the price of one US gallon (3.8 litres) normalized to four litres, the size of the Canadian container, with prices from the American Midwest Region. As it turns out, that Region is by far the cheapest for dairy among all official Regions as denoted by the US Bureau of Labor. Milk in big cities in Canada is approximately C$4.27 for four litres. Normalized to four litres and using the 27-year average exchange rate, milk in the American Midwest is the equivalent of C$3.86, cheaper than the product in Canada. However, when any of the other Regions measured by the US Bureau of Labor is treated similarly, the result favours Canada as these examples demonstrate:

Generally, critics of supply management oppose the system because it impinges on the 'free market' they purport functions elsewhere. As already noted, however, the Canadian supply management system has ensured a stable, fair system that mitigates the influence of financial institutions, whereas many farmers in these sectors in other countries are experiencing tremendous upheaval. In the United States, for example, farmer incomes have declined by 50 per cent over the five years from 2013. Producers quit their business or retrench in an effort to cut debt with a concomitant effect on dollars spent in the local community. For example, US dairy production has experienced a consistent surge over the

Table 12.1 Conversion of US milk prices to Canadian $

I	*US price of milk as reported by US Bureau of Labor, January 2017, 1 US gallon (3.8 litres)*	*US milk normalized for exchange rate and 4 litres in C$*
Mid-Atlantic Region	US$3.61	C$4.78
West Region	US$3.42	C$4.53
South Region	US$3.57	C$4.72

Source: US Department of Labor, US Bureau of Labor Statistics. Average retail food and energy prices, US Bureau of Labor Statistics. Average retail food and energy prices, US city average and (insert) region, 2017.

past half-decade while consumer demand has not gone up as much. The result has been a dramatic fall in dairy prices by about 40 per cent over the past four years, and in dairy farmer income. The US National Farmers Union has pointed out the obvious: 'U.S. dairy farmers are experiencing an extended period of very low milk prices which, unless corrected, will force thousands of farmers out of business' with a spill over effect on rural sustainability (Clayton, 2016). The emphasis on size and 'go big or go home' in the US, discussed below, is deleterious for the family farm which cannot compete with this emerging trend and the establishment of mega-farms owned or controlled by agribusiness. This has an immediate and detrimental effect on rural communities.

During the recent extreme downturn in global dairy prices from 2013–16, Canada's producers remained unaffected with a concomitant effect on rural stability unlike the situation in the US. When considering how various agricultural marketing systems function, a more comprehensive approach is needed, one that recognizes that the dynamics of the structures which are sustained and supported in agriculture have real-life implications for farmers and their rural communities. 'What a boon', the *Western Producer* opined, 'it would be to agriculture if efficient farmers could consistently meet their cost of production, realize a profit and provide safe food to consumers' (*Western Producer*, 2015). Further, as food supply chain critic, Sylvain Charlebois, has noted, 'For any organization to be sustainable, it needs to be profitable for everyone across the supply chain: farmers, processors and retailers' (Charlebois, 2016). With SM, it is; in non-supply managed sectors or in jurisdictions which eschew the model, it often is not.

Challenges and the financialization of agriculture

Another concern with respect to the trend favouring the financialization of agriculture is its implications for the vertical integration and corporatization of agriculture, where smaller farmers are forced from their lands and larger farm operations emerge. Supply management has limited the size and scale of farm amalgamation in the sectors it manages because it allows producers to earn a living with more modest-sized farm operations. As pointed out above, the average milking dairy herd in Canada is about 80 cows while for eggs, it is approximately 22,000 layers. In most Western countries, these numbers would be viewed derisively as insufficient. Richard Longworth, for example, has accurately pointed out that

> in the era of globalized agriculture, any American farmer who wants to succeed has to be a mega-farmer.... Each deals with giant global corporations under contracts that dictate not only price but almost every stage of the growing process.
>
> (Longworth, 2008, p. 63)

In the US, for example, the average farm number for dairy cows is 430, but those mid-range farms are in the process of being shuttered. Growth is happening in the 1,000+ range and about 80 percent of all dairy products produced in the United States come from farms of that size. Processing has followed a similar trajectory. Dean Foods, for example, controlled 36 per cent of the American fluid milk supply in 2015, marketed under 55 different regional brands. This concentration of agriculture and processing leads naturally to financialization of the sector given the requirements of 'bigness'.

With respect to layer barns, a recent Cal-Maine/Rose Acre facility in Bogota, Texas houses three million birds. Rose Acre is also constructing a facility near Bouse, Arizona that will host an equal number of birds. The state of Iowa alone has more than 54 million layers on multi-million bird farms, while Indiana and Ohio have about 61 million between them. In total, the US numbers 351 million egg-laying hens, with the great majority of them housed on very large farms located in only ten states. Indeed, with this geographic concentration has come concentration in birds – the average size of egg operations in the US has grown by more than 74 per cent over 15 years, rising from 399,000 in 1997 to more than 695,000 in 2012 (*Factory Farm Nation, 2015*, 2015). The Canadian system with its total layer numbers approaching 22 million is clearly much smaller, even more so than the usual 10:1 ratio that is normally used for US-Canada comparisons. Egg consumption per capita in the former is larger than in Canada, but not markedly so.[5] The US number is suggestive of agribusiness, where finance plays a role equal to or greater than that of the layer corporation in terms of guaranteeing sustainability. Cal-Maine, for example, is traded on the Nasdaq National Market, offering common stock and class A common stock. In 2015, it also accounted for sales of 12 billion eggs in the US totalling 23 per cent of the market.

Conclusion

Financialization of agriculture is a serious issue as Greta Krippner noted at the beginning of this chapter, one that will only become more pronounced with time if regulation in the farm sector continues to be seen as a dirty word. Krippner has defined financialization as 'the tendency for profit making in the economy to occur increasingly through financial channels rather than through productive activities' (Krippner, 2011, p. 4). Indeed, the activities of financial institutions are driving farms to become more responsive to the former's needs especially as the mantra of go big or go home continues to inform behaviour in the farm sector. In the United States, a family farmer is not able to build a 30,000 cow milking operation that addresses concerns about efficiency and economic return in the face of plummeting international prices, or destabilizing and quixotic price levels and fluctuations in the global market more generally.

Those buffeted by international forces require scale to survive, or so the modern mantra has it, which means corporate structures and access to levels of finance that are necessary to fund those operations.

Supply managed systems do not require the same inputs as do those participating in the so-called free market. There is no necessity to grow ever-bigger given the regulated, negotiated approach to pricing. As has been shown, their smaller size means that they are family operations and can survive on family labour. The concept or operationalization of financialization does not affect their farms in the way it does those larger operations. They are not dependent on those market forces that determine their viability, nor are they forced to sell the fruits of their labours at less than the cost of production.

This does not mean, however, that supply managed farmers are not efficient and effective. Quite the contrary. They are early users of new technology like robots and equipment designed to make farm life easier, as well as providing a quality product at a reasonable and competitive cost, as has been shown above. For those products not widely traded on international markets, like dairy and eggs, production controls are critical to a sustainable lifestyle for farmers and for the rural communities that depend on their continuing prosperity.

Supply management does represent an alternative vision to the neoliberal, deregulated and privatized agricultural world which has animated government policy since the late 1970s. Profits are made on these farms through productive activities and, while Canada's banks like the system for the stability it imposes in the sectors in which it operates, the economics of those agricultural sectors mean smaller is beautiful and more sustainable, while also representing a part of the narrative unfolding around Canadian food security and sovereignty.

Notes

1 Longworth noted the same phenomenon 20 years later: 'Why own the farm when you can own the farmers?' (p. 65).

2 These include Jennifer Clapp, Annette Desmarais, Ryan Isakson, André Magnan and Margaret Somerville, among a very few others.

3 This paragraph largely comes from Jostein Brobakk and Bruce Muirhead, 'A tempest in a teapot or a teapot in a tempest? Financialization of Saskatchewan farmland and rural restructuring'. Unpublished manuscript.

4 Ryan Cardwell, Chad Lawley and Di Xiang, (2015) 'Milked and feathered: The regressive welfare effects of Canada's supply management regime', *Canadian Public Policy*, vol 41 no 1, (March 2015), 1–14.

5 In the US, per capita egg consumption is about 270, while in Canada it approaches 255.

References

Bender, Lora, (2015) 'Ontario minister of ag reinforces supply management system', *Peterborough Examiner*, 31 March, available at: www.progressivedairycanada.com/news-topics/industry-news/ontario-minister-of-ag-reinforces-supply-management-system, accessed 2 February 2018

Cardwell, Ryan, Lawley, Chad and Xiang, Di (2015) 'Milked and feathered: The regressive welfare effects of Canada's supply management regime', *Canadian Public Policy*, vol 41 no 1 (March), 1–14

Canada, CDC. (2017) 'Cost of production based on 2016 survey data', November, available at www.cdc-ccl.gc.ca/CDC/userfiles/file/REPORT__PE__2016_COP_Indexed_to_Q3_2017_Booklet__November_2017.pdf, accessed 2 February 2018

CBC News (2017) 'Q&A: Supply management still works, says Ontario dairy farmer', 15 May, available at www.cbc.ca/news/canada/kitchener-waterloo/dairy-supply-management-bruce-sargent-1.4116159, accessed 26 March 2018

Charlebois, Sylvain (2016) 'For the food industry, fraud is the elephant in the room', *Globe and Mail*, 15 March, available at: www.theglobeandmail.com/report-on-business/rob-commentary/counterfeit-products-threatening-the-food-industrys-delicate-balance/article29220689/], accessed 19 March 2016

Clapp, Jennifer (2014) 'Financialization, distance and global food politics', *The Journal of Peasant Studies*, vol 41 no 5, 797–814, DOI: 10.1080/03066150.2013.875536

Clayton, Chris (2016) 'Ag policy blog: NFU asks Congress to boost dairy safety net', 16 December, available at: www.dtnpf.com/agriculture/web/ag/perspectives/blogs/ag-policy-blog/blog-post/2016/12/16/nfu-asks-congress-boost-dairy-safety-2?referrer=NLSnapshot, accessed 19 December 2016

Clements, Mark (2015) 'Who are the world's largest egg companies?' WATT AgNet, 15 January, available at: www.wattagnet.com/articles/20682-who-are-the-world-s-largest-egg-producers, accessed 17 January 2018

Congress Thumbs Its Nose at WTO and the Doha Round: US Federal and State Subsidies to Agriculture, Grey, Clark, Shih and Associates, 2018, available at www.greyclark.com/wp-content/uploads/2018/02/US-Subsidies-Post-2014-Farm-Bill-FEB-2018.pdf, accessed 28 March 2018

Factory Farm Nation: 2015. (2015) Food and Water Watch, May, available at: www.foodandwaterwatch.org/sites/default/files/factory-farm-nation-report-may-2015.pdf, accessed 28 March 2018

Fairbairn, Madeleine (2014), '"Like gold with yield": evolving intersections between farmland and finance", *The Journal of Peasant Studies*, vol 41, no 5, 777–795

Garner, E. and Campos, O. and Paula de la, A. (2014) *Identifying the "Family Farm": An Informal Discussion of the Concepts and Definitions*. ESA Working Paper No. 14–10. Rome, FAO. Gurian-Sherman, Doug, 'CAFOs: The Untold Costs of Confined Animal Feeding Operations', Union of Concerned Scientists, April 2008, available at: www.ucsusa.org/sites/default/files/legacy/assets/documents/food_and_agriculture/cafos-uncovered.pdf, accessed 10 March 2017

Hafemeister, Jason (2018) 'What's next for agriculture in the WTO?' USDA Foreign Agricultural Service, 5 January, available at www.fas.usda.gov/newsroom/what-s-next-agriculture-wto, accessed 15 March 2017

Hall Findlay, Martha with Dalke, Eric (2017) 'Supply Management: A win-win opportunity for reform', Canada West Foundation, 2017 2, available at: http://cwf.ca/wpcontent/uploads/2017/06/CWF_SupplyManagement_Report_JUNE2017.pdf, accessed 5 January 2018

Hamilton, Neil (1994) *Why Own the Farm If You Can Own the Farmer (and the Crop)?: Contract Production and Intellectual Property Protection of Grain Crops,* 73 Neb. L. Rev.

Hunter, Paul (2017) '"We don't blame you": Wisconsin farmers on Trump's blast at Canada's dairy industry', CBC News, 22 April, available at www.cbc.ca/news/world/dairy-farmers-wisconsin-trump-1.4081391, accessed 27 March 2018

Imhoff, Daniel (2012) 'Overhauling the Farm Bill: The real beneficiaries of subsidies', *The Atlantic,* 21 March, available at www.theatlantic.com/health/archive/2012/03/overhauling-the-farm-bill-the-real-beneficiaries-of-subsidies/254422/, accessed 27 March 2018

Isakson, Ryan (2013) 'The financialization of food: A political economy of the transformation of agro-food supply chains: A political economy', Conference Paper #9, International Conference, Yale University, 14–15 September, available at: www.tni.org/files/download/9_isakson_2013.pdf

Johnson, Kelsey (2017) 'Ontario won't "cede one inch" on supply management, says minister', *iPolitics,* 28 April, available at: https://ipolitics.ca/2017/04/28/ontario-wont-cede-one-inch-on-supply-management-says-minister/, accessed 2 February 2018

Krippner, Greta (2011) *Capitalizing on Crisis: The Political Origins of the Rise of Finance,* Boston, HUP

Lawrence, Geoff (2014) 'Book reviews', *The Journal of Peasant Studies,* vol 41, no 3, 421

Lee, Larry (2018) 'Wisconsin Farmers Union Supports Canadian-Style Dairy Supply Management', Brownfield Ag News for America, 5 February 2018, available at https://brownfieldagnews.com/news/wisconsin-farm-bureau-supports-canadian-style-dairy-supply-management/, accessed 28 March 2018

Longworth, Richard C. (2008) *Caught in the Middle: America's Heartland in an Era of Globalism,* New York, Bloomsbury

Magnan, André and Desmarais, Annette (2017) *Who is Buying the Farm? Farmland Investment Patterns in Saskatchewan, 2003–2017,* Canadian Centre for Policy Alternatives

Moreau, Alexandre (2017) 'Does agricultural prosperity require supply management?' Montreal Economic Institute, Augus, available at: www.iedm.org/sites/default/files/web/pub_files/lepoint1317_en.pdf, accessed 5 January 2017

National Farmers Union. (2016) 'Strengthening Supply Management – Defending Canadian control of our market space and advancing food sovereignty', January available at: www.nfu.ca/policy/strengthening-supply-management-defending-canadian-control-our-market-space-and-advancing-foo, accessed 26 March 2018

Nurse-Gupta, Jodey (forthcoming) 'Marketing Milk: The Ontario Milk Marketing Board and the Origins of Supply Management, *Journal of the Canadian Historical Association*

OMAFRA. (2017) 'Allocation of Ontario's supply managed commodities to processors', 19 May, available at: www.omafra.gov.on.ca/english/farmproducts/factsheets/allocation.htm

Palley, Thomas (2007) 'Financialization: What it is and why it matters', Paper presented at a conference on 'Finance-led Capitalism? Macroeconomic Effects of

Changes in the Financial Sector', sponsored by the Hans Boeckler Foundation and held in Berlin, Germany, 26–27 October, available at: www.levyinstitute.org/pubs/wp_525.pdf, accessed 15 May 2016

Panetta, Alexander (2017) 'The final NAFTA bombshell: U.S. demands Canada end supply management for dairy, poultry, eggs', *National Post* 16 October, available at: http://nationalpost.com/news/politics/nafta-u-s-demands-end-to-supply-management-in-canadian-dairy-poultry-eggs#comments-area, accessed 5 January 2018

PCO Cabinet Conclusions (1960), "Dairy support program for 1960–61," 5 April, available at: www.collectionscanada.gc.ca/databases/conclusions/001039-119.01-e.php?&sisn_id_nbr=19610&page_sequence_nbr=1&interval=20&&PHPSESSID=sldjvjr1b84u7ktnk7frv33k32, accessed 15 December 2014

Ross, Lois (2016) '"Gold with yield": A primer on land-grabbing', *Rabble*, 26 September, available at: http://rabble.ca/columnists/2016/09/gold-yield-primer-on-land-grabbing, accessed 17 January 2018

Skerrit, Jennifer (2016) 'Foreign traders moving in on Canadian wheat market', *The Globe and Mail*, 16 June, available at: www.theglobeandmail.com/report-on-business/economy/foreign-traders-moving-in-on-canadian-wheat-market/article30495508/, accessed 17 January 2018

Somerville, Margaret and Magnan, André (2015) '"Pinstripes on the prairies": Examining the financialization of farming systems in the Canadian prairie provinces', *Journal of Peasant Studies*, vol 42, no 1, 119–144

Statistics Canada. (2017) 'Farm and Farm Operator Data: Saskatchewan remains the breadbasket of Canada', 10 May, available at: www.statcan.gc.ca/pub/95-640-x/2016001/article/14807-eng.htm, accessed 15 January 2018

van Praet, Nicholas (2014) 'Land-grabber or farming's future? Quebec's Pangea says bigger is better', *Financial Post*, 12 May, available at: http://business.financialpost.com/entrepreneur/land-grabber-or-farmings-future-quebecs-pangea-says-bigger-is-better, accessed 27 March 2018

Welshans, Krissa (2017) 'Egg market conditions lead to losses for Cal-Maine Foods', *Feedstuffs*, 24 July, available at: www.feedstuffs.com/news/egg-market-conditions-lead-losses-cal-maine-foods, accessed 6 January 2018

Western Producer (2015) 'Editorial – Voters deserve answers on supply management,' 27 August, available at: www.producer.com/contributor/western-producer-editorial/page/9/

White, Ed (2017) 'Has post-CWB marketing freedom been good for farmers?', *The Western Producer*, 10 August, available at: www.producer.com/2017/08/has-post-cwb-marketing-freedom-been-good-for-farmers/, accessed 17 January 2018

13 The local counterparts of financialization

Organizational accomplishments of farm/land investments and sales

Jana Lindbloom

Introduction

The current era of financial capitalism is supposed to be distinct for the way finances circulate in the economy and the mechanisms they create to extract value from finance-linked operations (Erturk *et al.*, 2008a). Even though spare capital, investments, and reinvestments have always been part and parcel of capitalism (Weber, [1930] 2001), the role of financial operations in the present economy as well as the associated changes in organizational and stakeholders' relations co-produce shifts in experienced economic/social realities. They also alter the knowledge and skills needed in order to make sense of – and to address – the various challenges at hand (Krippner, 2005; Erturk *et al.*, 2008b). The new institutionalized mechanisms of managing financial transactions and flows – including innovative banking devices – assist capital in its restless search for new investment opportunities and for quick and increasing returns (Erturk *et al.*, 2008b; Russi, 2013; Švihlíková, 2014).

Recently, and certainly not for the first time in history (Ouma, 2016), agriculture has become an attractive site for financial capital ventures (Borras *et al.*, 2011; Cotula, 2012; McMichael, 2012; Gertel and Sippel, 2016; Clapp *et al.*, 2017). The current international land rush and ongoing farm/land large-scale acquisitions reveal yet another enactment of global finance in potential – and supposedly promising – opportunities for value extraction. The current pattern of financialization has paralleled global economic divides and has thus drawn scholarly attention toward countries of the global South, prompting some authors to deploy the term 'post-colonialism' to describe what is occurring (Robertson and Pinstrup-Andersen, 2010). However, other regions have also received attention, including the post-socialist countries, since their farming regions also attract financial investments (Visser and Spoor, 2011; Visser *et al.*, 2012; Mamonova, 2015). Finance-led farm/land acquisitions appear to be spreading swiftly, seemingly portraying financial investors (whether individual or institutional) in the role of villain and the poor(er) target farming communities in the role of victim.

Important, and often neglected questions, are: How does financialization and farm/land acquisition actually happen? Which actors and which factors enable, facilitate, mediate, or obstruct the relation between farms and finance? It has been pointed out that academic reporting on land grabbing and financialization tends to resort to 'black-boxed' accounts, leaving the concrete processes at site unexplained (Ouma, 2014 and see Chapter 5, this volume). This chapter aims to fill in this gap and contribute to our understanding of how the financialization of agriculture actually happens on the ground, how it is organized and practically accomplished, and how it impacts rural communities. The analysis focuses on the local counterparts of international (but also domestic) finance-investment undertakings. It examines the case of Slovakia, a country in which the current increase in farm/land demand has also substantially reshaped farm/land ownership and use.[1] Drawing on the processual (organizational) approach (Czarniawska, 2004, 2013), it analyses and discusses how the involvement of new outsider investors is practically organized and how the associated property rights are procedurally transferred and rearranged.

The enrolment of a financial investor is not a simple achievement and requires the mobilization of many intra-organizational and inter-organizational activities. This analysis focuses on determining the particular circumstances and factors which either permit and support, or curb and hinder, the entry of an outsider financial investor. At the same time, it also examines the factors which farmers take into an account when deciding whether to affiliate an outsider investor with the farm and what kind of investor/new owner to choose. It provides a close exploration of lived encounters, applied strategies, and puzzling dilemmas which local actors deal with when facing financial capital, as well as the opportunities and challenges it brings.

Agriculture in post-socialist Slovakia

In Slovakia there are three types of farms. First, there are large agribusiness enterprises – legal entities (limited liability companies or joint-stock companies), established via privatization of largely former socialist state farms along with some collective farms. Then, there are small individual or family farms established after 1989. The last type is cooperatives – former socialist collective farms, with privately-owned land plots. Their members are people who had worked on them before 1989 or who were accepted as members after this time. The shareholder structure is organized according to a formula enacted in post-1989 legislation.

Cooperatives have dominated Slovak agriculture, both in terms of the number of farms and their size. Their special feature is that associated members from the local community held property shares. Cooperatives played a significant role in villages' (or even towns') economy, offering employment and public services (Blaas, 1995; Buchta, 2007). All these aspects have made cooperative (and former collective) enterprises a topic of wide public interest (Torsello, 2003; Lindbloom, 2014; Pospech, 2014, p. 102).

Cooperatives (as well as state farms) were large and they stayed large even after the transformation of the 1990s. Slovak farms are among the biggest in the EU and, to a significant extent, it is precisely their large size which makes them sites worthy of pursuing substantial entrepreneurship. In the post-socialist era, there have been basically two main waves of new business interests and shifts concerning (former) cooperatives. The first happened in the late 1990s and early 2000s, which will be referred to as 'insider investments' and the second has taken place since the late 2000s, consisting of 'outsider investments', which is the focus of this analysis.

Insider versus outsider investments

Agricultural cooperatives entered the post-socialist period as enterprises with extremely dispersed share ownership. There were around two million people who had shares in a cooperative farm – either as a member or as a non-member.[2] This situation was regarded as untenable in the long run and unfavourable to the cooperatives themselves. In order to facilitate the consolidation of farms' property structures, new legislation was passed which changed the property shares into shareholder bonds that would allow for and facilitate not only the determination of the shares' nominal value but also the negotiation of the shares' market value.

The transfers of ownership and other property rights has been happening for some time (in cases of many farms since around 2000s, in a few cases even earlier) and has been considered contentious since the beginning. One of the main disputes has involved the relatively low price of shares (the bonds).[3] Especially in prospering coops, members actively working there (particularly members working at higher, managerial positions) have been able to acquire a significant number of shares, usually aiming for and obtaining the majority (Lindbloom, 2014).

Even though these members investments, in most cases classifiable as management buyouts, were often considered unfair, the fact that the leading members were local people and they managed to navigate the enterprise through the tough perils and challenges of the post-socialist transformation outbalanced, to a certain extent, the usual grudges for 'privatizing' the common local cooperative farm. In general, however, despite being criticized, this first wave of investments has nonetheless been less controversial and less publicly discussed than the subsequent and current rush of outsider investors which has swept across Slovakia in the last decade.

There have been two basic types of 'new' investors in the fields of Slovak agriculture: those involved in farming and those focusing merely on land acquisition. Either can be domestic, foreign, or a hybrid of the two. Investments in purchasing farms are regularly followed by investments in acquiring land, with some investments in land being turned into a 'farm grab' as well. These new investors, unlike the known members of the local community or region, are 'outsiders' in relation to both the farm and the village. This category includes farmers from other regions of Slovakia, farmers or farm representatives from other countries managing invested foreign capital, and especially various legal entities, such as financial groups, real-estate agencies, or companies with diverse investment portfolios.

Unlike the previous wave of investments, the current wave differs not only in the number and variety of incoming 'outsiders' but also due to a lack of transparency regarding the identity of the actual investors:

> Nobody knows how much is owned by whom.... The only thing we know is the name of the company or where the firm is located. But we don't know who is behind it, whose money it is.
>
> (High-ranked leader of Slovak agricultural association)

Investor anonymity (with its anticipated or ascribed unpredictability and disloyalty) has stirred public concern. Media articles, a petition, and new legislation have occurred as a result of public uncertainty and suspicion.

In the 1990s and early 2000s, agriculture was not recognized as an area worthy of investment. This was primarily a result of the fragmentation of land ownership, the risky nature of farming, and insufficient subsidization. Agriculture was seen as an area with uncertain prospects and the value of the hectare was vanishing (Verdery, 2003).

The regular payments from the EU CAP funds changed the situation. Financial support for a farm(er) is based on the hectares of cultivated agricultural land. That is, every hectare is an important source of income. Land has become crucial not only for its productive capacity but also as a base for subsidy provision. Moreover, the 2008 financial and food crisis has also turned the spotlight on the long-term quality of land as a financial value keeper, hedge against inflation, along with its value-producing capabilities (Fairbairn, 2014). This triple role has made agricultural land a desired commodity – an object of thoughtful and strategic calculation. Farmers are trying to secure their land and – on the grounds of potential lease withdrawals – seek to gain ownership instead of a rent contract for the land they farm.

The growing interest in land and farms, the extent of property and ownership transfers, as well as the concern about the expanding accumulation of land and farms in the possession of foreign farmers or various financial groups/entrepreneurs (that is companies whose primary area of activity is not agriculture), have altered the options for the current farmers and the prospects for the potential new ones. What is little known are the actual strategies of investors and the exact processes of property transfer. Despite the many voices of concern and disapproval, the sales of both the farms and land continue to thrive.

Four pathways

This section explores 'how the incorporation of farmland and agriculture practically unfolds' (Ouma, 2016, p. 83) in relation to Slovakia. The perspective 'from below' allows the identification of the intra-organizational and inter-organizational circumstances, decisions and considerations which farmers (especially the representatives of cooperative farms) have to take into account when they desire, or feel pressure, to allow an outsider investor to become part of their enterprise. It is argued, below, that there are presently four pathways for investors to access Slovak farmland.

Entry of capital into a farm experiencing financial hardship

The first pathway through which an outsider investor can enter a farm business (usually a cooperative) is by providing capital to an indebted enterprise. Despite the generally improved financial situation of most agricultural operations, achieving and maintaining a positive balance sheet is no straightforward matter. There are many farms that became quite indebted in the 1990s and during the first half of the 2000s. Falling prices for meat, fluctuating prices of grains, or failed harvests (due to weather extremes) frequently made investments into inputs in production a critical issue. In the past, farms unable to pay the debts to their creditors went bankrupt. Currently, however, farms in financial difficulties do not cease to exist but are often taken over either by one of the creditors or by another finance-provider. There were several cases in which the farm's main representative (the Chair of a cooperative) explained the takeover of a farm as a means of solving the farm's financial difficulties:

> The firm lends the farm money.... And if the harvest is bad and prices are low, they cannot pay back. What can they do in such situation? They can only negotiate. And when they see that there is no other alternative, finally they say: Ok, buy us.
>
> (Farmer, high-ranked leader of Slovak agricultural association)

In these cases, farm enterprises allowed for an entry to either a company which had been their long-term business partner (usually the production inputs provider), a financial group, or a group of foreign farmers/financiers. Whether it is a company that used to be in a (different) relationship with the farm before or it is a company which has newly engaged with the farm, there are several diverse mechanisms or options for the involvement of a financier in a farm. One fairly common form of intermingling is a sometimes more – sometimes less – a clear arrangement between an official leaseholder/landlord and a leaseholder:

> ... an investor from [a foreign country] came and he offered to buy their pasture land. They needed money so they sold their pastures to him and now they are leasing this land from him. And this is the mechanism by which it is done. Poor people need money so they often take this wrong step and then they rent the land that used to be theirs.
> (Environmental and agricultural activist)

Agricultural cooperatives in Slovakia still have many members (between 50 and 100, or sometimes even more) and many of them are the farm's retired employees. If there is no majority owner (or a small group of members who together reach a majority of votes), then the decision-making about whether to incorporate an outsider investor (as well as which one, and how) rests on the shoulders of all members. However, due to significant disparities regarding education, experience, and position, the Chair and the members holding managerial positions play the most crucial role in both the decision and the process. Thus, the member(s) in leading positions open the gate and prepare the ground. Such a large and consequential move requires difficult, demanding and laborious interaction and negotiation. It also requires a gradual, step-by-step, process of convincing the ordinary members that this is the best choice given the circumstances.

Direct purchase – a succession option

The second possibility is direct purchase. There are many large farms in Slovakia which are either in private ownership or in semi-private ownership, in a kind of hybrid form. 'Semi-private' enterprises are (formally) cooperative farms in which there is either an individual person or a small group of members or, more recently, a legal person who owns the majority of shares in the cooperative. With regard to property transfer, there is not much difference between these cases.

The transaction is always primarily a matter of agreement between the seller and the buyer even though, to a certain extent, the opinions and preferences of the other co-owners or co-members are normally taken into consideration. However, due to the high market value of a majority share in a large farm, it is extremely hard for the other stakeholders to suggest other viable alternatives. That is, it is unlikely they could devise a better deal than the one on the table.

> They were five of them who owned the farm. Three of them wanted to retire. The other two had no chance to pay their shares. They were given an offer from the Danes.... It was a very nice package. They

> didn't hesitate for a second! … It is always a matter of money. Who gives more [is chosen]. There is no doubt about that.
>
> (Slovak farmer, a friend of the sellers)

With regard to the timing of the dramatic increase in property transfers, it is important to bear in mind the temporal length between the transformative era of the 1990s and the current times of the 2010s. It has been around 20–25 years, so many people who were economically active and put their efforts into reaching and consolidating property in the late 1990s and 2000s (see Lindbloom, 2014) are either retiring or getting close to their retirement age. Thus, since they were not only owners but especially active managers of a farm, the issue of succession has become pressing for many of them. If none of their children displays an interest in inheriting and actively working the farm, the parents-shareholders decide to sell their shares.

Even though the wish to sell their shares has been understandable to both the co-owners and the (local and national) public, the choice of a trading partner has often been contested, and is ambiguous in practice. The interviews suggest that the optimal new owner would be someone of local origin and, ideally, someone who is willing to actually work on the farm. This preferred option seems to clash with the fact that ordinary farmers (those not linked to any outsider financier) or people interested in starting farming, rarely have sufficient capital to purchase a share in a large, developed, mechanized agricultural enterprise. The price, therefore, would have to be far below its current market value. Hence, the sellers prefer and accept alternative offers from outsider investors – either large Slovak agribusiness companies (usually with sub-firms either horizontally or vertically integrated), or foreign companies specializing in farming. The latter are companies officially based in Slovakia but with foreigners as their co-owners, agents or managers, or foreign financiers.

If someone owns a majority share in a large farm, whether a private farm or a cooperative, they regularly receive offers of purchase. From my research, all of the majority owners state they would have no trouble selling their share. They say they know about people, firms or farms that would be eager to buy. Even though the potential buyers are interested in farms in any economic condition, all the farmers included in this research who sold their shares via direct transaction on agreement had shares/bonds in farms that were doing very well.

Covert, hostile takeover

When examining the practice of 'unwelcome seizing', it is important to emphasize that it can be undertaken via two routes, that is, by purchasing farm shares or by grabbing land. A covert, hostile takeover strategy is essentially different from the first two pathways even though, in actual practice, they can be interconnected. A covert, hostile takeover is employed without the consent of the main shareholders of the farm. This is particularly true with regard to land grabbing. Investors (or people/firms acting on their behalf) can contact the land owners and acquire farmland directly. Farmers are informed only after the transaction has been accomplished – for example, when signing a rent contract with the new land owner.

As for the pursuit of farms themselves, this scenario happens only in cooperative operations and predominantly in those that approve trading with shares/bonds but do not yet have a majority owner. Cooperatives with shareholders willing to sell their bonds open up an attractive territory for investment by allowing for the entry of an outsider. Understandably, however, no outsider can really orient itself in the ownership structure unless provided with the assistance of someone from inside. Thus, a gate opener plays a crucial role. Such practice usually involves a great deal of initial secrecy on the side of a member (or other local person) who – with capital from an incoming financier – rapidly buys a significant number of shares/bonds.

Opening trading with bonds/shares permits acquiring a bigger share. The availability or unavailability of finance is the prime factor determining one's likelihood of obtaining more bonds/shares. And, as the interviewees point out, any sudden, unexpected, action of a member to purchase many bonds/shares, or an offer to notably enlarge his/her member's deposit, is viewed with a high degree of suspicion. Especially in the present day, when there are many financial groups or entrepreneurs trying to enter farms, a purchasing outburst might be the sign of an existing alliance between a member (or sometimes, initially, even a non-member) and an investing financier. Hence, these covert, non-transparent pacts are being reported as perhaps the most daunting threat currently affecting cooperatives as quotes from two different cooperative members show: 'The bonds were seemingly bought by the managers [cooperative members]. But you never know who is behind them' (cooperative member).

INTERVIEWER: So in your opinion, whenever someone is doing well and greatly extends their shares, there has to be someone behind them?

COOPERATIVE MEMBER: Not always. There are some people who got it due to their skill. But many are just dummy shareholders and they are doing it for someone else.

Various individuals and legal persons are pursuing shares in farms, some as majority shareholders, but others seeking to eventually own the entire asset. In a cooperative, the only way to enact shareholders' influence is to be a member. If one does not hold membership status it is necessary to act through a legal member. In the most common scenario, a member purchases shares/bonds on behalf of an outsider – until the latter can acquire a significant amount or, if possible, the majority of shares. Another recently applied strategy is to endow an allied member with finances which s/he invests in the farm as his/her additional deposit which increases his/her share in the company's assets. Once becoming a majority owner, the person can subsequently approve the financier as a new member, allowing the transfer of purchased shares to the new member's account. This practice, while despite being formally legal, ravages trust among cooperative members as well as members of the local community.

> And he [the Chair of a cooperative] suggested that a member can insert an additional deposit in the farm but that it should be minimally two hundred thousand euro. Two hundred thousand! At this moment it was obvious because how can an ordinary man have two hundred thousand?
>
> (Cooperative member)

The fact that an outsider can be disguised as an insider member makes such a situation perplexing, upsetting and blurred. Yet another uncertainty is the unknown identity of the people who actually provide the capital. The names of the actual purchasers usually remain undisclosed. First, it is the prevailing practice that investors do not work on the transactions themselves but hire individuals or firms who act on their behalf. In the case of foreign investors, in particular, there are usually multi-layered intermediaries between the farm-members, bond-owners and the original finance-providers. One legal person can be linked to another, and even their representatives can be unknown. As Folkman *et al.* (2007) argue, intermediaries play a crucial role in current capitalism, and closer studies of their actual work on site would indeed help our understanding of the changes of relations between investors and firms. Second, in the current form of financial capitalism, capital can be provided by a variety of entities – banks, networks of financial groups, or various funds (Blackburn, 2006; Erturk *et al.*, 2008b).

To make the portrayal of pathways for intrusive invasion complete, it is necessary to underline land as a highly permeable gateway. Land ownership, like the cooperative shareholder structure, has been fragmented into tiny parcels. Like the cooperative shares/bonds, the heirs to land parcels do not have as strong an attachment to them as their parents or grandparents did. Most of them have moved out of the local community. Usually, they are just the co-owners of the inherited parcel and frequently, they are in need of money. Thus, they are often prepared to sell it, and they benefit from competition among buyers when it inflates prices.

From the investor's point of view, land in Slovakia – especially in the fertile low-land regions – is considered a good investment. Compared to prices in the Western Europe, arable land in Central Europe is much cheaper and prices are expected to rise, thus bringing capital gain once the land is sold in the future. In Slovakia, one of the main procedures of enhancing the value of land is to unite fragmented parcels into one consolidated property which can be later sold for a higher price.

With regard to agricultural subsidies, despite the fact that only farmers – and not the owners – are direct beneficiaries of CAP, buying more land parcels is a sound strategy for generating income. An acquisition of significant acreage allows the owners to pressure the farmer for more rent:

> We looked into the agricultural cooperatives and discovered the amount of finances that flow in there. We explored the amount of subsidies, their possibilities and so on. And we found out that the rent they would give us was laughable. It was only a small chunk of what they were getting.... Your leverage is quite different when you have three hundred hectares compared to just three. If they lost us as their client, they would have to give up their business altogether.... Of course, they came to settle on an agreement.... They had to give up part of their profit.
>
> (A real estate agent working on behalf of foreign investors)

Moreover, since large farms in Slovakia mostly operate on rented land, having an ownership of many hectares empowers the owner not only in rent negotiations but also in potential discussions over the ownership of the farm itself. Basically, there is no crop farm (and few animal husbandry possibilities) without land. If new land owners decide to rent their hectares out to another farmer or if they establish their own farm/firm that would cultivate the parcels, this threatens the existing farmer. Therefore, having a farm with an outsider investor is usually viewed as a better alternative than having a substantially diminished agricultural operation or no land at all.

Invited counter-investor

This fourth variant of investors' enrolment might shed a critical light on the mixed and interactive nature of farm-investor relations. On the one hand, it is undoubtedly an asymmetric power play, an interaction of unequal participants. Ordinary local dwellers (for example, cooperative members) cannot financially compete with the dominating agribusiness companies, foreign farmers (with their backing investors), or foreign or domestic financial groups. Finances matter. Moreover, pressing circumstances may highlight some options as more reasonable and perhaps even inescapable. On the other hand, this process cannot be simply interpreted as a movement initiated and accomplished by the investors and financiers alone. They do not seize the businesses ruthlessly, regardless of a united local opposition. On the contrary, as already partially shown, they generally rely on a local person to 'open the gate' on their behalf. And increasingly, local farms themselves play an active role.

A close analysis of the development in Slovakia reveals that the entry of an outsider investor in a region often results in a spill-over effect. Basically, the encroachment of a strong, expanding investor in one farm makes it probable that this investor – or other outsider investors – will establish their presence in neighbouring farms. In other words, once an outsider investor sets foot in one farm, other farms in the region are very likely to be targeted, as well. To a certain extent, farms owned by (or linked to) an outsider financier seem increasingly to come in clusters.

It could be argued that one of the core empowering drivers might be the different entrepreneurial culture which the outsider investors frequently display. In previous periods, even in cases when trading with shares opened or when farms started to buy the land from owners, there had been one unwritten and – at that time – virtually unspoken rule which was nevertheless strongly respected if one wanted to keep good relations with the other farmers. When purchasing shares or land, the farmer was expected to buy the land s/he farmed, not looking beyond the farm gate. That is, despite it being formally legal, transgression of neighbouring farm's property borders had been considered a strong violation of the informal market morals and norms of the community (Jayyusi, 1993; Mandel and Humphrey, 2002).

The new outsider investors, however, treat cooperative bonds and land as an open market where demand, supply, and price decide what is on offer. To a certain extent, they are not the only ones who advocate this. They have many bond and land owners as allies who do not express their support vocally, but through their sales. The owners welcome the opportunity to have more (prospective) buyers, especially when their competition generates rising prices.[4]

> Now [after the alliance with a financier] we buy land even in the neighbouring village. They have good soil over there, so it is a good strategy. I suppose that colleagues in the neighbouring farms do not like us very much.... But what can I say? A higher offer [always] trumps.
>
> (Chair of a cooperative)

Land- or farm-grabbing can result in bitter conflict between competing legal and moral rights. It can also result in a somewhat paradoxical move: encouraging a counter-investor to outbid the unwanted invader. The struggle has no easy solution. On the one hand, there are the owners of land (or bonds) who appreciate that there is a solvent buyer, and the price matters to them the most. On the other hand, there are the member-workers who would like to achieve or keep their control over the enterprise. In such a situation, finding a trusted investor has the potential to settle the conflicting interests and address the demands of both parties:

> At that time the neighbouring farm started to interfere in our cooperative. They had their own 500 hectares but started being attracted by having more.... So they started to buy our land.... We would not be able to manage it alone. And just at that time, completely unrelatedly, these guys came [foreign investor and a Slovak intermediary firm]. We immediately agreed ... because without land you have nothing. So now they are buying the land.... For us it was a rescue.... At that time, we saw the investor as a rescue[r].
>
> (Chair of a cooperative)

> Several of us put our heads together and decided we have to do something to save the cooperative.... We declared that the cooperative will buy land and shares from those who want to sell but they [the invading company] would keep raising the price. Then we basically ran out of money and we had to ask him [the invited counter-investor] to come.
>
> (Member of a cooperative)

Unlike the invading investor, the invited counter-investor is expected to be a better alternative primarily in terms of the employees' and shareholders' interests and visions. If the farmers conclude that admitting an outsider financier is an inevitable step, they prefer someone who is willing to negotiate the conditions, one who promises to act in a considerate manner. They are fully aware that allowing an investor to get a shareholder majority, or the land, means giving up control over the farm. In these situations, keeping the farm staff employed (via assurance not to downsize) and maintaining the salaries at current levels are among the core requests when settling conditions with an investor.

There is, however, one crucial prerequisite. One has to know who they actually are and, ideally, one has to be able to meet them in person. Based on the interviews, it seems that this characteristic truly matters to farmers. Instead of permitting an unknown, often untraceable, firm to enter their farm, they generally suppose and decide that a known, embodied and preferably local financier would be a lesser evil. They hope for someone who might care for the ordinary employees, members and shareholders, who might have a long-term interest in keeping the farm, and who might play a supportive role in the local community, when needed.

Discussion

Krippner (2005) defines financialization as the rising importance of the financial sector in economy and as an increase of firms' revenues from financial activities such as interest, dividends and capital gains. This chapter deals with the latter and shows how farms in Slovakia have been a target of financial investors. Also, it has been revealed that farms with an outsider investor are apt to act in an expansionary fashion – as financiers of other agricultural organizations. Besides the gains from agricultural production, investing in Slovak farms (also due to their size and via EU membership) bestows the venture with a variety of financial opportunities. These include: shareholder dividends; interest from loans; a buy-and-hold strategy and opting for land ownership as value storage; higher rents for owned land in other farms; the augmented value of acquired enterprises in terms of the economies of scale; and, subsidies. Yet, despite these possible advantages, transforming land into an investible and profitable asset requires the concurrence of numerous factors and mechanisms, with the prospect of success being more fragile than it seems (Li, 2014; Visser, 2017).

For the financialization of agriculture to happen, mobile, abstract capital has to connect with fixed, tangible farmland. As many studies have pointed out, the process is very complex (Clapp *et al.*, 2017). On one hand, domestic investors still play an important role (Deininger, 2011, p. 218) and a significant share of the activities can be characterized as private investments with a seemingly simple and transparent structure. Many investments (also in Slovakia), however, can be categorized as private equity farms/firms (Daniel, 2012) and these, as well as other types of funds involved in the current financialization of agriculture, are accomplished and mediated via a multi-level network of intermediaries (Sud, 2014; Ouma, 2016), that is, organizations acting on behalf of third parties.

According to Czarniawska (2004, 2013) it is much more ontologically accurate and methodologically beneficial to consider and approach organizations (that is farms and financial companies) not as static entities but as an outcome of the processes of organizing. The focus on acts of organizing which reach beyond presumed organizational boundaries can help us to understand the dynamics as well as depict the relationships between agriculture and finance as a practical accomplishment situated in a specific geographical and historical context (Ouma, 2014, 2016).

The case of the financialization of agriculture in Slovakia shows the advantages of pursuing a non-static approach in which intra-organizational processes substantially intermingle with inter-organizational processes and can be interpreted only in relation to one another. For instance, a Chair and other managerial representatives of a cooperative would gradually connect to – or curb connection with – the other members with a financial company by acting on various alternative strategies. Such strategies include agreeing on a loan, conveying lands, intervening in share transactions, or persuading the members (not) to accept a new member with a large deposit. Any action, nevertheless, would inevitably entail careful, balanced, accounting, keeping an eye on a neighbouring farm's affairs, engaging in intra-organizational precautionary measures, contacting lawyers as well as landowners, scouting out possibilities, noticing and interpreting signals, and comparing one's own personal expectations and age with the prevailing institutional expectations.

The determination of triggering events and decisive moments from 'below' (Czarniawska, 2004) can prove as elusive as detecting the identity of the original source of capital that is altering the farm. Was it the bad harvest? Was it the consolidation of shares? Was it the unchallenged fragmentation of shares? A neighbour's ruthless action? Brussel's subsidies and policies? The financial crisis? Outsiders' push and drive? Angst not to miss out on a convenient opportunity? Not trusting others? Trusting others too much? Making sense of the situation and reorganizing are intrinsically intertwined (Weick, 1995; Weick *et al.*, 2005).

Financial actors and local owners represent counterparts in the financialization process. Yet, they must interact. Outsiders, including foreign investors, regard farms and lands in Central and Eastern Europe as having the potential to increase their yield, and as places with scope for land consolidation (Deininger, 2011; interviews). The target countries, on the other hand, ponder whether the investors intend a long-term, socially and environmentally considerate endeavour or aim solely for short-term gains (Zoomers, 2010). They consider whether the investments will contribute to a country's development or to the worsening of local people's livelihoods (De Schutter, 2011), and to what extent capitalization will exacerbate financial and power inequalities, and whether or not the investment is simply speculative trickery (Visser *et al.*, 2012).

Indeed, structural asymmetries and investors' sway play a fundamental role both in international transactions and in domestic investments. Virtually all four identified pathways in the case of Slovakia confirm that outsider investors have the capacity to conquer any potential local competitor whether it be a financially struggling or prosperous farm. At the same time – and this is where institutional theory (Powell and DiMaggio, 1991; Hofreiter, 2011; Scott, 2014) can help to illuminate some cultural aspects which broaden inequality – outsiders devoid of local cultural expectations can, and do, utilize practices which have been normatively not admitted in the repertoire of (virtuous) local farmers (e.g. purchasing the land of other farms).

However, at the same time, besides dwelling on generalizations at the national level, it is crucial to examine localized processes in order to uncover the plurality of circumstances, opinions or strategies which contribute to the unfolding transformation of agricultural and rural settings. Studies by Mamonova (2015), Sippel *et al.* (2017), and – from a somewhat different angle – Hall *et al.* (2015), as well as this chapter, point out that affected stakeholders deal with and accommodate social change differently. Surely, intermediaries represent one of the best incorporated actors in the new game. As for the local farmers and land owners, due to their different social positions and roles (Šubrt, 2017), they are not necessarily allies (Mamonova, 2015). While landowners might profit from the enrolment of an outsider investor in the local setting (as the price of land and rent are likely to increase), local farmers (who compete with financiers) might resist their admission or expansion. Or, as Sippel *et al.* (2017) as well as this research specify, they might accept the outsiders' presence under certain conditions or even accommodate their preferences in light of the opportunities the financial actors provide.

When local people categorize outsider investors, one of the criteria applied in distinction-making is their visibility/invisibility and their transparency/non-transparency. Sippel *et al.* (2017) note that having a positive presence in the local community facilitates acceptance. Also, the possibility of negotiating with the 'real' investor – and not just an agent – can act to diminish suspicion. Both Desmarais *et al.* (2015) and Sippel *et al.* (2017) observe that the non-transparency of land deals heavily undermines trust in local communities. In Slovakia, two particular aspects stood out in the interviews: the role of a (disloyal) gate opener who facilitates a deal, and the untraceable nature of the invested capital. Both generate an atmosphere of uncertainty, suspicion and, not infrequently, anger or resignation.

As the case of Slovakia precisely displays, 'the major factors that organizations must take into account are other organizations' (see DiMaggio and Powell, 1983, p. 150). What happens in one farm is of utmost importance to other farms and therefore, methodologically, it is necessary to analyse and interpret them not as discrete cases of certain actions or reactions but together as movements in interaction. In this vein, and given the widespread financialization of agriculture worldwide, it might also be helpful to remember the concept of institutional isomorphism (DiMaggio and Powell, 1983; Powell and DiMaggio, 1991), the observation that organizations within one field are becoming increasingly similar. Undoubtedly, competitive pressures partially explain this, but the perspective 'from below' (at least in the case of Slovakia) indicates that there is more to the story. As already argued, the spill-over effect in a locality and the act of inviting a counter-investor epitomizes a rather startling passage which does not quite fit the usual narrative.

Inviting a counter-investor does not primarily emerge from the internal financial urgency of the farm, nor does it stem from the wish to make the most of a good sale. Its meaning is ingrained in inter-organizational processes, in inter-farm and inter-investor interaction. Basically, the aim of becoming an investor-backed farm is not only to literally purchase the shares and lands so that there is nothing left for the (possible) competing buyer. In fact, to accomplish such a strategy in its entirety would not be financially and practically viable. The meaning of such a move is, to a great extent, figurative. Within its own field, the farm sends a symbolic warning that it is equipped to defend its fields, to fight back, or even to launch its own conquest, if needed or desired. Hence, besides the conflict over the actual land, it is also a struggle for a legitimate, respectful position in the organizational field where a farm linked to a finance-backed investor might become an institutionalized norm. However, paradoxically, it may be that the farm's status in this inter-organizational network will be redeemed only by sacrificing its internal autonomy.

Conclusion

Local dwellers, farmers and land owners constitute a crucial link in the organizational accomplishment of the financialization of agriculture. While at times it might seem that trans-local, trans-sectoral and trans-national processes are simply imposed on target countries and regions, the analyses of the encounters uncover a plurality of opinions, actions and reactions as well as complexity, procedurality, relationality, interaction and even contingency which such an enrolment involves. Slovakia offers a case which, like any other example, has its geographic and historical specificities that must be taken into account not only in the interpretation of investment, but especially in the process of establishing farm-finance networks themselves. This study distinguishes the peculiarities of local circumstances because that is precisely what finance-actors, their representatives, and their intermediaries, are compelled to recognize and manage.

Despite the particularities of Slovakia (cooperatives, previous managerial takeovers, membership in the EU, farming on rented land, and land fragmentation), it nevertheless shares many characteristics identified elsewhere, such as power asymmetries, a variety of pathways, the connection of financialization to farm succession, differentiation among various financial investors, non-transparency and secrecy with regard to farm/land transactions, opportunity grabbing, a structured impact on local communities, the erosion of trust, and concerns about the jobs and livelihoods of rural inhabitants.

This study has explored the dynamics of establishing the farmland/finance network – including the considerations, decisions, and situated practical activities that co-produce the reordering of agricultural relations. On the one hand, the identified four pathways represent the most common circumstances and decisive moments which render the enrolment of an outsider financier likely. On the other hand, when observed carefully, they display the interactional and inter-organizational features which further clarify the speed and extent of investor-allied farm/land transactions. They point out that the processes of organizing agriculture as a site of not only commodity production but (more significantly) of financial extraction increasingly transgresses individual farms and blurs the notion of organizational borders. Therefore, treating the reactions and actions from below not as discrete (sub)cases but approaching them as interacting and connected constituents of an organizational field can assist in the explanation of contemporary farmland/finance developments.

Acknowledgement

The research was carried out as a part of the project No. 2/0184/15 of the VEGA grant agency.

Notes

1 The study is a part of ongoing research into the post-socialist transformation of Slovak agriculture carried out since 2010. There have been more than 80 in-depth interviews conducted with various stakeholders (primarily the (ex)chairs and members of cooperative farms) of which at least 20 discussed, in detail, new investors, farm sales and large-scale land acquisitions.

2 The category of non-members refers to shareholders who, at the time of the official transformation in 1992 (or earlier – in the case of retirees), had not worked at the cooperative and thus were not its members.

3 There were several reasons for the relatively low market value of the shares. The main one was the (initial) lack of competing buyers. At the same time, since the majority of the shareholders had also been (tiny or more major) landowners, the cooperatives considered paying the rent both as their priority and sufficient compensation to landowners-shareholders. Also, since the individual shares were mostly small, for most people, it has been hard both to exert an influence in the cooperative and to negotiate higher price when selling the bonds.

4 The alliance between the outsider investors and the local owners, however, is a contested one. Not only the farmers but also local authorities (e.g. mayors) and other public figures (e.g. government or parliamentary representatives) voice their concerns and disapproval. While farmers mostly depict the financial groups as the main danger to the contribution which agriculture provides to society, other stakeholders portray foreign land buyers as the main threat to the very existence of relatively tiny Slovakia. Yet, there is a paradox: while audible public opinion is quite negative, the sales themselves flourish.

References

Blaas, G. (1995) 'Agriculture in Slovakia – a forced re-structuring', *Sociológia*, vol 27, no 7–8, pp. 134–141

Blackburn, R. (2006) 'Finance and the fourth dimension', *New Left Review*, vol 39, pp. 39–70

Borras, S.M. Jr., Hall, R., Scoones, I., White, B. and Wolford, W. (2011) 'Towards a better understanding of global land grabbing: An editorial introduction', *Journal of Peasant Studies*, vol 38, no 2, pp. 209–216

Buchta, S. (2007) 'Možnosti a perspektívy riešenia vidieckej zamestnanosti', *Sociológia*, vol 39, no 2, pp. 119–135

Clapp, J., Isakson, S.R. and Visser, O. (2017) 'The complex dynamics of agriculture as a financial asset: Introduction to symposium', *Agriculture and Human Values*, vol 34, no 1, pp. 179–183

Cotula, L. (2012) 'The international political economy of the global land rush: A critical appraisal of trends, scale, geography and drivers', *Journal of Peasant Studies*, vol 39, no 3–4, pp. 649–680

Czarniawska, B. (2004) 'On time, space, and action nets', *Organization*, vol 11, no 6, pp. 773–791

Czarniawska, B. (2013) 'Organizations as Obstacles to Organizing', in D. Robichaud and F. Cooren (eds) *Organization and Organizing: Materiality, Agency and Discourse*, Routledge, New York, pp. 3–22

Daniel, S. (2012) 'Situating private equity capital in the land grab debate', *Journal of Peasant Studies*, vol 39, no 3–4, pp. 703–729

Deininger, K. (2011) 'Challenges posed by the new wave of farmland investment', *Journal of Peasant Studies*, vol 38, no 2, pp. 217–247

De Schutter, O. (2011) 'How not to think of land-grabbing: Three critiques of large-scale investments in farmland', *Journal of Peasant Studies*, vol 38, no 2, pp. 249–279

Desmarais, A., Qualman, D., Magnan, A. and Wiebe, N. (2015) 'Land grabbing and land concentration: Mapping changing patterns of farmland ownership in three rural municipalities in Saskatchewan, Canada', *Canadian Food Studies*, vol 2, no 1, pp. 16–47

DiMaggio, P. and Powell, W. (1983) 'The iron cage revisited: Institutional isomorphism and collective rationality in organizational gields', *American Sociological Review*, vol 48, no 2, pp. 147–160

Erturk, I., Froud, J., Johal, S., Leaver, A. and Williams, K. (2008a) *Financialization at Work: Key Texts and Commentary*, Routledge, London

Erturk, I., Froud, J., Johal, S., Leaver, A. and Williams, K. (2008b) 'General introduction: Financialization, coupon pool and conjuncture', in I. Erturk, J. Froud, S. Johal, A. Leaver and K. Williams (eds) *Financialization At Work: Key Texts and Commentary*, Routledge, London, pp. 1–43

Fairbairn, M. (2014) ' "Like gold with yield": Evolving intersections between farmland and finance', *Journal of Peasant Studies*, vol 41, no 5, pp. 777–795

Folkman, P., Froud, J., Sukhdev, J. and Williams, K. (2007) 'Working for themselves: Capital market intermediaries and present day capitalism', *Business History*, vol 49, no 4, pp. 552–572

Gertel, J. and Sippel, S. (2016) 'The financialization of agriculture and food', in M. Shucksmith and D. Brown (eds) *Routledge International Handbook of Rural Studies*, Routledge, London, pp. 215–226

Hall, R., Edelman, M., Borras, S.M. Jr., Scoones, I., White, B. and Wolford, W. (2015) 'Resistance, acquiescence or incorporation? An introduction to land grabbing and political reactions "from below" ', *Journal of Peasant Studies*, vol 42, no 3–4, pp. 467–488

Hofreiter, R. (2011) 'Nová ekonomická sociológia a štruktúrny kontext trhu', *Sociológia*, vol 43, no 5, pp. 528–549

Jayyusi, L. (1993) 'Values and moral judgement: Communicative praxis as moral order', in G. Button (ed.) *Ethnomethodology and the Human Sciences*, Cambridge University Press, Cambridge, pp. 227–251

Krippner, G. (2005) 'The financialization of the American economy', *Socio-Economic Review*, vol 3, pp. 173–208

Li, T. (2014) 'What is land? Assembling a resource for global investment', *Transactions of the Institute of British Geographers*, vol 39, no 4, pp. 589–602

Lindbloom, J. (2014) 'From workers' cooperative to hired-labor farm: Accommodating to global market pressures in Slovak post-socialist agriculture', in A. Bonanno and J. Cavalcanti (eds) *Labor Relations in Globalized Food*, Emerald Group Publishing Limited, UK, pp. 115–141

McMichael, P. (2012) 'The land grab and corporate food regime restructuring', *Journal of Peasant Studies*, vol 39, no 3–4, pp. 681–701

Mamonova, N. (2015) 'Resistance or adaptation? Ukrainian peasants' responses to large-scale land acquisitions', *Journal of Peasant Studies*, vol 42, no 3–4, pp. 607–634

Mandel, R. and Humphrey, C. (eds) (2002) *Markets and Moralities. Ethnographies of Postsocialism*, Oxford, Berg

Ouma, S. (2014) 'Situating global finance in the land rush debate: A critical review', *Geoforum*, 57, pp. 162–166

Ouma, S. (2016) 'From financialization to operations of capital: Historicizing and disentangling the finance-farmland-nexus', *Geoforum*, 72, pp. 82–93

Pospech, P. (2014) 'Discursive no man's land: Analysing the discourse of the rural in the transitional Czech Republic', *Journal of Rural Studies*, vol 34, pp. 96–107

Powell, W. and DiMaggio, P. (eds) (1991) *The New Institutionalism in Organizational Analysis*, The University of Chicago Press, Chicago

Robertson, R. and Pinstrup-Andersen, P. (2010) 'Global land acquisition: Neo-colonialism or development opportunity?', *Food Security*, vol 2, no 3, pp. 271–283

Russi, L. (2013) *Hungry Capital: The Financialization of Food*, Zero Books, Winchester, UK

Scott, W. (2014) *Institutions and Organizations: Ideas, Interests, and Identities*. 4th ed., Sage, London

Sippel, S.R., Larder, N. and Lawrence, G. (2017) 'Grounding the financialization of farmland: Perspectives on financial actors as new land owners in rural Australia', *Agriculture and Human Values*, vol 34, no 2, pp. 251–265

Statistics Canada. (2017) 'Farm and farm operator data: Saskatchewan remains the breadbasket of Canada', 10 May. Accessed on 15 January 2018 at www.statcan.gc.ca/pub/95-640-x/2016001/article/14807-eng.htm

Šubrt, J. (2017) 'Homo sociologicus: Co se skrývá za tématem sociálních rolí', *Sociológia*, vol 49, no 2, pp. 203–230

Sud, N. (2014) 'The men in the middle: A missing dimension in global land deals?', *Journal of Peasant Studies*, vol 41, no 4, pp. 593–612

Švihlíková, I. (2014) *Přelom: Od velké recese k velké transformaci*, Inaque.sk, Bratislava

Torsello, D. (2003) *Trust, Property and Social Change in a Southern Slovakian Village*, Lit Verlag, Münster

Verdery, K. (2003) *The Vanishing Hectare: Property and Value in Postsocialist Transylvania*, Cornell University Press, Ithaca

Visser, O. (2017) 'Running out of farmland? Investment discourses, unstable land values and the sluggishness of asset making', *Agriculture and Human Values*, vol 34, no 1, pp. 185–198

Visser, O. and Spoor, M. (2011) 'Land grabbing in post-Soviet Eurasia: The world's largest agricultural land reserves at stake', *Journal of Peasant Studies*, vol 38, no 2, pp. 299–323

Visser, O., Mamonova, N. and Spoor, M. (2012) 'Oligarchs, megafarms and land reserves: Understanding land grabbing in Russia', *Journal of Peasant Studies*, vol 39, no 3–4, pp. 899–931

Weber, M. ([1930], 2001) *The Protestant Ethic and the Spirit of Capitalism*, Routledge, London

Weick, K. (1995) *Sensemaking in Organizations*, Sage, Thousand Oaks

Weick, K., Sutcliffe, K. and Obstfeld, D. (2005) 'Organizing and the process of sensemaking', *Organization Science*, vol 16, no 4, pp. 409–421

Zoomers, A. (2010) 'Globalisation and the foreignisation of space: Seven processes driving the current global land grab', *Journal of Peasant Studies*, vol 37, no 2, pp. 429–447

14 Financialization and the Swedish Land Acquisition Act

Elin Slätmo

Introduction

Research on financialization shows that financial actors are playing an increasingly active role in the ownership and control of food and farmland (Isakson, 2014; Clapp, 2016). Increased dominance of financial motives, financial actors, financial markets and financial institutions in access to, and control over, farmland has been reported in global South contexts such as Tanzania and Cambodia (Cotula *et al.*, 2009; Robertson and Pinstrup-Andersen, 2010; Fairbairn, 2015; Bergius *et al.*, 2017) and in global North contexts such as Europe, Australia and Canada (Franco and Borras, 2013; Magnan, 2015; Sippel *et al.*, 2017).

One expression of this dominance is the influence of financial logics in how problems and solutions are framed in policy. As in many other countries, the agricultural sector in Sweden is perceived as having problems with growth and access to capital (FAO, 2012; Swedish Board of Agriculture, 2015; Statistics Sweden, 2016). Solutions to the perceived problem of lack of capital in the agricultural sector were put forward in a report following a state investigation in 2015 (SOU, 2015). Among the range of different measures and changes to laws and regulations suggested in this report, a fundamental alteration in the Swedish Land Acquisition Act (SFS, 1979) was proposed to increase corporate ownership of agricultural land. The state report does not explicitly discuss who would gain from such policy changes or the possible negative effects for existing (referred to here as 'active') farmers. Therefore, in this study, Swedish state documents, empirics from encounters with active Swedish farmers at two fairs, and material from the Swedish press, are analysed through the lens of research on rights to land and financialization of food and farm land, to investigate possible effects of the suggested policy changes in Sweden.

Corporate involvement and financial investment in farmland

Financial investment in farmland involves investment by capital-rich nations and/or affluent corporations/individuals in agricultural land in countries where the climatic conditions for food production are good and where the economic risks are perceived as low (Cotula *et al.*, 2009; Hallam, 2009; Sippel *et al.*, 2017). Importantly, drawing on the cases of Australia and Canada, Sippel *et al.* (2017) and Magnan (2015) show that global North contexts such as Sweden can also be attractive to investors as it is assumed to have 'low-risk environments' in terms of stable government, well-established investment laws, and strong private property rights.

Researchers such as McMichael (2012) and Ouma (2014) report that the standard narrative which has evolved in investment brochures and industry intelligence in the aftermath of the 2008–11 financial crisis is that investment in farmland guarantees stable returns on capital and that, due to climate change and the scarcity of land, the value is likely to grow. Moreover, investment in agricultural land is driven by a range of other factors, including the rise in global food prices, concerns about food security in nations with growing populations, and general acceptance of biofuels as a viable alternative source of energy (Robertson and Pinstrup-Andersen, 2010; Cotula, 2012; Magnan, 2015).

Positive aspects of financialization of farmland

Some studies describe potential positive effects of an increase in corporate-owned farming and capital investment in agricultural land. However, these effects seem to rely on the host government's ability to steer and regulate the financial actors in terms of tax systems, lease terms, contract obligations, environmental regulations and so forth. For instance, Cotula (2012), Hallam (2009) and Robertson and Pinstrup-Andersen (2010) argue that contractual obligations concerning, for instance, infrastructure and service investments in rural areas should be a primary concern for host governments considering land investment and sales to corporate parties. Those studies also acknowledge that an increased influence of financial actors can lead to 'spill-over effects' in terms of capital, knowledge and technology transfer to local, contract-based, farmers.

Negative aspects of financialization of farmland

The positive effects reported above must be coupled with insights on the risks and potential negative effects related to farmland investments. Researchers such as Clapp (2014, 2016) and Isakson (2014) have demonstrated that increased financialization can have negative effects on food workers and small farmers, as financial logics tend to prioritize the interests of actors in powerful positions. Furthermore, a review by Robertson and Pinstrup-Andersen (2010) highlights that investment in farmland by financial actors carries risks of food insecurity, irreversible environmental degradation due to more short-term management, and loss of livelihood and farming knowledge, when financial actors gain more control. There is also instability caused by conflict over rights to exploit natural resources. Moreover, research on foreign investment in agricultural land in global South contexts show that, from the perspective of the local population, large-scale land investments are deeply problematic for self-sufficiency and rights to land (Cotula, 2012; GRAIN, 2012; Olsson *et al.*, 2012; Clapp, 2014; Bergius *et al.*, 2017). Extensive control of land by other countries or corporations can raise questions about political interference and influence over territory (Hallam, 2009).

In order to avoid severe negative effects such as local conflicts, failed projects, and investment withdrawal (Ouma, 2014; Magnan, 2015), it is of fundamental importance to acknowledge that, for many farmers and, particularly, Indigenous people, cultural identity and livelihood are directly tied to the land (Quinn and Halfacree, 2014). Therefore, land that is valued and designated as 'degraded' or 'for development' by public authorities might still have other emotional, cultural heritage or religious values for local populations. These values are not directly translatable into the economic logic of financial actors, which is why the integration of different types of 'knowledge' is important to shape legitimate and sustainable decision-making processes (Daugstad *et al.*, 2006; Grubbström and Sooväli-Sepping, 2012; Howley *et al.*, 2015; Suess-Reyes and Fuetsch, 2016).

Nuancing the picture of the financialization of farmland

In a study investigating how increased financialization of farmland takes shape at the local level in rural Australia, Sippel *et al.* (2017) report that rural residents and farmers perceive and respond to financial actors with a mix of acceptance, accommodation and unease. Those authors observed that some farmers appreciate the increased interest by financial actors in farmland, as that can allow them to continue their farming through increased access to capital, while other farmers view the changing access to land resources as a threat to the continuation of family farming. In previous studies on farming in Sandnes in Norway and Hållnäs in Sweden, I found that such divergent responses and perspectives are also evident in the Scandinavian context (Slätmo, 2014, 2016). Importantly, as described in the next section, none of the agricultural land in Sweden is currently directly owned by foreign nations or foreign corporations. However, foreign and domestic corporations and other financial actors are active and dominant in the food and agricultural sector, from the purchase of seeds, machinery and diesel to the selling and marketing of food. This is, in part, due to the market-oriented economic reforms that have been put in place over the past 20 years (Kristensen and Grundfelder, 2018).

Context: farming and land ownership in Sweden

In 2015, around 7 per cent of Swedish territory was designated as agricultural land by public authorities (Swedish Board of Agriculture, 2015). The Swedish agricultural sector is dominated by individual land ownership and family farms. Figures from Statistics Sweden (2013) show that in 2010 some 90 per cent of Swedish farmland was privately owned by individuals. The rest was divided between Swedish corporations (3 per cent), estates (1 per cent), other/unknown (3 per cent), municipalities and regions (2 per cent), the state (1 per cent) and the church (0.1 per cent). The relative proportions of these different owner categories have remained relatively stable over time, although private land ownership has increased slightly in the past ten years, while state and municipal ownership has decreased (Statistics Sweden, 2013).

This structure is positive for access by individuals to the basic prerequisite to produce food (that is, agricultural land). However, on a national level only about 30 per cent of agricultural land in Sweden was fully owned by active farmers in 2010, which means that farming in the Swedish agricultural sector is dominated by partly- or wholly-leased land. A large proportion of the Swedish food production, therefore, rests with farmers who do not own the land upon which they are dependent for their survival and livelihood (Statistics Sweden, 2013, p. 48).

The standard narrative regarding the Swedish agricultural sector is that there have been increasing difficulties in achieving profitability because of aspects such as increased competition with farmers in other EU member states, a relatively short growing season, relatively strict regulations on environmental protection, animal welfare, and, high labour costs. And, as a consequence, the number of farms in Sweden has decreased rapidly during the past 50 years and the remaining farms have become larger and more specialized (for example, in cereal, dairy and broiler production). In addition, about one-third of farmers combine farming with other activities such as contracting, forestry or tourism. Only about 1.5 per cent of the total Swedish labour force is employed in agriculture (Swedish Board of Agriculture, 2015).

Figures from Statistics Sweden (2016) and the Swedish Board of Agriculture (2015) show that the value of production was 51 billion Swedish Crowns (5.7 billion euro) in 2014, which corresponds to about 0.5 per cent of Swedish GDP. Approximately half of this value came from crop production, most importantly cereals, and half from livestock production (about half of which was associated with dairy production). Livestock production, which in recent years has predominantly been dairy production, has struggled to achieve profitability, while cereal production on the southern plains has shown more stable economic results.

During the 1900s and 2000s, the agricultural sector in Sweden focused upon production for the domestic market (Antonsson and Jansson, 2011).

In a report, following a state investigation, (SOU, 2015) it was recommended that the Swedish agriculture sector could be altered in regard to a wide range of changes to environment, animal welfare and land use regulations. These suggested changes are perceived as needed to increase food production rates and economic growth, as well as enhance the competitiveness of Swedish agriculture on the global market:

> The investigation states that labour costs, taxes on means of production, regulations for animal welfare and plant protection, applications according to the Swedish Environmental Code and the practical use of regulations, have great impact on the agricultural and horticultural competitiveness.
>
> (SOU, 2015, p. 12)

This problem formulation of 'competitiveness' can be regarded as an indication of an attempt to create a 'new' financial, neoliberal and export driven logic for the Swedish agricultural sector (Lambin and Meyfroidt, 2011; Clapp, 2014). Importantly, however, the policy changes that are required for such a shift in production focus in Sweden are likely to undermine the values that are used for marketing for the domestic market (for example, environmental and animal welfare regulations, as well as the relative strong labour rights, in relation to EU-standards).

Methods

The methods used for analysing agricultural land ownership in this study were the interrogation of Swedish state documents, interviews and interactions with active farmers, and an analysis of articles in the Swedish media. The state documents included in the study were the Swedish Land Acquisition Act (SFS, 1979), the amendments to the law (SFS, 2010) and the state report *Attractive, Innovative and Sustainable – Strategy for a Competitive Agricultural and Horticultural Sector* (SOU, 2015).

The interviews and interactions with active farmers took place in June and July 2016, when I was engaged in presenting and discussing research on the future of Swedish agriculture at the Borgeby agricultural show in Skåne and the political fair in Almedalen, Gotland. During these events, I interacted with a range of different stakeholders involved in agriculture: national, regional and local politicians, public authorities, bank officials and other financial actors, lawyers, representatives of non-government organizations (NGOs) and civil society, farming and other press, researchers, consultants and 'ordinary' rural and urban residents, among whom were many active farmers. The Borgeby show (June 2016) was held on former arable fields outside Lund and attracted some 22,100 visitors. The farming focus of the show was apparent given the high number of representatives from the farming industry, as well as those selling seeds, fertilizers and machinery. There were also small demonstration plots of different crops, as well as seminars by researchers, consultants and companies. During my two days of making presentations (under the heading 'Future farming in Sweden'), I talked to approximately 30 farmers, mostly from southern Sweden. I did not discuss the issues of finance and land ownership with all of them, but discussed social, economic and environmental issues that were important in their farming lives. These other issues included soil governance, plant protein for feed and food, mechanisms for interaction with consumers in urban areas, and EU policy after Brexit.

The annual one-week political fair at Almedalen in July is a well-known democratic arena in Swedish society that attracts a large number of representatives from political parties, NGOs, financial actors, researchers and citizens to meet, discuss, and mingle in the medieval town of Visby. In 2016 the event had approximately 40,000 participants and visitors. Unlike the annual Borgeby show which has a clear farming and 'rural' focus, Almedalen is more of an 'urban' arena, both in terms of location and content. I undertook research in Almedalen for two of the seven days of the event in 2016. During and after my seminar on the future of Swedish farming, I interacted with approximately 50 people, around eight of whom were active farmers from Gotland. We talked about the issues that were important to them for their farm livelihoods, such as the importance of closing the nutrient gap between urban and rural areas and new strategies for selling local products. We also discussed the proposed changes to the Land Acquisition Act that the state report (SOU, 2015) had suggested.

To complement the material, I interviewed officials at the regional authority in Halland in September 2017 and ran a search in the media database Mediaarkivet retriever on 26 September 2017, concentrating on the Swedish printed press (850 sources) and internet press (293 sources). The search terms used were: Jordförvärvslag* (in English: Land Acquisition Act), and articles also including one of the Swedish words: jordförvärv*, jordbruks*, mark*, storbolag, aktiebolag, konkurrenskraft* (land acquisition, agricultural, land, corporation, limited company, competitiveness). The search period was 1 January 2015 to 25 September 2017, allowing coverage of the period from when the suggested policy changes were announced, until recently.

The search generated 276 hits, 202 articles, many in the farming press but also in other national and regional newspapers, government documents and the websites of NGOs and political parties. The search results showed that the issue has been increasingly reported and debated in the search period (Figure 14.1).

I printed out, sorted and read the material with the purpose of identifying differing views on the proposed policy changes regarding who should be allowed to own agricultural land in Sweden. Articles from 2015 and early 2016 mentioned the suggested policy change, while articles from

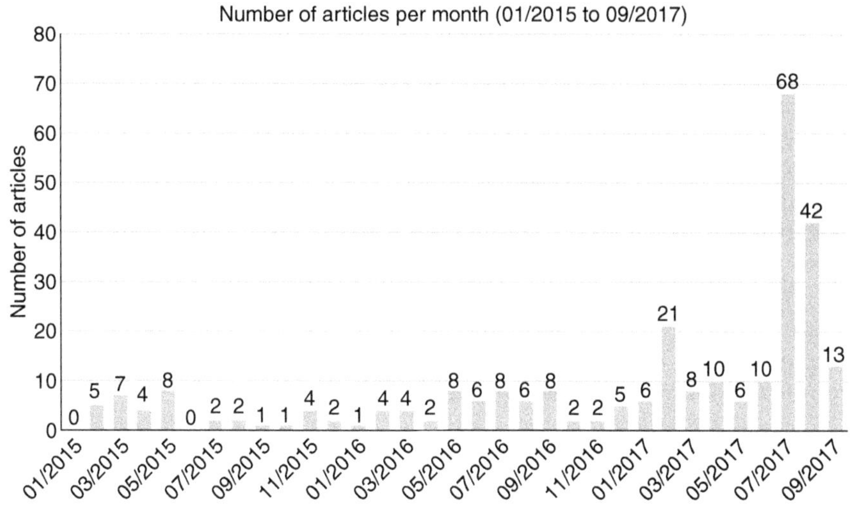

Figure 14.1 Number of articles per month discussing the Land Acquisition Act in Sweden, during the search period January 2015 to September 2017.

Source: Mediaarkivet retriever, 26 September 2017.

2016 onwards more clearly discussed the pros and cons. There were also more opinion pieces for and against the suggestion to change the legislation. The material also included reports on different sales and purchases and how the current version of the law had been enacted.

In total, 113 individual pieces were included in the final analysis of the press material (Table 14.1). With the aim of capturing some voices from active farmers, 36 articles and opinion pieces by, and about, private

Table 14.1 Final list of documents and articles included in the study

Type of document	*Number of articles (total 113)*
Journalist writing/article	53
Government documents (propositions and protocols)	7
Opinion piece from national politicians	11
Opinion piece from Swedish Environmental Protection Agency (Naturskyddsföreningen)	5
Opinion piece from Hela Sverige ska leva	1
Opinion piece from individual forest and farm owners	10
Opinion piece from National Farmers' Union (Lantbrukarnas Riksförbund)	12
Opinion piece from La via Campensina Sweden (NOrdBruk)	3
Opinion piece from the Swedish Smallholders' Organisation (Förbundet Sveriges småbrukare)	11

landowners and farmers' organizations were scrutinized. It should be noted that the discrepancy in numbers between these 113 articles and total hits (276) is partly a result of publication on multiple sources. The collected empirical material from state documents, interviews and media was analysed in order to identify who would gain from the proposed policy changes and the possible negative effects active farmers would experience by such changes. In order to capture differing views on financialization – and on agricultural and farmland ownership – in Sweden, the empirical material was analysed both for content and for underlying meanings (Jørgensen and Phillips, 2002; Silverman, 2010). The results are presented in the following sections, starting with a description of the current Land Acquisition Act in Sweden. An analysis of the proposed policy changes and responses from active farmers is then presented. In a final section, the results are summarized and discussed.

Results and analysis

The Land Acquisition Act to ensure continued family farming and rural development in Sweden

The current Swedish Land Acquisition Act was introduced in 1979 to control who has the right to own land. In particular, the law was intended to protect small/family farms from being purchased by forest corporations and to ensure that farmland would be owned by people living in close proximity to the land. In fact, the law states specifically that owners should live in peripheral rural areas.

Among other things, the law (SFS, 1979) prescribes that corporate entities (largely, corporations) must apply for a permit to purchase a farm unit. The permit is issued by the regional authorities or the relevant state body in cases where it is necessary for the state to interfere. Such a permit is only issued in special circumstances, for instance if a farm is rated at too high a value for cultural heritage and at the same time is too expensive for a single farmer or business owner to run and care for. According to SOU (2015, p. 65), around 250 permits were issued in the year 2013, mainly to 'limited companies' in the forestry business.

In a book on forest investments, af Kleen (2010) shows that this type of permit has been given to large estates that have been in the possession of Swedish noble families for generations, as a 'solution' to keep the farm-based estates intact. When discussing the matter with the relevant officials at regional authorities in the county of Halland in September 2017, it was reported that, although applications from corporations seeking to buy agricultural land have increased in recent years, permits are never issued. However, the regional authorities guide the affairs of land ownership with the purpose of keeping a balance between different land ownership categories. In practice, this means that if a corporation wants to buy farmland, it can do so if it first sells other land, whether agricultural or forest, to a private individual.

It should be noted that there have been some recent changes in the law stating, for instance, that if agricultural land is designated a 'development area' in spatial planning, no permit is needed for a corporation to purchase that agricultural unit (SFS, 2010). This is because it is no longer categorized as agricultural land by the relevant authorities.

As mentioned above, in 2015 a report on a state investigation concluded that, in order for the Swedish agricultural sector to fulfil the goal of competitiveness, a change in those allowed to own agricultural land would be necessary (SOU, 2015). The report proposed that the Swedish Land Acquisition Act should be changed in order to increase the possibilities for limited corporations to purchase agricultural land in order for efficient purchases for big and capital-intensive corporations (SOU, 2015, pp. 12, 64–66, 159–160). At the time of writing, the Swedish government had tasked the Swedish University of Agricultural Sciences with investigating the details regarding the changes in the Land Acquisition law.

The results from the media analysis of the media between 2015 and 2017 showed that some political parties, state officials, corporations and major forest owners agreed on the need to change the Swedish Land Acquisition Act. Others – political parties, land and forest owners and a majority of NGOs – are clearly against such a change. In the following sections the motives behind the suggested policy change and some of the responses are elaborated.

Motives for the proposed changes in the law that regulates farmland ownership

The official report from 2015 argues that the current version of the Swedish Land Acquisition Law hinders competitiveness. Going through the report it is argued that changes in the business structure of the Swedish agricultural sector and an increase in capital intensity are viewed as reasons to question the current land ownership regulations (SOU, 2015, pp. 159–160). Further, the report states that the increase in capital-intensive farming that has taken place in the sector is a legitimate reason to further increase the corporate structure of ownership, as it is assumed that this is linked to growing financialization. From the perspective pursued in the report this increase in corporate activity is positive; it is the strategy proposed to meet the goal of a sustainable, attractive and innovative agricultural and horticultural sector in Sweden, one based upon economic growth and value creation (SOU, 2015, pp. 12, 159–160). In an article in farm press the state official and social democrat Rolf Annerberg, who was the main author of the official report, explain the need to change the law: 'The capital market is functioning well (but) the Land Acquisition Act restricts capital investment and sales of agricultural land and comes with administrative costs' (Jordbruksaktuellt, 2015). Furthermore, the Swedish minister responsible for agriculture and rural areas, the social democrat Sven-Erik Bucht, argued for a change in the law, stating that the development towards increased capital-intensive farming is desired: 'Shall we, through legislation, limit positive development in primary production?' (Ektander, 2015).

In analysing the state report and the results from the media coverage, this 'financial' or 'free market' logic seems to be the only legitimate base for the suggested policy changes. There are no other arguments or perspectives applied. Importantly, it should be noted that the official report might have reached other conclusions, concerning both problems and solutions, if it had started with a more 'balanced' perspective on farming and agriculture. Agriculture is important both for the livelihood of private farmers and for public access to healthy and affordable foods and high-quality farm land (Slätmo, 2017). This latter, societal, perspective on farming and agriculture, is generally lacking in the official report suggesting the policy changes (SOU, 2015).

Although it can be agreed that economic aspects are important, there are risks that a single focus will neglect environmental, cultural and societal aspects of development. It is also not well-elaborated in the state report which economic developments are desired. As pointed out by Robertson and Pinstrup-Andersen (2010, p. 272) in their review article on large-scale purchases of agricultural land, 'it's important to not conflate corporate investment and economic development'.

Another argument used by the report to justify a need to change the law is that it complicates farm succession and sales of farmland to big and capital intensive-farms (SOU, 2015, p. 12). What is not mentioned is that this 'complication' could be needed to safeguard the current ownership structure of farmland. This raises the question of what active farmers might think about the issue and therefore I assessed this with the audience at the agricultural show and political fair in 2016.

Responses to the suggested policy change from Swedish farmers

When asked directly whether the shift in ownership or succession would be smoother if land were in corporate ownership, the farmers at Almedalen said that would not be the case. Instead, other economic aspects, such as securing long-term financial support in seasons when production levels were low, were perceived as more important to continue farming and to enable farm succession. Similarly, discussions with farmers at Borgeby revealed that other issues were more important for them, such as how to get banks to lend them money to invest in such things as machinery and farm buildings, along with the lack of legal expertise on issues concerning land and business ownership, especially in relation to farm succession.

Other non-economic issues were also mentioned in the discussions with the farmers. As revealed in studies on the agency of farmers, there are complex underpinnings of decisions to continue farming and agricultural activities and land use cannot be understood by applying economic models alone (Howley *et al.*, 2014; Slätmo, 2016). Such studies report that, besides increasing profit, internal drivers for continued farming are: a real interest in working with soil and animals; consideration of past and coming generations; and, a societal responsibility to produce food (Raggi *et al.*, 2013; Slätmo, 2014; Saunders, 2016).

As the farmers with whom I interacted at Almedalen and Borgeby represent only a small fraction of the total number of farmers in Sweden (around 60,000–70,000), these results are not generalizable. Rather, they can be taken as an indication that other issues, rather than corporations being prevented from owning agricultural land, could be more important to active farmers. To place this indication in a broader empirical setting the results from the media search were read to learn what other active farmers said about the proposed changes of the law. In a joint opinion-piece on the land ownership structure, the Swedish National Farmers' Union (NFU) and its youth division stated that the Swedish Land Acquisition Law should be left in its current form (Jonsson *et al.*, 2016). Acknowledging what the NFU sees as the purpose of the law – to ensure local land ownership that enables a vital countryside, to prevent corporate entities from owning agricultural land, and to create possibilities for long-term and responsible management of soils and forests – changes to the law were not viewed as a priority by the NFU (also see NFU, 2017a; Odell, 2017; Sandberg, 2017; Törner, 2017). The head of the Swedish NFU, the Swedish NFU forest division and the NFU youth division jointly expressed their unwillingness to change the law: 'Long-term sustainable business in rural areas is created through increased economic viability, innovation and access to capital – not by letting corporations buy land more freely. Thanks for the thought, but no thanks!' (Jonsson *et al.*, 2016, p. 2). Following this, the deputy head of the NFU, Åsa Odell, reported that the Land Acquisition Act has been a recurring issue at member meetings, and that both the challenges and the benefits with this law are clear to those who live and run businesses in rural areas: 'We have repeatedly voted on the issue at NFU member meetings and have decided that the positives of the current legislation weigh heavier than the negatives' (NFU, 2017b). Importantly, among the farmers and land owners debating the issue in the media 2015–17, there are also calls for a change in the Act. However, the strongest positive voices (in terms of number of articles) are not those of farmers but of private forest owners in possession of much land. One of the strongest advocates pointed out the need for a change in the law for landowners and rural entrepreneurs to be able to decide over their land and resources without state interference:

> Rural inhabitants of today are well aware and there is no need for the state to prohibit forms of ownership. There will probably be few that will change their [form] of ownership. But for those who want to, a law that is 110 years old should not be a hindrance. There are no arguments for prohibition of corporate ownership.
>
> (Öster, 2016, p. 13)

During February and March 2017, a heated debate between the minister in charge and two representatives from the Swedish Smallholders' Organization (SSO) took place in the media. The SSO refers to the official report (SOU, 2015) as 'a wolf in sheep's clothing' and a 'sneaky Trojan horse' (Wangsten and Eriksson, 2017a, 2017b, 2017c, 2017d). Among other things, the SSO claims that it is thanks to the Land Acquisition Act that Swedish rural areas have a relatively high diversity of both small and large farms, and refer to the risk of land grabbing if the law were to be abolished:

> Like a Trojan horse, the sneaky plan has been embedded in positive words about competitiveness.... Farmers' most important resource, farmland, will soon be so expensive that normal farmers will not have a chance to buy it. Unfortunately, such a development seems to be a deliberate strategy.... The few farmers that continue to exist can therefore be slaves to capital – as leaseholders on their former farms.
>
> (Wangsten and Eriksson, 2017a, p. 54)

In his response to this critique, Minister Bucht stated that there is no 'sneaky' plan and that the Land Acquisition Act is outdated if the goal is to increase Swedish food production rates. He also stated that any changes in the legislation will not be done readily:

> It is important to me to clarify that the government doesn't readily change the legislation on land acquisition and leases. There are many different values and factors that need to be addressed. Sweden has many valuable natural resources that make it possible to increase food production, which gives more jobs, sustainable growth and a vibrant countryside.
>
> (Bucht, 2017, p. 54)

In addition to the NFU and SSO, the board of the Swedish branch of the global peasant movement La Via Campesina, called NOrdBruk, is clearly against the proposed change to the Land Acquisition Act (Karnehed, 2016). In an article, the board of NOrdBruk states the need for Sweden to leave the European Union in order for food sovereignty to be realized, as well as to stop land grabbing and the impoverishment of rural areas:

> We are worried that too many forces which claim to fight for Swedish agriculture, especially small and middle-sized mixed farming, don't realize the depth of this fight – what the work for food sovereignty in Sweden really means.… We want farming organizations and politicians to stop the misappropriation of land and natural resources caused by the current legislation, not only for the sake of farmers but also for rural areas, a cornerstone in societies, which are currently impoverished in Sweden, Europe and other continents.
>
> (NOrdBruk, 2017, p. 6)

Concluding discussion: financialization and Swedish agriculture

This study shows the increased presence of financialization, where the economic sphere is the main arena for policy and decision-making of Swedish food and farming. To have a voice and be heard in this arena the use of 'economic language' can be one strategy. However, using such language – which considers financial motives as the only legitimate basis for decisions – comes with the risk of excluding other motives and knowledges, especially those of active farmers.

An important finding of the present analysis is that economics/finance is one of the dominant narratives for how Swedish agriculture and food should be advanced. However, the Swedish state documents, personal interactions with farmers and the media coverage all highlight different economic problems. In the material analysed, somewhat different aspects of financialization were perceived as 'problems' that needed to be 'solved'. According to the official state report (SOU, 2015), lack of capital is the main problem in 'the sector'. However, it is not clear who they see operates in this rather anonymous sector, or which type of capital is perceived to be lacking. The farmers I consulted, those at the fairs and those writing in the newspapers, also mentioned economic issues as prominent concerns. Lack of capital was mentioned as a problem by some, but these farmers did not necessarily consider this to be related to the land ownership structure and therefore they see no need in changing the Land Acquisition Act. Instead, they believe that actors such as banks and lawyers need to step up and meet farmers' needs, to ensure long-term management and food production in times of succession or in years when the yields are low. They consider it important to strengthen the possibilities to buy and sell the farm in several steps. This would enable newcomers to enter and/or strengthen farm succession. The Swedish tax system was viewed as an issue that could be changed to ease succession, for instance by enabling the sale of a farm in several steps, but only being taxed once (Jonsson *et al.*, 2016; NFU, 2017a). Moreover, in times of expansion and development of the farm, they see a need for increased possibilities to access capital via financial actors in banks. The Swedish Smallholders'

Organization and especially La Via Campesina's NOrdBruk were more radical in their critique of the official report than the NFU. These small farmer organizations protested more strongly against any change to the Land Acquisition Act. They explicitly referred to the economic difficulty for 'normal farmers' to compete with corporations in the purchase of farmland, and acknowledged that food sovereignty is difficult to ensure even within the current system.

These results indicate that the different views on economics and finance arise due to differing temporal and spatial scales applied to understand the situation and what is at stake. The economic reality for active farmers spans several temporal scales from everyday life to annual (such as harvesting) to the perspective of past and coming generations (Head *et al.*, 2012; Phillips, 2013). In contrast, the economic reality in the official report (SOU, 2015) tends to be decoupled from the physical material sphere of the reality (that is, the land). It advocates increased financialization in terms of a desire to increase the capital-intensive farming as a goal along with further market-oriented economic reforms, and wants to expand the legislation for corporate ownership as a means of achieving those goals.

This study indicates increased financialization and the economic sphere of reality as the main arena for policy and decision-making of Swedish food and farming. Using only financial motives as the only legitimate basis for decisions, however, comes with the risk of excluding other motives and knowledges. What then can be concluded regarding who would gain from the proposed policy changes to the Swedish Land Acquisition Act, and what are the possible negative effects for active farmers?

The analysis also revealed that farmers are not a homogeneous group of 'victims' of a financialized food system (Slätmo, 2014, 2016; Sippel *et al.*, 2017). Financial logic is, at least partly, incorporated into the mindset of Swedish farmers. Even so, listening to the Swedish farmers' perceived needs, financial actors involved in food production need to further incorporate dimensions of responsibility for long-term accessibility to nutritious food in their basis for decisions and compliances (Isakson, 2014; Clapp, 2016; Luyendijk, 2016). From the farmers' perspective this is fundamental to ensure production and access to the resources (especially land and capital) needed to produce sufficient food now and in the future. This would require politicians to ensure that financial and corporate actors, with the help of measures such as education, spatial plans, regulations, subsidies and sanctions, act according to the political visions and goals, such as the global Sustainable Development Goals, to which they have subscribed.

At the time of writing, the Swedish Land Acquisition Act is under further investigation. As argued in this study, it is important to include insights from active farmers in Sweden and about the financialization of food and agricultural land from other geographical contexts in such an investigation. Increased financialization, in terms of foreign capital investments in agriculture and farmland, comes with possibilities of positive 'spill-over effects' in terms of knowledge and technology transfer to contract-based farmers and possibilities to use measures such as contract obligations to enhance infrastructural investments in rural areas (Hallam, 2009; Robertson and Pinstrup-Andersen, 2010). For this to be achieved, the host government needs to be actively engaged, to maintain control over the prerequisites for human survival and well-being, such as soils and agricultural land, and to avoid increased conflict over the rights to land and environmental resources (Cotula et al., 2009; Hallam, 2009; Cotula, 2012; Clapp, 2014; Fairbairn, 2015; Bergius *et al.*, 2017).

Previous research also reports negative effects of corporate involvement and financial investment in farmland. These effects include risks of food insecurity, loss of livelihood and farming knowledge, and instability caused by conflict over rights to exploit natural resources (Robertson and Pinstrup-Andersen, 2010; Ouma, 2014; Fairbairn, 2015; Magnan, 2015; Clapp, 2016). Indeed, foreign investments in land can be deeply problematic for local inhabitants' food sovereignty and active farmers' access and rights to land (Cotula, 2012; Franco and Borras, 2013; Clapp, 2014). Based on this body of research and the results obtained in the present study, it can be deduced that, from the perspectives of active Swedish farmers, the proposition to relax the terms of the Swedish Land Acquisition Act and allow corporate ownership to become normal is both politically naïve and socially contested.

Acknowledgements

Thanks to the farmers and others with whom I interacted during visits to Borgeby and Almedalen in 2016, to officials in the county of Halland, and to the editors of this volume.

References

af Kleen, B. (2010) *Jorden de ärvde*, Månpocket and Bonnierpocket, Stockholm, Sweden

Antonsson, H. and Jansson, U. (2011) *Jordbruk och skogsbruk i Sverige sedan år 1900 – en kartografisk beskrivning*, vol 53. Stockholm: Sveriges Nationalatlas and Norstedts.

Bergius, M., Benjaminsen, T.-A. and Widgren, M. (2017) 'Green economy, Scandinavian investments and agricultural modernization in Tanzania', *Journal of Peasant Studies*, pp. 1–28, DOI: doi.org/10.1080/03066150.2016.1260554

Bucht, S.-E. (2017) 'Ministern: Nej, vi har ingen lömsk plan i livsmedelsstrategin!', *Land*, 17 March 2017

Clapp, J. (2014) 'Financialization, distance and global food politics', *Journal of Peasant Studies*, vol 41, no 5, pp. 797–814

Clapp, J. (2016) *Food*, Second Edition, Polity, Cambridge

Cotula, L. (2012) 'The international political economy of the global land rush: A critical appraisal of trends, scale, geography and drivers', *Journal of Peasant Studies*, vol 39, no 3–4, pp. 649–680

Cotula, L., Vermeulen, S., Leonard, R. and Keeley, J. (2009) *Land Grab or Development Opportunity? Agricultural Investment and International Land Deals in Africa*, IIED, FAO and IFAD, London and Rome

Daugstad, K., Rønningen, K. and Skar, B. (2006) 'Agriculture as an upholder of cultural heritage? Conceptualizations and value judgements – A Norwegian perspective in international context', *Journal of Rural Studies*, vol 22, no 1, pp. 67–81

Ektander, V. (2015) Bucht vill låta aktiebolag köpa mark, *Jordbruksaktuellt*, 5 May 2015

Fairbairn, M. (2015) 'Foreignization, financialization and land grab regulation', *Journal of Agrarian Change*, vol 15, no 4, pp. 581–591

FAO. (2012) *The State of Food and Agriculture: Investing in Agriculture for a Better Future*, available at: www.fao.org/docrep/017/i3028e/i3028e00.htm, accessed 22 June 2017

Franco, J. and Borras, S. (2013) *Land Concentration, Land Grabbing and People's Struggles in Europe*, Transnational Institute (TNI) for European Coordination Via Campesina and Hands off the Land network, The Netherlands

GRAIN. (2012) 'Genetic Resources Action International (GRAIN) releases data set with over 400 global land grabs', available at: www.grain.org/article/entries/4479-grain-releases-data-set-with-over-400-global-land-grabs, accessed 22 June 2017

Grubbström, A. and Sooväli-Sepping, H. (2012) 'Estonian family farms in transition: A study of intangible assets and gender issues in generational succession', *Journal of Historical Geography*, vol 38, no 3, pp. 329–339

Hallam, D. (2009) 'International investments in agricultural production', in M. Kugelman and S. Levenstein (eds) *Land Grab? The Race for the World's Farmland*, Woodrow Wilson International Center for Scholars, Asia Program, Washington DC., pp. 27–38

Head, L., Atchison, J. and Gates, A. (2012) *Ingrained: A Human Bio-geography of Wheat*, Ashgate, UK

Howley, P., Dillon, E. and Hennessy, T. (2014) '"It's not all about the money": Understanding farmers' labor allocation choices', *Agriculture and Human Values*, vol 31, no 2, pp. 261–271

Howley, P., Buckley, C., Donoghuec, C. and Ryanc, M. (2015) 'Explaining the economic "irrationality" of farmers' land use behavior: The role of productivist attitudes and non-pecuniary benefits', *Ecological Economics*, vol 109, pp. 186–193

Isakson, S. (2014) 'Food and finance: The financial transformation of agro-food supply chains', *Journal of Peasant Studies*, vol 41, no 5, pp. 749–775

Jonsson, H., Hammar, S. and Björid, N. (2016) 'Jordförvärvslagen bör behållas intakt', *ATL*, 23 June 2016

Jordbruksaktuellt. (2015) 'Konkurrenskraftsutredningen sammanfattad', *Jordbruksaktuellt*, 6 March 2015

Jørgensen, M. and Phillips, L. (2002) *Discourse Analysis as Theory and Method*, Sage, London

Karnehed, S. (2016) 'Småbrukare kritiska till ny liivsmedelsstrategi', *Landets Fria*, 21 September 2016

Kristensen, I. and Grundfelder, J. (2018) 'Foreign direct investment: Trend and patterns of FDI inflows', in J. Grunfelder, L. Rispling and G. Norlén (eds) *State of the Nordic Region 2018*, Nordic Council of Ministers, Copenhagen, Denmark, pp. 132–142

Lambin, E. and Meyfroidt, P. (2011) 'Global land use change, economic globalization, and the looming land scarcity', *Proceedings of the National Academy of Sciences of the United States of America*, vol 108, pp. 3465–3472

Luyendijk, J. (2016) *Among the Bankers: A Journey into the Heart of Finance*, Melville House, New York

McMichael, P. (2012) 'The land grab and corporate food regime restructuring', *Journal of Peasant Studies*, vol 39, no 3–4, pp. 681–701

Magnan, A. (2015) 'The financialization of agri-food in Canada and Australia: Corporate farmland and farm ownership in the grains and oilseed sector', *Journal of Rural Studies*, vol 41, pp. 1–12

NFU. (2017a) 'Palle Borgström LRF: Självägande bönder bäst för landsbygden', *Lantbrukarnas Riksförbund*, 13 July 2017

NFU. (2017b) 'Jordförvärvslagen: Landsbygdsföretagarna har redan gjort analysen', *Lantbrukarnas Riksförbund*, 26 June 2017

NOrdBruk. (2017) 'Jordförvärvslagen får inte ändras!', *Miljömagasinet*, 15 September 2017

Odell, Å. (2017) 'Stark koppling mellan ägande och brukande', *Gotlands Tidningar*, 16 August 2017

Olsson, L., Tambang, Y., Faran, T. and Jerneck, A. (2012) *Internationell Handel med Jordbruksmark – Ett Modernt Baggböleri*, LUCSUS, Lund

Öster, L. (2016) 'Fel med bolagsförbud och bosättningskrav', *ATL*, 14 June 2016

Ouma, S. (2014) 'Situating global finance in the land rush debate: A critical review', *Geoforum*, vol 57, pp. 162–166

Phillips, C. (2013) *Saving More Than Seeds: Practices and Politics of Seed Saving*, Routledge, London

Quinn, C. and Halfacree, A. (2014) 'Place matters: An investigation of farmers' attachment to their land', *Human Ecology Review*, vol 20, no 2, pp. 117–132

Raggi, M., Sardonini, L. and Viaggi, D. (2013) 'The effects of the Common Agricultural Policy on exit strategies and land re-allocation', *Land Use Policy*, vol 31, pp. 114–125

Robertson, B. and Pinstrup-Andersen, P. (2010) 'Global land acquisition: Neo-colonialism or development opportunity?' *Food Security*, vol 2, no 3, pp. 271–283

Sandberg, A. (2017) 'Palle Borgström LRF: Självägande bönder bäst för landsbygden', *Aftonkuriren*, 14 July 2017

Saunders, F. (2016) 'Complex shades of green: Gradually changing notions of the "good farmer" in a Swedish context', *Sociologia Ruralis*, vol 56, no 3, pp. 391–407

SFS. (1979) *Jordförvärvslag (1979:230)*, Svensk författningssamling

SFS. (2010) *Lag om ändring i jordförvärvslagen (1979:230)*, Svensk författningssamling

Silverman, D. (2010) *Doing Qualitative Research. A Practical Handbook*, Third Edition, SAGE, London

Sippel, S.R., Larder, N. and Lawrence, G. (2017) 'Grounding the financialization of farmland: Perspectives on financial actors as new land owners in rural Australia', *Agriculture and Human Values*, vol 34, no 2, pp. 251–265

Slätmo, E. (2014) 'Jordbruksmark i förändring. Drivkrafter bakom och förutsättningar för offentlig styrning i Sverige och Norge', PhD thesis, University of Gothenburg, Sweden

Slätmo, E. (2016) 'Challenges in agricultural land management: A Scandinavian perspective on contextual variations and farmers' room to manoeuvre', in L. Head, K. Saltzman, G. Setten and M. Stenseke (eds), *Nature, Temporality and Environmental Management: Scandinavian and Australian Perspectives on Peoples and Landscapes*, Routledge, Oxon, pp. 169–185

Slätmo, E. (2017) 'Preservation of agricultural land as an issue of societal importance, *Rural Landscapes: Society, Environment, History*, vol 40, no 1, pp. 1–12

SOU. (2015) *Attraktiv, innovativ och hållbar – strategi för en konkurrenskraftig jordbruks- och trädgårdsnäring*, available at: www.regeringen.se/49bbaa/contentassets/2062cad823a54061bfa7f9028e8a71b3/attraktiv-innovativ-och-hallbar-strategi-for-en-konkurrenskraftig-jordbruks-och-tradgardsnaring-sou-201515, accessed 4 May 2015

Statistics Sweden. (2013) *Land Use in Sweden*, Sixth Edition, Stockholm, Sweden

Statistics Sweden. (2016) *Economic Accounts for Agriculture – Second Estimate for 2015*, Örebro, Sweden

Suess-Reyes, J. and Fuetsch, E. (2016) 'The future of family farming: A literature review on innovative, sustainable and succession-oriented strategies', *Journal of Rural Studies*, vol 47, pp. 117–140

Swedish Board of Agriculture. (2015) *Yearbook of Agricultural Statistics 2014, Including Food Statistics*, Jönköping, Sweden

Törner, M. (2017) 'Palle Borgström om åtgärdspaket 4', *Land Lantbruk*, 14 July 2017

Wangsten, F. and Eriksson, M. (2017a) 'Nya livsmedelsstrategin är en lömsk trojansk häst', *Landets Fria*, 8 February 2017

Wangsten, F. and Eriksson, M. (2017b) 'Ny livsmedelsstrategi – en ulv i fårakläder', *ATL*, 2 April 2017

Wangsten, F. and Eriksson, M. (2017c) 'Här är lömska planen som bäddats in i livsmedelsstrategin', *Land*, 3 March 2017

Wangsten, F. and Eriksson, M. (2017d) 'Sälj inte ut vår svenska åkermark, Bucht!', *Land*, 24 March 2017

15 Institutional investment in a multi-functional forested landscape

Neoliberalism and pragmatism in contemporary land conservation

Steven Wolf

Introduction

Applied to rural land, financialization is a reference to expanded roles for investors and investment logic in shaping land use, land access, and who captures the benefit flows. Financialization implies displacement of traditional users of land, reduced capacity for self-determination by local people, and diminished prospects for environmental conservation. As expressed in the call for papers that gave rise to this volume, private equity firms and developers purchasing land for investment threaten 'family farmers'. Parallel to Goldschmidt's (1947) analysis of agribusiness and absentee landlords in California agriculture, financialization is understood as a process of distancing and disembedding arising from changes in property relations. These changes in ownership and access undermine accountability mechanisms, social norms, and long-range commitments to specific places that traditionally govern relations between production, environment, and community. In essence, financialization is a process of neoliberal abstraction, and this abstraction is contested because it threatens social and ecological relations (Krippner, 2005; Clapp, 2014; Fairbairn, 2014). Polyani's analysis of the construction of land as a 'fictitious commodity' centres similarly on a process of abstraction and resulting social conflict.

The markers and implications of financialization can be specified in the following terms:

1 New enclosures that displace local people, disrupt traditional patterns of access, and marginalize traditional livelihoods
2 Accelerated extraction of natural resources and ecological degradation
3 Absence of transparency, informed consent, and accountability mechanisms in land transactions (the 'grab' in land grabs).

A number of authors (Russi, 2013; Lawrence, 2015; Schmidt, 2016) emphasize elements of this established critique, identifying financialization applied to agri-food as implicated in large-scale land acquisitions, in the purchasing and asset stripping of companies in the food sector, and in speculative activities surrounding agricultural commodities and farmlands. These authors offer a programmatic message regarding a need to complement existing macro-structural analysis of financialization with more attention to empirical specificity and historical significance. Similarly, in outlining a research agenda on 'green-grabbing' – land appropriation justified through reference to ecological risks – Fairhead *et al.* (2012, p. 241) caution that '(s)ome of the critical literature on neoliberalizing nature adopts a rather uniform position, assuming a singular hegemonic project, failing sometimes to analyse the consequences for diverse, differentiated and contingent settings'. Through grounded applied research, it is possible to leverage the concept of financialization to develop a more multi-dimensional appreciation of the ways in which finance capital is implicated in contemporary land governance and rural transformations.

This chapter analyses a specific real estate deal in which institutional investors have played a key role. This transaction generated conflict by redistributing access rights, fuelling tensions between local and non-local actors, posing ecological risks, and raising questions about the allocation of costs and benefits of public investments and the adequacy of procedural controls governing such investments. At the same time, the changes in property relations resulting from this high profile transaction yielded significant and varied public goods (e.g. ecological conservation, recreational access, maintenance of jobs in the timber trade, and cultural continuity) in response to problem definitions of a wide range of stakeholders. Environmental conservation interests, those supporting traditional livelihoods (logging) and community economic development, along with advocates for diverse forms of recreational access, found something to cheer about in this complex deal that unfolded over the period 2007–16. In parallel, the deal had many detractors and has been a focus of protest, law suits, and policy critique by State officials.

Labels such as neoliberal natures (Büscher *et al.*, 2012), land grabbing (Wolford *et al.*, 2013), green-grabbing (Fairhead *et al.*, 2012), integrated conservation and development project (ICDP) (Newmark and Hough, 2000), and sustainable rural development, fail to capture the dynamics and outcomes of this land deal. On the one hand, the deal focused on land conservation and ecological imperatives. On the other, the deal was about maintenance of logging and paper manufacturing in a de-industrializing region. The institutional investors were, presumably, focused on profit-making and not on social problem-solving. Analysis of the deal highlights blending and blurring (Hodge and Adams, 2012) of categories, strategies, and roles of actors engaged in contemporary land governance in circumstances characterized by tension, and varying degrees of parity, among political and economic drivers of 'old' economy (agriculture, or in this case logging), 'new' economy (service, amenity or consumption-oriented development) and environmental conservation.

Critiques of multi-functionality and post-productivism highlight the value of avoiding claims of epochal shifts, as material and institutional landscapes of rural regions in advanced industrial societies are more usefully characterized as ambiguous (Potter and Tilzey, 2007; Wolf and Klein, 2007).

The admixtures of strategies and roles of actors, and the lack of correspondence with existing academic categories, highlights the ad hoc and pragmatic character of action in these governance settings. By definition, pragmatism represents a willingness to work with what is available and to build on openings that present themselves (Healy, 2009). To be pragmatic is to reject orthodoxy and to cultivate a capacity to live with ambiguity and contradictions as part of navigating shifting terrain. While demanding maintenance of a critical stance and a commitment to ongoing search for more attractive options, pragmatism requires engagement. Pragmatists do not invest much time in reflecting on imagined alternative dimensions. They insist on addressing problems at hand, and they are comfortable relying on creative approaches – *bricolage* (Cleaver, 2001) – as necessary. Pragmatism can, of course, institutionalize incrementalism and reformism, and crowd out imaginative radical visions, but it can also be a vector of new development pathways. These pathways can enable transformative change over time. These potential changes include a cementing of neoliberal trajectories (Büscher *et al.*, 2012; Hodge and Adams, 2012), but there is also the prospect of embedding land ownership, land management, and land transfers in social and legal contexts. Embeddedness refers here to capacity for expression of local interests, scientifically informed risk assessments, and outputs of democratic processes in socioecological relations and development trajectories.

This chapter addresses financialization applied to the Northern Forest, a 10.5 million ha forest stretching across Northeastern States of the US and Southeastern Canadian provinces (see Figure 15.1). Based on a case study and a specific framing of socioecological problems that threaten forest landscapes and communities in the Northern Forest, the chapter analyses financialization empirically. The case study centres upon a high profile real-estate transaction in the Adirondack Park located in upstate New York in which The Nature Conservancy, New York State government, local communities, private equity firms from New York City, and a Danish pension fund play central roles. The land is situated in the centre of the Adirondack Park, the largest protected area in the US outside of Alaska. The sale of this land – the single largest parcel of privately owned land in the Adirondack Park – resulted in the biggest expansion of the Park in over 100 years including maintenance of commitments to logging and paper manufacturing in a region characterized by long-term decline of primary industry. In this sense, the property claims that institutional investors advanced in the forest are linked to problem definitions and strategic responses enacted by public agencies, civil society organizations, and local

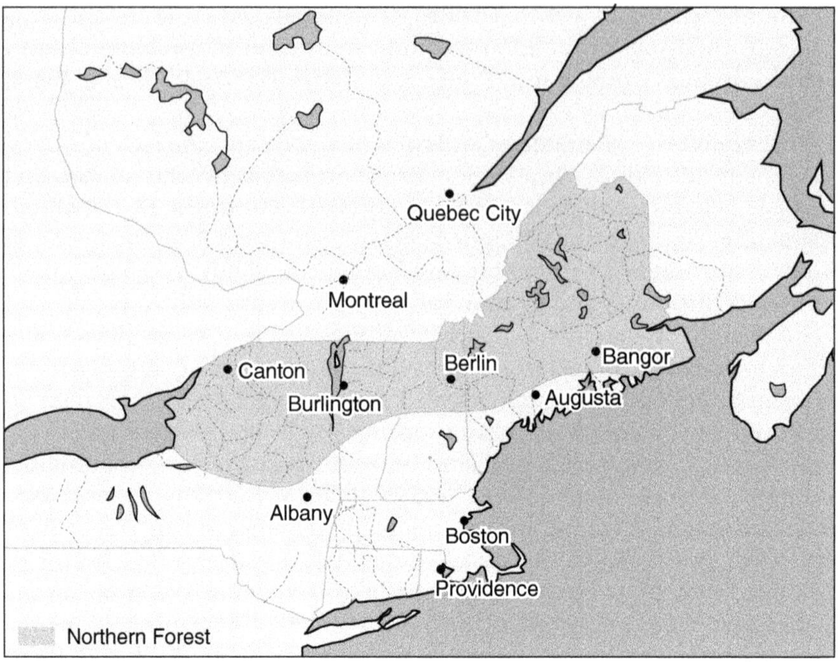

Figure 15.1 The Northern Forest.

Source: Northern Forest Center, 2003, available at: http://northernforest.org/techmap.htm

communities. Based on the argument that institutional investment played a constructive role in mitigating socioeconomic and ecological risks identified by public agencies, environmental NGOs and rural communities in this particular landscape, the chapter situates institutional investment in a broader analysis of environmental governance practice. The chapter seeks to complicate simple characterizations of institutional investment in rural landscapes as a consistent set of practices linked to a consistent set of outcomes. Finally, the chapter aims to advance an analytic strategy in which the meaning and significance of financialization is understood through attention to tensions and interdependencies between institutional investors and other actors making property claims in specific material and institutional landscapes.

Socioecological dynamics of the Northern Forest

The Northern Forest is subject to many stresses as a function of ongoing socioeconomic and environmental change. Two dynamics warrant particular attention. The first is subdivision of ownership (that is, parcelization) and associated processes of landscape fragmentation. The second is a major shift in land tenure arising from entry of institutional investors as a new class of owners of US forestland.

Like most of the forests in the eastern US (and in contrast to the forests of the western US) the Northern Forest is largely privately owned (around 80 per cent privately owned). Private forests, particularly non-industrial private forests, are subject to parcelization, a process of subdivision of ownership over time that expands the number of owners and decreases the average size of forest parcels owned by individuals (Best and Wayburn, 2001). Increasing demand for second homes and access to rural amenities creates incentives for subdivision in a rural region proximate to large populations and great wealth. Parcelization also arises as a function of inheritance (that of dividing land among surviving children). Parcelization changes how land and timber is managed, resulting in ecological, social, and economic changes at larger geographic scales. Applied to ecological considerations and prospects for maintenance of a viable forest products industry, forestland parcelization is problematic due to its relationship to forest fragmentation (Schulte *et al.*, 2008). Fragmentation – loss of integrity – arises from land use conversion (most usually for second-home development), and introduction of roads, fences, and other constructed barriers. Processes of parcelization and fragmentation disrupt ecological processes, degrade the value of wildlife habitat, and accelerate the introduction of invasive species (Schulte *et al.*, 2008). The Centre for Large Landscape Conservation and the Practitioners Network for Large Landscape Conservation are two current initiatives in the US that highlight the significance of scale – and risks attached to parcelization and fragmentation – within current debates on biodiversity and conservation more generally. Apart from ecological considerations, parcelized and fragmented forests also pose political and economic challenges for forest-based industry. The quality of timber, cost of timber, and maintenance of a workforce, of infrastructure, and of a service sector capable of supporting the harvesting, transportation and processing of timber, are all negatively affected by the 'suburbanization' of forests.

As Petersen (1998) colourfully phrased the current situation in the Northern Forest, the most dangerous critter in the woods is not a person with a chainsaw but a person with a lawnmower. This quip captures a profound political realignment in the Northern Forest. Concerns regarding ecology (biodiversity), environment (aesthetics and recreational opportunities), maintenance of cultural identity (preserving the relevance of a history of logging in a region increasingly focused on environmental amenities), and survival of a flagging forest products industry (such as conserving jobs in a sector subject to decades of automation and pressure from global market integration) have unified strange political bedfellows (Neugarten *et al.*, 2012). Environmentalists, community economic development interests, and the wood products industry identify relentless processes of parcelization and fragmentation as existential threats. Following a high profile sale of a larger tract of timber in 1988, concerns came to a head. The US Congress funded the Northern Forest Lands Council (NFLC), a group chartered in 1990 to make public policy recommendations aimed at reinforcing traditional forestland uses and long-term stewardship of the forest resource (Wolf and Klein, 2007). Applied to the Adirondack Park, linking a commitment to logging with a commitment to conservation highlights how parcelization and fragmentation present political common ground in a famously divisive context. Conflict between local communities and Park administration, and between logging, recreational, and environmental interests has defined the regional politics for many generations (Terrie, 2008).

The second important dynamic in the Northern Forest is a change in patterns of ownership. Since the late 1980s, vertically integrated forest products companies – owners of roughly 25 per cent of US privately owned forestland – began selling their land holdings (Block and Sample, 2001). Nearly ten million ha of forest land in the Northern Forest was sold between 1980 and 2005 (Hagan *et al.*, 2005), reflecting a nationwide pattern (Bliss *et al.*, 2010). In this sell-off, land owned by publicly traded and private industrial forestland owners – paper and cardboard manufacturers – was sold to a range of new actors including Timber Investment Management Organizations (TIMO), Real Estate Investment Trusts (REIT), timber barons and, in rare cases, land trusts. At this time, the sell-off is complete and no forest product companies hold substantial amounts of land (Stein, 2010). In the US, TIMOs have come to own more than 9.3 million ha of US timberland compared to fewer than 11,180 ha as of 1987. In 2015 this investment represents US$57 billion as compared to US$1 billion in 1989 (Table 15.1). Eighty to 90 per cent of the timberland

Table 15.1 Largest private owners of timberland in USA, January 2013

Top U.S. Timberland owners and managers as of January 2013

Firm	*Type*	*US acres*
Plum Creek	Public REIT	6,379,000
Weyerhaeuser	Public REIT	6,000,000
American Forest Management	TIMO/forest manager	4,700,000
Hancock Timber Resource Group (HTRG)	TIMO/forest manager	3,397,500
The Forestland Group (TFG)	TIMO/forest manager	3,250,946
The Campbell Group	TIMO/forest manager	2,985,300
Resource Management Service (RMS)	TIMO/forest manager	2,654,500
Wagner Forest Management*	TIMO/forest manager	2,640,000
Rayonier	Public REIT	2,389,048
Forest Investment Associates	TIMO/forest manager	2,371,295

Source: 2013 Forisk Timberland Owner List (Forisk Blog, 2013)

Note

* Includes acres in Canada

investment is based in the US, Canada, Australia and New Zealand. The investment industry reports an expectation for expanded investment in forestland in Latin America, Asia, Africa and Europe (Jacobius, 2015), but such pronouncements can be understood as promotion rather than forecasting (Fairbairn, 2014).

The explanation for the US timberland sell off is not thoroughly clear. In their analysis of financialization of US timberland, Gunnoe and Gellert (2011) emphasized the rise of a shareholder value imperative as a key motivating factor for paper and lumber manufacturers. At a more macro-structural level, analysts emphasize: the Tax Reform Act of 1986, which made it unattractive for vertically integrated timber companies to own land relative to institutional investors; mergers and acquisitions in the globalizing forest products industry, which prompted firms to liquidate land holdings to pay down debts; and, institutional investors' discovery of forestland as an attractive element of diversified portfolios (Block and Sample, 2001; Hagan *et al.*, 2005). As part of a diversification strategy, timberland is regarded as a safe investment and as a risk-buffering asset (for farmland, see Fairbairn, 2014). Timberland values are thought to be negatively correlated with other asset values and, in this sense, it can act as a hedge relative to market downturns and to inflation. As is the case of agriculture, the drivers, dynamics, and implications of entry of institutional investors in forestry – manifest across varied contexts and scales – remain an important focus of research.

TIMOs expect to derive roughly one-third of investment returns from the sale of timber and two-thirds from appreciation in land values. The wave of institutional investment in the Northern Forest is reflected in timberland prices, as land was selling at two to eight times the estimated timber value ten years ago. These prices reflect speculative pressure rather than potential for the development of silviculture. When opportunities exist, TIMOs subdivide land and sell off parcels for housing or commercial development in keeping with a logic of 'highest and best use'. Widespread land use conversion is, however, not a reasonable expectation for significant parcels of land, as these forested tracts are very large and often very remote. Within the Adirondack Park, while growth management controls constrain potential for widespread land use conversion, it is possible to imagine private purchases of large estates contributing to landscape fragmentation. In this context, anxiety regarding institutional investment on ecosystems in the Northern Forest stems from varied concerns including fragmentation, retreating from maintenance of commitments to eco-certification, perceptions that harvesting is dictated by short-run timber prices rather than long-term forest management planning, and lack of clarity about future land use (Hagen *et al.*, 2005).

In a small number of cases, institutional investment in forestland and environmental conservation efforts have been integrated in a manner that goes beyond blocking the suburbanization of forests. In these situations, investors purchase the land and the attendant right to harvest timber, but the development rights – the rights to subdivide the land and convert to non-forest uses – are separated from the ownership rights and are purchased by a land trust or a public agency. These 'working forest conservation easements' (Wolf and Klein, 2007; Neugarten *et al.*, 2012) allow the investors to buy the land – and secure the flow of revenues from timber harvesting – at a reduced price. At the same time, they advance public interests including conservation of biodiversity, recreational access, and defense against the suburbanization of the forest. At least one TIMO, Lyme Timber, pursues these types of land deals as a standard part of its business model.[1] These types of property arrangements highlight interdependencies that characterize multi-functional landscapes and potentially confuse problem statements that structure critical analysis of both green-grabbing and financialization.

Deal of the century:[2] institutional investors in the forest of environmental governance

This case study is based on a review of publicly available documents over a period of ten years, interviews with participants and analysts of the deal, and a research partnership with representatives of the Adirondack Chapter of The Nature Conservancy of New York State (see Neugarten *et al.*, 2012, and Kraft, 2014 for methodological details and expanded references).

> Finch, Pruyn Paper (FPP) commenced business in 1865. The firm had a paper mill in Glens Falls and owned substantial timberland in the nearby Adirondack Park. The 2.4 million ha Adirondack park was established at the end of the 19th century. This park is larger than Yosemite or Yellowstone National Parks. It is unusual by US standards in that the land is roughly half publicly owned and half privately owned. Private land in the park is subject to a variety of controls that limit development opportunities. The Park is celebrated by some as a model of the integration of environmental conservation and economic development interests. It is also subject to sustained criticism as a playground established by elites at the expense of the rights and economic prospects of local communities experiencing chronic rural poverty.
>
> (Terrie, 2008)

In line with the entire paper industry in that part of the country, a sector that shed 38 per cent of its jobs between 2000 and 2003, FPP struggled through the end of the twentieth century and in the early part of the new century. In April 2007, FPP – the mill and 67,000 ha of forestland – was purchased by two private equity firms working in partnership – Atlas

Holdings LLC and Blue Wolf Capital (here after referred to as Atlas). At the time of purchase, Atlas operated four paper mills and nine cardboard packaging plants, and they expressed an intention to hold on to the business rather than restructure it and immediately put it up for sale. The forestland sold by FPP was the single largest privately owned parcel in the Adirondack Park. This parcel was located in the heart of the park in the region known as the High Peaks (Figure 15.2). This land supports rare and endangered plants and animals and plays a vital role in regulating water flows in the region. FPP managed their forests under Forest Stewardship Council eco-certification standards and was widely regarded to be a good steward of the land.

In June 2007, The Nature Conservancy (TNC), the top fundraiser among environmental NGOs and one of the largest landowners in the US, purchased the land formerly owned by FPP from Atlas and Blue Wolf for US$110 million. While the resources of this international NGO were important elements of this purchase, TNC is a substantially decentralized organization and the Adirondack chapter of TNC of New York (NY) State was the engine behind this purchase. TNC secured finances for the land deal from the Open Space Institute, a not-for-profit organization dedicated to land conservation, and John Hancock, a commercial lender. At the time of the purchase, TNC had conserved over 225,000 ha in the Adirondacks, half of which via purchase of working forest conservation easements. The purchase of land from Atlas and Blue Wolf was the largest land deal that TNC had ever advanced in NY State.

TNC has significant experience in purchasing land and selling it to the state or federal government to be managed as parkland. This pattern of acting as an intermediary reflects an inability of government agencies to move fast enough to participate in land markets, and it also provides a buffer for government agencies which might be subject to criticism for competing with private interests and undermining economic development prospects of rural territories. For TNC this pattern of acting as an intermediary allows them to apply ecological inventory tools to identify and protect the most ecologically sensitive sites in these large landscapes, and it allows them to recycle their capital and thereby protect more area. TNC purchased the former FFP land with the expectation of selling a portion to the State of NY to be incorporated into the Adirondack Park and to place a working forest conservation easement on all lands not purchased by the State. TNC did not intend to own the land long term. In July, just one month after the TNC purchased the land, the State government signalled some degree of uncertainty regarding its ability to purchase the land and the timing of any future purchases. At this time TNC was conducting a detailed ecological inventory of the land, paying US$1.1 million in annual local taxes, and servicing debt on the US$110 million purchase price. TNC was also capturing revenue from active logging operations on the land.

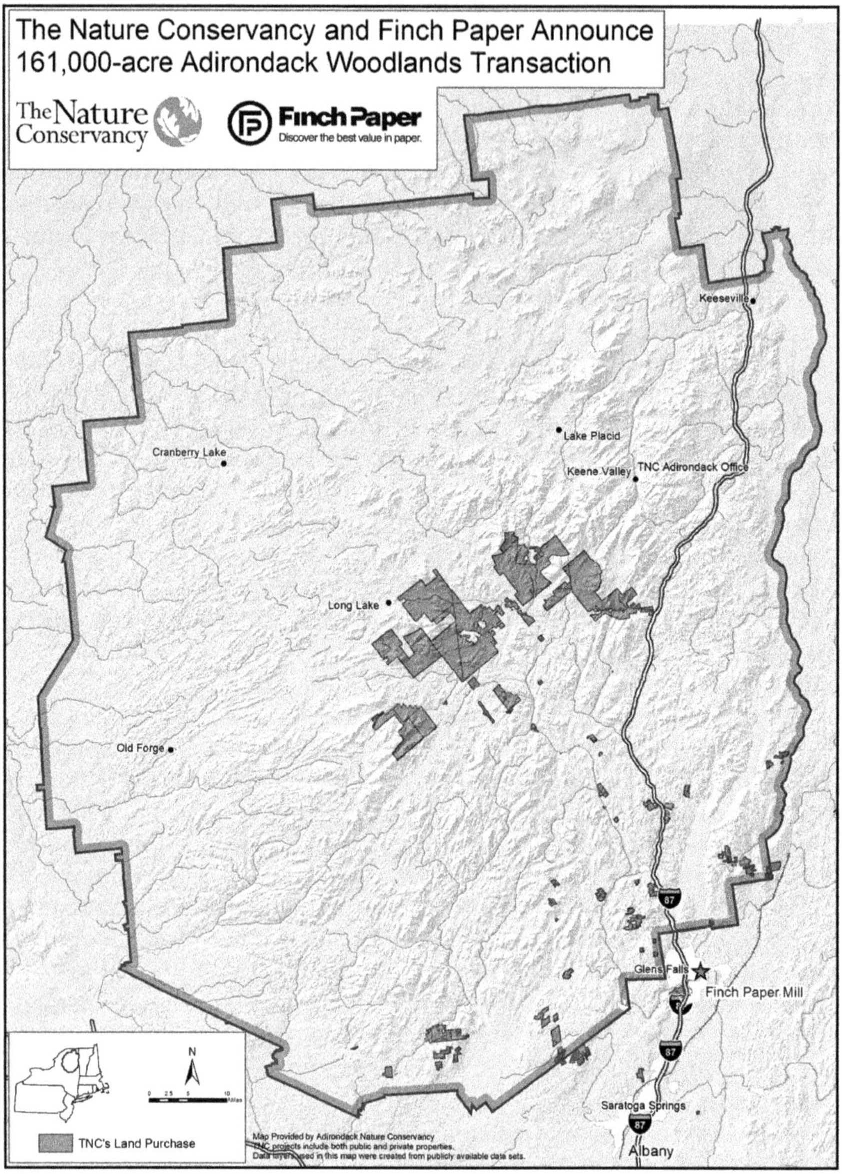

Figure 15.2 The former Finch, Pruyn lands, purchased by The Nature Conservancy (TNC) in 2007.

Source: Craig Cheeseman, The Nature Conservancy.

The State of NY has a history of purchasing land to create parks and preserves. The State established an Environmental Protection Fund in 1993 supported by a Real Estate Transfer Tax. Roughly 50 per cent of this tax revenue has been used to purchase land as part of the Open Space Initiative, a State-wide strategic conservation planning process. In 2007 there was concern regarding revenues to the fund. Also, the State expressed concern about existing money owed to the Open Space Institute and to TNC for prior land purchases. The global financial crisis of 2007–8 is an important backdrop to these events. This acute economic crisis lay on top of the long history of conflict surrounding the establishment and administration of the Adirondack Park, mounting evidence of ecological risks at global and regional scale, decline of the region's paper industry, and the sell-off of the last of the large industrial timber estates.

In February 2008, prior to Governor Patterson replacing the outgoing Governor Spitzer, the State agreed to buy 23,350 ha of timberland and buy conservation easements on an additional 29,800 ha. In NY, local communities hold a veto power over local land purchases advanced under the Open Space Initiative. The land proposed for purchase was located in the towns of Newcomb, Indian Lake, North Hudson, Minerva, and Long Lake. The Adirondack chapter of the TNC of NY organized a process of consultation and negotiation with community leaders. Based on this dialogue and planning process, the towns agreed to the purchase of land by the State. The deal included the sale of 772 ha of land to these communities to be developed for local projects such as the expansion of tourism and recreational facilities and affordable housing. While this land is a small fraction of the total, because of development restrictions in the Adirondack Park this land has significant value. TNC regarded the outreach they conducted with local communities to be a very important and very positive element of the larger process.

During this same period, TNC engaged local 'hunt clubs', which were some of the most vocal critics of TNC and the State in relation to the sale of the former FPP land. Hunt clubs are locally important institutions that structure recreational access to privately owned forest. FPP had maintained long-term leases with roughly 44 recreation clubs devoted to hunting, fishing and camping on their land. Some of these clubs have memberships of more than 100 families and have existed for three generations (Earl, 2007). Some hunt club members are local families. Other families travelled across the country to enjoy the forest of the Adirondacks. As a product of negotiations, TNC entered into leases with 11 clubs who had traditionally accessed land on which the State intended to buy a conservation easement. TNC entered into ten-year leases with 20 other clubs that traditionally accessed land that would be converted to forest preserve under purchase by the State. The expectation was that club members' exclusive access would cease with the expiration of these leases at the end of the ten-year period. The land would be open to the public at that point.[3]

In January 2008, TNC announced the sale of 38,000 ha to ATP, a Danish Pension Fund. Under the terms of the sale, the land was encumbered by a conservation easement designed by TNC to protect riparian zones and high elevation lands. In addition, the land was subject to a fibre supply agreement. This agreement maintained the historical linkage between the land and the mill by providing an annual first-right-of-refusal for a set volume of timber to Atlas, owner of the former FP paper mill. Finally, the easement required that the land was to be managed under a third-party verified Sustainable Forest Management System. ATP hired RMK Timberland (a TIMO) to manage the land and timber operations. At the time of this sale, TNC expressed confidence that the State would honour its commitment to buy timberland and easements according to the plan under which they purchased the land.

In January 2010, citing budget deficits, Governor Patterson proposed a moratorium on all State land purchases. This moratorium affected the pledge to purchase the former FPP land from TNC in addition to other previous agreements made with TNC and other conservation organizations elsewhere in the State. The legislature eventually rejected the moratorium in May and appropriated funding for State conservation programming, but this policy debate and a drop in the level of funding relative to historical levels introduced additional uncertainty about the fate of the former FPP lands and future agreements between the State and conservation organizations.

The relationship between TNC and the State of NY became more complicated in April 2010 when the Attorney General of NY, now Governor Andrew Cuomo, opened an investigation into payments made by the State to TNC for land purchased in 2008. The investigation arose from allegations that the State over-payed TNC for land it purchased and then sold three years later to the State at a 57 per cent profit (Dicker, 2010). The Attorney General's office concluded that the relevant procedures for conducting valuation and making payments were appropriate and that no misconduct had occurred by TNC or by the State, but as with the proposed moratorium, this episode highlighted the fragility of this model of land conservation.

In December 2010, almost two years after TNC purchased the land, the State of NY purchased the conservation easement TNC placed on the ATP land for US$30 million. From the perspective of TNC and those who supported the original plan for dispensation of the FPP lands, this was an isolated success. As of June 2011, the State had still not made good on its agreement to purchase the 26,000 ha of land from TNC. At this time, local critics of this land deal became more vocal in their efforts to derail the State's purchase of the land. Local residents raised questions about fairness of outcomes under the proposed sale and about the procedures through which the deal was constructed. People claimed to have been pressured into agreeing to the plan by TNC and by local officials who negotiated on their behalf. Lawsuits and protests were launched. Reports indicate that the people losing their access to hunting camps played a central role in these resistance efforts. In May 2012, local people began circulating a petition urging the State to buy a conservation easement on the 26,000 ha in question rather than purchase it outright. The petition argued:

> So far, our voices have fallen on deaf ears. We are the towns and villages of the Adirondacks, the businesses, families, outdoorsmen and women, conservationists, naturalists, environmental stewards, concerned constituents, and taxpayers. We want our State to do the right thing for our economy, for our environment, and for our access to nature.[4]

In 2012, NY State confirmed a commitment to purchase 27,920 ha from TNC for US$49.8 million in a phased five-year contract beginning with US$13 million for 11,180 ha in 2013. By March 2014, two acquisition phases had been completed representing a total of 12,160 ha. In May 2016, nine years after the initial purchase of the land by TNC, the final transaction was completed. The State manages this land and organizes recreational access and infrastructure based on detailed plans developed by the State's Department of Environmental Conservation.

Discussion

The FPP land deal is many things to many people, as represented by its champions and its detractors. The redistribution of property rights responded positively to ecological concerns and to the displacement of people and communities in the wake of socioeconomic change. If the appropriate counterfactuals include Atlas selling off the land for trophy homes and private hunting camps (parcelization and fragmentation), an unrestricted sale to timber investors,[5] or TNC and the State transferring all the land to parkland, the deal that did materialize has had many advantages in relation to the historical socioecological problems of the region. While the outcome in place at present looks to be a success, these arrangements present a challenge for analysts interested in making sense of land governance and how we might think productively about the meaning of financialization as a concept and a set of practices. The point of this analysis is not to introduce evidence of positive outcomes linked to processes of financialization. Rather, the point is to emphasize the diversity of forms and functions institutional investment can assume in land governance. Recognizing the pragmatic and ad hoc character of the strategies of actors in land governance, including institutional investors, we move beyond homogenous categories and deterministic assumptions. By paying attention to grounded practices in land governance in settings like the Adirondacks, we can understand how state and non-state actors enable and constrain institutional investment and the effects it has on people and places.

In framing this volume, the editors emphasized how site-specific engagement with financialization presents opportunities to address ways in which state re-regulation and social movement resistance are setting roadblocks or speed bumps in the path of financialization. In this framing of the analytical challenge, financialization is cast as an undesirable destination, and social regulation is understood in terms of defence. The case study presented in this chapter highlights the potential for landscape governance through which public agencies, civil society organizations, local communities, and institutional investors are brought into an alignment that pushes back against ecological disorganization (in particular, landscape fragmentation) and sociocultural displacement. In this sense, the analytical challenge can be framed in terms of research that allows us to understand how geographic, ecological, historical, and institutional characteristics of landscapes shape expressions of financialization.

The case study allows us to gain some perspective on the three critiques or markers of financialization identified in the introduction to this chapter. With respect to institutional investment (as an agent of financialization), the view that such investment displaced local people and traditional livelihoods is thoroughly ambiguous. ATP's investment extends the future of large-scale commercial logging in the region. Applied to the issue of economic displacement, ATP's presence contributes positively to conservation of traditional livelihoods. Importantly, by purchasing a conservation easement and reducing the purchase price of the land, the State of NY can be seen as subsidizing ATP and contributing to the longevity of commercial logging in a context of long-term decline of primary industry. Similarly, as architects of the deal, TNC contributed to efforts to avoid the suburbanization of the forest and thereby maintained the dominance of the chainsaw over the lawnmower. It may be possible to go further by suggesting that investments made by TNC and the State of NY served to subdivide the former FPP lands in such a way that the most economically productive timberland was sold to ATP,[6] thereby strengthening the business case for commercial logging. The fibre supply agreement on the land owned by ATP symbolically, financially, and materially embeds papermaking by private equity investors and logging by institutional investors in the social and ecological history of the region.

At the same time, members of local hunt clubs in the region have been displaced by TNC's actions and by ATP, triggering significant and sustained protest. More generally, shifting one-third of the privately owned former FPP land into the public domain, and ceasing logging operations on this land, has fuelled long-running resentment linked to the creation of the Park, infringement of property rights, and a perceived loss of local sovereignty. The moratorium on parkland acquisition proposed by the Governor (which was rejected by the legislature) and the investigation into sweetheart deals between TNC and the State of NY (which resulted in a finding of no impropriety) reflect political sensitivities arising from historical and contemporary displacement in the name of environmental conservation.

The extent to which the redistribution of property rights in this land deal upended or conserved socioecological relations is a complex question. Attributing displacement to institutional investment should be based on arguments that account for a broad range of trends and drivers. Landscapes are subject to multidimensional and multiscalar dynamics that redistribute costs and benefit flows. Automation, deindustrialization, global market integration, the rural amenity-turn, and environmental change, have been alienating people from traditional livelihoods in this region for a long time. Risks attributable to the parcelization of forests and the sell-off of industrially owned timber land should be assessed in relation to these macro drivers of change. In this context, the analytical challenge is to assess the significance of institutional investment against this complex and ambiguous backdrop.

With respect to the ecological implications of this land deal, at one level it contains elements that connect to the narrative of 'green grabbing' (Fairhead *et al.*, 2012). TNC, through its local chapter, was able to navigate considerable political and economic barriers and conserve a significant land resource. TNC's ability to sell property interests to an institutional investor and to the State of NY allowed it to advance an important conservation objective. The Adirondack Park experienced its largest expansion in over 100 years and environmental controls are in place on lands subject to continued logging. While it took much longer than expected, TNC did eventually sell off the conservation land to the State, which presents opportunities to deploy its capital in other locations to advance conservation. At the same time, there is substantial ambiguity attached to environmental outcomes of the deal. A significant portion of the land incorporated into the public domain is open to public hunting and two thirds of the former FPP land continues to be logged commercially. Data are not available to comment on if and how logging practices might change under ATP ownership relative to FPP ownership. The 'brown' (that is, anti-ecological) dimension of this deal is further reflected in the status of the Finch Paper mill in Glens Falls having been identified as the number one polluter in the State of NY (see Doolittle, 2010). Most of the pollution is nitrate pollution to waterways attributable to ammonia used in processing wood chips. This industrial pollution seems to run contrary to the ecological objectives that motivated TNC and the State's engagement with the land. This pollution is an additional cost that should be considered when assessing the policies and investments that promote maintenance of traditional livelihoods.

Displacement is an historical process linked closely to development dynamics and social inequality. It is clear that social, economic, and environmental change cannot be resisted categorically. Applied to land grabbing through financial channels, social and ecological concerns arise from the exercise of power in the absence of informed social debate and appropriate application of existing institutional controls. In this sense, the issue is transparency and commitments to deliberative processes and the rule of law. As a function of having timely access to information and access to platforms through which they can make themselves heard, a broad range of parties can participate in decisions about public investment and public controls on private actors and private property. On this basis, changes in property rights gain legitimacy and governance is strengthened.

The sale of FPP lands was the subject of significant attention and contestation because of their singular economic, social and ecological significance in the region. While this land had special importance, changes in ownership of large parcels was an established concern. The sell-off of the land owned by the paper mill was part of a well-recognized pattern in the Northern Forest. This issue was recognized by the US Congress and Governors of Northeastern States, and it led to the creation of the Northern Forest Land Council in 1991. A central recommendation of the NFLC final report was public–private cooperation in the construction of 'working forests', very much in line with what happened to the FPP lands (Wolf and Klein, 2007). This NFLC report grew out of a consultation and planning process. Similarly, in 2008, approximately one year after TNC purchased the land from Atlas, TNC and the State of NY issued a detailed plan for the dispensation and management of the former FPP lands. The report was the subject of public meetings, public comment, and media coverage. While communications took place, there is no normative basis against which it is possible to assess the adequacy of the information provided to the public, and assessment of the quality of this information: Was the information timely? Was it accessible? Was it comprehensive? Was it biased?

The issue of informed consent holds an important status in the land grabbing literature and in efforts to regulate institutional investment in land through promulgation of norms of responsible conduct (Silva-Castaneda, 2016). Local communities in NY State are formally empowered to veto the use of public funds to purchase land, as structured in this land deal. TNC has repeatedly emphasized its commitment to dialogue with local communities as part of an effort to allay concerns about displacement and impropriety. The sale of land to communities for local development purposes and the decision not to exercise their right to veto the land purchase by the State point directly to a process of consent. That said, protests and lawsuits continue. Some local people bitterly contest the procedures and the outcomes of the consultation and consent process.

The accountability mechanisms in place in multifunctional landscapes are the focus of ongoing research (Kraft and Wolf, 2018). This research addresses the broad range of instruments and logics of accountability, and it grapples with the challenge of distinguishing contexts in which there is meaningful potential for imposition of sanctions, contexts in which interdependence makes sanctions very unlikely, and contexts in which actors strategically advertise accountability in the absence of potential for sanctions. Applied to the assessment of risks to communities and ecosystems posed by financialization, the core accountability issue is the capacity of communities, civil society, and the state to discipline investors and their enablers, and the resulting land management. More generally, the challenge in this complex, multi-functional landscape is to support information flows and to maintain productive tensions that present material consequences to parties that fail to meet their obligations or violate grants of trust. ATP, TNC and State agencies that occupy central roles in this land deal are subject to significant scrutiny. They have made public commitments, and they have entered into a variety of private contracts and property relations that carry formal obligations (see Kraft, 2014 for a description of the relevant accountability mechanisms in place). Parallel to assessment of transparency, there is no objective basis through which to evaluate the strength or weakness of existing accountability relations.

Conclusion

In reflecting upon institutions that enable and constrain distancing and disembedding associated with financialization (and green-grabbing), property rights are the central concern. Accountability controls, information flows, protests, and the capacity of citizens to advance self-protection through state interventions warrant close attention, as property rights are social relations that reflect changing notions of legitimacy and shifting balances of power. In addition to analysis of governance processes that mediate how property rights are expressed and how they change, research must address changes in ownership (Jacobs, 1998; Best and Wayburn, 2001). The FPP deal subdivided land to support specific land uses (i.e. ecological preservation, recreation, timber supply, and real estate development) and to finance environmental conservation. At a finer scale, it subdivided property rights in ways that made it possible to accommodate multiple interests on single parcels. The former and the latter represent complementary strategies to advance multi-functionality. It is possible to argue that the distribution of property rights, along with the governance controls, effectively protects against the social and ecological risks posed by institutional investors. It is also possible to argue that finance capital has found a way, in an institutionally dense and politically fractious landscape, to extract value and buffer itself from economic and political risks. In other words, just as we might conclude the case points to the potential for effective social regulation of financialization of rural land, we must also reflect on the plasticity of finance and its capacity to capitalize on defensive investments made by civil society and the state.

This empirical case highlights new land conservation strategies that have emerged to maintain the socioecological integrity of landscapes. These strategies reflect the neoliberal context and a pragmatic response to the inability of state agencies, science, and communities to mount credible responses to contemporary socioecological problems (Bonanno and Wolf, 2017; Hodge and Adams, 2012). These problem-focused efforts seek to integrate economic, social, and ecological demands on forests. They seek to respond to claims of local sovereignty and self-determination and at the same time yield global public goods. They are premised on investment and engagement by public agencies, institutional investors, non-governmental organizations, and local communities. As novel and creative governance arrangements that bridge interests, scales, and actors, these pragmatic strategies are premised on institutional entrepreneurship, new combinations of resources, and justifications (Kraft and Wolf, 2018; Lounsbury and Glynn, 2001). The potential to reproduce, adapt, and scale-up these models is not known, as these initiatives are unfolding in real time. Further, it is not clear how ecological and social characteristics of this specific territory mediate generalizability of this case.

This study raises questions about the unit of analysis in the study of financialization of rural land. Applied to this case study, for certain questions (what environmental controls on logging are in place, what terms structure recreational access, and what information is disclosed?) a focus on the land owned by institutional investors is warranted. But the case highlights the interplay between a wave of institutional investment in timber land and efforts to design multi-functional landscapes to respond to shifting social, economic, and ecological problem definitions. In terms of making the deal work financially and politically, TNC and the State of NY needed a commercial partner interested in owning timberland. Attention to the interaction and interdependence of this diverse set of actors within this larger landscape emerges as a productive focus relative to a more narrow analysis directed strictly toward the motives and behaviours of institutional investors.

In reflecting upon how institutional investment plays a role in the construction of a multi-functional landscape in a region characterized by conflict along multiple dimensions, it is possible to point to a practical contribution of social science critiques of land grabs and green grabs and neo-colonialization of the rural. The varied land uses and access provisions; the integrated response to ecological, environmental, and diverse economic demands at multiple scales; the planning and consultative processes; the contractual preservation of the link between the land and the mill (through the inclusion of a fibre supply agreement); and, accountability structures in place, mirror many of the objections and concerns highlighted in the critical literature. No specific mechanism is identified through which public and private sector actors in this land deal have come to internalize new commitments derived from critiques of past land deals, but there does seem to be heightened sensitivity and substantive engagement with problems discussed in the academic and grey literature. The practices documented above highlight how the critical conversation about land grabbing is reflected in the practices and justifications of public officials and a global conservation NGO. In this sense, this case study points to the importance of deepening and broadening this conversation.

Notes

1 www.lymetimber.com/about-us.html
2 See '"Deal Of Century" to Open 70,000 Adirondack Acres to Public Hunting', available at: www.hunting.be/wp-content/uploads/Deal-Of-Century-to-Open-70000-Adirondack-Acres-to-Public-Hunting-Outdoor-Life.pdf, accessed 18 June 2017.
3 The hunt clubs made the provocative argument that recreational use of the land would decline (relative to maintenance of the status quo) under a plan to make the land publicly accessible.
4 See www.change.org/p/andrew-cuomo-change-the-finch-land-acquisition-to-a-conservation-easement-purchase, accessed 18 June 2017.
5 Presumably, Atlas chose to sell the land to TNC because they produced a superior, or at least competitive, offer relative to other interested parties.
6 It is possible that the land conserved by TNC and State of NY investments featured the highest ecological values, but this correlate is not necessary to the argument offered about ATP's calculations.

References

Best, C. and Wayburn, L. (2001) *America's Private Forests: Status and Stewardship*, Island Press, Washington, DC

Bliss, J., Kelly, E., Abrams, J., Bailey, C. and Dyer, J. (2010) 'Disintegration of the US industrial forest estate: Dynamics, trajectories, and questions', *Small-scale Forestry*, vol 9, no 1, pp. 53–66

Block, N. and Sample, A. (2001) 'Industrial timberland divestitures and investments: Opportunities and challenges in forestland conservation', Pinchot Institute, Washington, DC

Bonanno, A. and Wolf, S. (2017) 'The new frontier of resistance in agri-food', in A. Bonanno and S. Wolf (eds) *Resistance to the Neoliberal Agri-Food Regime: A Critical Analysis*, Routledge, London

Büscher, B., Sullivan, S., Neves, K., Igoe, J. and Brockington, D. (2012) 'Towards a synthesized critique of neoliberal biodiversity conservation', *Capitalism, Nature, Socialism*, vol 23, no 2, pp. 4–30

Clapp, J. (2014) 'Financialization, distance and global food politics', *Journal of Peasant Studies*, vol 41, no 5, pp. 797–814

Cleaver, F. (2001) 'Institutional bricolage, conflict and cooperation in Usangu, Tanzania', *IDS Bulletin*, vol 32, no 4, pp. 26–35

Dicker, F. (2010) 'NY gave environmental org. absurd US$3.7m profit for forest', *NY Post*, available at: http://nypost.com/2010/04/05/ny-gave-environmental-org-absurd-3-7m-profit-for-forest/, accessed 18 June 2017

Doolittle, W. (2010) 'Finch tops federal list for chemical releases', available at: http://poststar.com/news/local/finch-tops-federal-list-for-chemical-releases/article_84256ea8-0a1c-11e0-8640-001cc4c03286.html, accessed 18 June 2017
Earl, G. (2007) 'Gooley Club's last stand', Adirondack Explorer, available at: www.adirondackexplorer.org/stories/gooley-clubs-last-stand, accessed 22 April 2017
Fairbairn, M. (2014) '"Like gold with yield": Evolving intersections between farmland and finance', *Journal of Peasant Studies*, vol 41, no 5, pp. 777–795
Fairhead, J., Leach, M. and Scoones, I. (2012) 'Green grabbing: A new appropriation of nature?' *Journal of Peasant Studies*, vol 39, no 2, pp. 237–261
Forisk Blog. (2013) '2013 US timberland and ownership: descriptive statistics', available at: http://forisk.com/blog/2013/03/21/2013-us-timberland-ownership-descriptive-statistics/, accessed 29 June 2018
Goldschmidt, W. (1947) *As You Sow: Three Studies in the Social Consequences of Agribusiness*, Allanheld, Osmun and Co, Montclair, NJ
Gunnoe, A. and Gellert, P. (2011) 'Financialization, shareholder value, and the transformation of timberland ownership in the US', *Critical Sociology*, vol 37, no 3, pp. 265–284
Hagan, J., Irland, L. and Whitman, A. (2005) *Changing Timberland Ownership in the Northern Forest and Implications for Biodiversity, Report # MCCS-FCP- 2005–1*, Manomet Centre for Conservation Sciences, Maine
Healy, P. (2009) 'The pragmatic tradition in planning thought', *Journal of Planning Education and Research*, vol 28, no 3, pp. 277–292
Hodge, I. and Adams, W. (2012) 'Neoliberalisation, rural land trusts and institutional blending', *Geoforum*, vol 43, no 3, pp. 472–482
Jacobius, A. (2015) 'Timber growing on investors as managers look at new markets', available at: www.pionline.com/article/20151005/PRINT/310059969/timber-growing-on-investors-as-managers-look-at-new-markets, accessed 18 June 2017
Jacobs, H. (ed.) (1998) *Who Owns America? Social Conflict over Property Rights*, University of Wisconsin Press, Madison
Kraft, B. (2014) 'Making sense of environmental governance: An empirical assessment of accountability in working landscapes'. MS thesis, Cornell University, Ithaca, NY
Kraft, B. and Wolf, S. (2018) 'Through the lens of accountability: Analyzing legitimacy in environmental governance', *Organization and Environment*, vol 38, no 1, pp. 70–92
Krippner, G. (2005) 'The financialization of the American economy', *Socio-Economic Review*, vol 3, no 2, pp. 173–208
Lawrence, G. (2015) 'Defending financialization', *Dialogues in Human Geography*, vol 5, no 2, pp. 201–205
Lounsbury, M. and Glynn, M. (2001) 'Cultural entrepreneurship: Stories, legitimacy, and the acquisition of resources', *Strategic Management Journal*, vol 22, no 6–7, pp. 545–564
Neugarten, R., Wolf, S. and Stedman, R. (2012) 'Forest fights and forest fights: Working forests as a strategy for reducing tensions in New York State', *Society and Natural Resources*, vol 25, no 12, pp. 1205–1220
Newmark, W. and Hough, J. (2000) 'Conserving wildlife in Africa: Integrated conservation and development projects and beyond', *BioScience*, vol 50, no 7, pp. 585–592

Petersen, J. (1998) 'War in the North Country: The battle for control of the Northeast's New Forest', *Evergreen*, September, pp. 24–43

Potter, C. and Tilzey, M. (2007) 'Agricultural multi-functionality, environmental sustainability and the WTO: Resistance or accommodation to the neoliberal project for agriculture?' *Geoforum*, vol 38, pp. 290–1303

Russi, L. (2013) *Hungry Capital: The Financialization of Food*, Zero Books, Winchester, UK

Schmidt, T. (2016) *The Political Economy of Food and Finance*, Routledge, London

Schulte, L., Rickenbach, M. and Merrick, L. (2008) 'Ecological and economic benefits of cross-boundary coordination among private forest landowners', *Landscape Ecology*, vol 23, no 4 pp. 481–496

Silva-Castaneda, L. (2016) 'In the shadow of benchmarks: Normative and ontological issues in the governance of land', *Environment and Planning A*, vol 48, no 4, pp. 681–698

Stein, P. (2010) 'Conservation easements', *Silviculture Magazine*, Winter, pp. 6–8

Terrie, P. (2008) *Contested Terrain: A New History of Nature and People in the Adirondacks*, The Adirondack Museum and Syracuse University Press, Syracuse, NY

Wolf, S. and Klein, J. (2007) 'Enter the working forest: Discourse analysis in the Northern Forest', *Geoforum*, vol 38, no 5, pp. 985–998

Wolford, W., Borras, S., Hall, R., Scoones, I. and White, B. (2013) 'Governing global land deals: The role of the state in the rush for land', *Journal of Peasant Studies*, vol 44, no 2, pp. 189–210.

Index

Page numbers in **bold** denote tables, those in *italics* denote figures.

For Product Safety Concerns and Information please contact our EU representative GPSR@taylorandfrancis.com
Taylor & Francis Verlag GmbH, Kaufingerstraße 24, 80331 München, Germany

www.ingramcontent.com/pod-product-compliance
Ingram Content Group UK Ltd.
Pitfield, Milton Keynes, MK11 3LW, UK
UKHW022009180926
475701UK00003B/280

* 9 7 8 0 3 6 7 5 8 6 2 7 0 *